# SUPERCONDUCTORS - MATERIALS, PROPERTIES AND APPLICATIONS

# SUPERCONDUCTORS - MATERIALS, PROPERTIES AND APPLICATIONS

*Editor*

**Mergen Balik**

scitus academics

**Superconductors - Materials, Properties and Applications**
Edited by **Mergen Balik**

ISBN: 978-1-68117-607-9
Library of Congress Control Number: 2016946462


SCITUS Academics LLC,
616, Corporate Way, Suite 2, 4766,
Valley Cottage, NY 10989

www.scitusacademics.com



Printed in United States of America on Acid Free Paper

# PREFACE

Superconductors are materials that conduct electricity with no resistance. This means that, unlike the more familiar conductors such as copper or steel, a superconductor can carry a current indefinitely without losing any energy. They also have several other very important properties, such as the fact that no magnetic field can exist within a superconductor. Superconductors already have drastically changed the world of medicine with the advent of MRI machines, which have meant a reduction in exploratory surgery. Power utilities, electronics companies, the military, transportation, and theoretical physics have all benefited strongly from the discovery of these materials.

*Superconductors - Materials, Properties and Applications* deals with various aspects of superconductivity, both theoretical and experimental. Some of the technological applications of superconductivity include: the production of sensitive magnetometers based on SQUIDs; fast digital circuits (including those based on Josephson junctions and rapid single flux quantum technology); powerful superconducting electromagnets used in maglev trains; Magnetic Resonance Imaging (MRI) and Nuclear magnetic resonance (NMR) machines, magnetic confinement fusion reactors, and the beam-steering and focusing magnets used in particle accelerators; low-loss power cables, etc.

# TABLE OF CONTENTS

# SUPERCONDUCTIVITY IN LAYERED ORGANIC METALS

**Jochen Wosnitza**

Hochfeld-Magnetlabor Dresden, Helmholtz-Zentrum Dresden-Rossendorf, 01314 Dresden, Germany

## ABSTRACT

In this short review, I will give an overview on the current understanding of the superconductivity in quasi-two-dimensional organic metals. Thereby, I will focus on charge-transfer salts based on*bis* (ethylenedithio) tetrathiafulvalene (BEDT-TTF or ET for short). In these materials, strong electronic correlations are clearly evident, resulting in unique phase diagrams. The layered crystallographic structure leads to highly anisotropic electronic as well as superconducting properties. The corresponding very high orbital critical field for in-plane magnetic-field alignment allows for the occurrence of the Fulde–Ferrell– Larkin–Ovchinnikov state as evidenced by thermodynamic measurements. The experimental picture on the nature of the superconducting state is still controversial with evidence both for unconventional as well as for BCS-like superconductivity.

## KEYWORDS

organic superconductors; electronic structure; superconducting state; Fulde–Ferrell–Larkin–Ovchinnikov state

# 1. INTRODUCTION

More than 30 years ago the first organic superconductor, $(TMTSF)_2 PF_6$, was discovered by D. Jérome *et al.* [1]. Materials based on TMTSF (tetramethyltetraselenafulvalene) with a monovalent anion are electronically quasi-one-dimensional (1D) materials that show very rich pressure-temperature-magnetic field phase diagrams with formidable ground-state properties. For some time a possible triplet superconducting state was discussed, but recent experimental data prove the singlet nature with presumed *d*-wave symmetry. More information on these topical issues can be found in a number of reviews [2–4].

Here, I want to focus on the quasi-two-dimensional (2D) organic superconductors. Among these, materials with BEDT-TTF (a.k.a. ET, *bis* (ethylenedithio) tetrathiafulvalene) as organic building block (Figure 1, left side) are the most investigated class with superconducting transition temperatures, $T_c$, of up to about 13 K [2,4,5]. The most common stoichiometry of these 2D organic metals is $(BEDT\text{-}TTF)_2X$, where $X$ represents a monovalent anion such as $I_3^-$, $Cu(NCS)_2^-$, $Cu[N(CN)_2]Br^-$, or $SF_5CH_2CF_2SO_3^-$. Charge transfer from the neutral BEDT-TTF molecule to the anion leads to partially filled molecular bands, resulting in metallic conductivity. Apart from BEDT-TTF, the organic molecule BETS (bisethylenedithio-tetraselenafulvalene), where the inner four sulfur atoms of BEDT-TTF are replaced by selenium, can be utilized as building block for a remarkable class of organic superconductors [6]. For example, charge-transfer salts containing $Fe^{3+}$ anions may lead to competing magnetic and superconducting interactions, reentrant superconductivity upon cooling, and magnetic-field-induced superconductivity.

Besides superconductivity, the organic metals show a large variety of different ground states such as antiferromagnetism, spin-Peierls, spin-density-wave, and charge-density-wave states. These states are accessible by tuning the structure, counter anion, magnetic field, temperature, and pressure. The study of these fertile phase diagrams has led to new theoretical concepts; however, a solid understanding of some of these states still remains a challenge. This includes the normal metallic state which often appears to be not explicable by conventional Fermi-liquid theory.

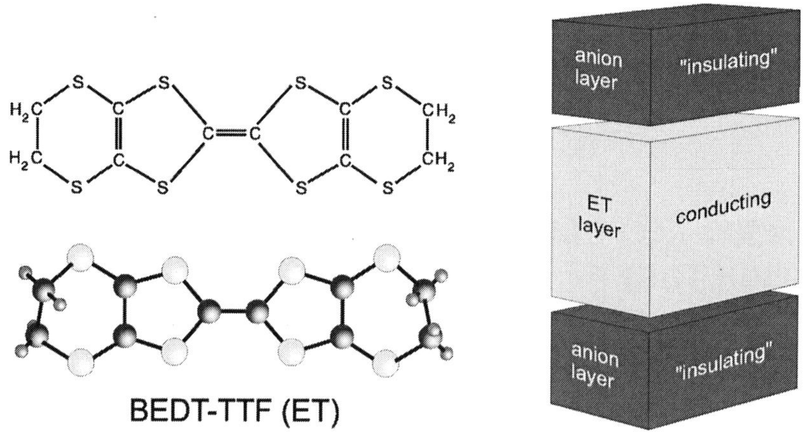


**Figure 1.** (left) The molecular structure of*bis* (ethylenedithio) tetrathiafulvalene (BEDT-TTF or ET); (right) Schematic arrangement of the ET and anion layers in the quasi-two-dimensional organic metals.

Many superconducting properties of the 2D organic superconductors, such as the $H$–$T$ phase diagram and the vortex dynamics, resemble the behavior found in the cuprate high-$T_c$ superconductors [7–11]. In contrast to the latter, however, much less is known about the nature of the superconducting state in organic metals. Clear-cut experimental proof for unconventional non-BCS-like superconductivity, such as phase-sensitive experimental evidence, is missing. For the organic superconductors, where singlet pairing has been established, the experimental picture on the symmetry of the order parameter is still diverse (see [4,5,12–15] for recent reviews). Some results hint at line nodes in the superconducting energy gap, *i.e.*, *d*-wave symmetry, whereas other experiments, in particular, specific-heat data, suggest the existence of a nodeless energy gap, fully consistent with BCS-like *s*-wave symmetry.

A further unsolved question is the mechanism responsible for the Cooper-pair coupling. While there is some belief in the existence of a non-phonon-mediated coupling mechanism, as appears to be operative in unconventional superconductors such as the cuprates, $Sr_2RuO_4$, and heavy-fermion compounds, this issue remains unclear and heavily debated for the organic superconductors. The proximity of antiferromagnetic or spin-density-wave states to the superconducting state spurred the notion that antiferromagnetic fluctuations are responsible for the coupling. On the other hand, a number of experiments prove the existence of clear interrelations between phonons and superconductivity.

## 2. CRYSTALLOGRAPHIC AND ELECTRONIC STRUCTURE

Mostly, the term "organic materials" implies biological substances or plastics, *i.e.*, polymeric carbon-based chains. However, the organic metals discussed here are charge-transfer salts of mostly high crystalline quality. They usually are grown by electrocrystallization from appropriate solutions, the details of which are reviewed in depth in Reference [16]. During this process the oxidation of the donor molecules occurs together with the crystal growth at inert electrodes. The 2D organic charge-transfer salts typically grow as single-crystalline plates that weigh up to a few milligrams.

A number of different crystal structures exists for the ET-based charge-transfer salts. They mainly differ in the packing motifs of the ET molecules and are labeled by different Greek letters [2,16]. The common feature, however, is the layered crystal structure depicted in the right panel of Figure 1. In the conducting ET layers, there are sufficiently close contacts between the π orbitals of the sulfur atoms of neighboring ET molecules. This leads to 2D molecular bands with rather small in-plane anisotropies (typically within a factor of two). The anions, sandwiched between the highly conducting ET layers, are responsible for the charge transfer that leads to partially filled molecular bands, enabling metallic conductivity. The electrical resistivity perpendicular to the layers is larger by about 3–6 orders of magnitude than for electronic transport within the ET layers. One should note, however, that the exact resistive anisotropy is difficult to determine since any dislocation disturbing the in-plane transport will immediately add a large out-of-plane component to the measured resistance [4,14].

Although the layered structure occurs in all ET-based superconductors, the varying in-plane packing motifs of the organic molecules lead to different 2D electronic structures. By utilizing the rather simple extended Hückel tight-binding approximation, the in-plane electronic band structures can be calculated. Four examples of the resulting energy dispersions and of the in-plane Fermi surfaces are shown in Figure 2. These calculations, in most cases, agree extraordinarily well with the experimentally determined Fermi surfaces, although the large number of atoms per unit cell necessitates quite crude simplifications in the tight-binding approximations. More sophisticated band-structure calculations either led to incorrect Fermi-surface topologies due to oversimplification or finally confirmed all in all the Hückel results [17]. It is remarkable that in spite of the complex crystallographic structure the electronic structure is astonishingly simple. Indeed, the in-plane Fermi surfaces often resemble free-electron topologies.

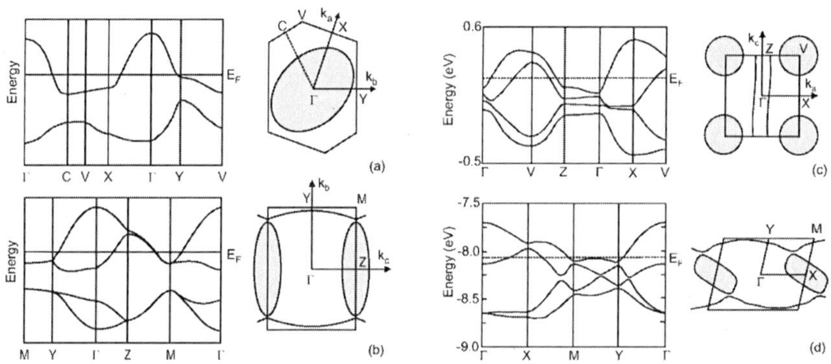

**Figure 2.** Calculated band structures and Fermi surfaces for (**a**) $\beta$-phase (after Reference [18]), (**b**) $\kappa$-phase [21,22], (**c**) $\alpha$-phase [23], and (**d**) $\beta''$-phase BEDT-TTF salts [25].

For the $\beta$ phase, one electron per unit cell is transferred to the anion leaving the highest occupied molecular orbital half filled. The nearly isotropic in-plane overlap integrals lead to an almost circular free-electron-like Fermi surface (Figure 2a) [18]. The validity of this result has been verified by measurements of magnetic quantum oscillations and angular-dependent magneto resistance oscillations [17,19]. Indeed, only one closed orbit with the predicted area (shaded area in Figure 2a) and only slightly more cornered shape [19,20] could be reconstructed from the data.

The Fermi surfaces of the other crystalline phases with pairs of dimerized ET molecules are almost as simple as for the $\beta$ phase. The dimerization doubles the unit cell, resulting in two transferred electrons on two molecular orbitals crossing the Fermi surface as visualized in Figure 2b–d. The resulting Fermi surfaces again resemble the free-electron picture. Since the dimerization is halving the first Brillouin zone, the almost free-electron-like Fermi-surface ellipse cuts the first Brillouin zone. Consequently, gaps open at the zone boundary due to Bragg reflections. These gaps are smaller for the $\kappa$-phase materials [21,22], but larger for the $\alpha$-phase salts. Besides the shaded closed orbits, wavy open bands appear (Figure 2c) [23]. These 1D bands are prone to a density-wave transition which indeed is realized in some of the $\alpha$-phase organic salts [24].

Although for most organic metals the calculated Fermi surfaces are favorable in line with experimental data, the measured Fermi surface in $\beta''$-(BEDT-TTF)$_2$SF$_5$CH$_2$CF$_2$SO$_3$ differs noticeably from the prediction (Figure 2d) [25]. The closed orbits are found to be only about 1/3 of the calculated orbits and so far no direct evidence for the open bands has been reported. Another drawback of the extended Hückel band-structure calculations is the negligence of many-body effects such as electron-electron and electron-phonon interactions. Consequently, the calculated effective masses are often much smaller than the measured values.

# 3. PROPERTIES OF THE SUPERCONDUCTING STATE

Before discussing the still controversial issues on the nature of the superconducting state in the organic metals, I will present some principal superconducting properties of the 2D organic metals. These properties resemble in many respect the behavior observed also in the cuprate superconductors. The 2D organic metals are strongly type-II superconductors with lower critical fields, $\mu_0 H_{c1}$, of only a few mT and upper critical fields, $\mu_0 H_{c2}$, in the Tesla range. These fields depend strongly on the direction of the applied field. For $H$ applied within the ET planes, $H_{c2}$ often exceeds the Pauli paramagnetic limit [26,27] and evidence for the Fulde–Ferrell–Larkin–Ovchinnikov state may be found, as will be discussed below. The moderately low upper critical fields allow, in contrast to the cuprate high-$T_c$ superconductors, to access conveniently the whole superconducting phase diagram and to study the rich physics of the vortex phase [4,5]. In the following, some selected experimental features observed for the organic superconductors are presented.

## 3.1. Principal Properties

As a typical example, Figure 3 shows the superconducting phase diagram of the well-studied material $\kappa$-(BEDT-TTF)$_2$Cu[N(CN)$_2$]Br for fields applied perpendicular to the ET planes. Shown are the temperature dependences of $\mu_0 H_{c1}$, $\mu_0 H_{c2}$, and of the irreversibility field, $\mu_0 H_{irr}$ derived from DC-magnetization, $M_{dc}$, and AC-susceptibility, $\chi_{ac}$, measurements [28–30]. The upper critical fields (closed circles) were obtained from a scaling analysis of the magnetization data which due to superconducting fluctuations are strongly rounded in fields [28]. The lower critical fields (closed diamonds) were extracted from the vanishing magnetic relaxation, *i.e.*, the fields where no longer a time dependence of the magnetization is detected [29]. From Figure 3, the very large range of the Shubnikov state, characteristic for strongly type-II superconductors, is evident (note the logarithmic scale). The extrapolated critical fields at $T = 0$ are $(12 \pm 2)$ T for the upper critical field and $(3 \pm 1)$ mT for the lower critical field [28,29]. This results in a thermodynamic critical field of about 100 mT, an in-plane coherence length $\xi_{||} \approx 5$ nm, and a Ginzburg–Landau parameter of $\kappa \approx 100$.

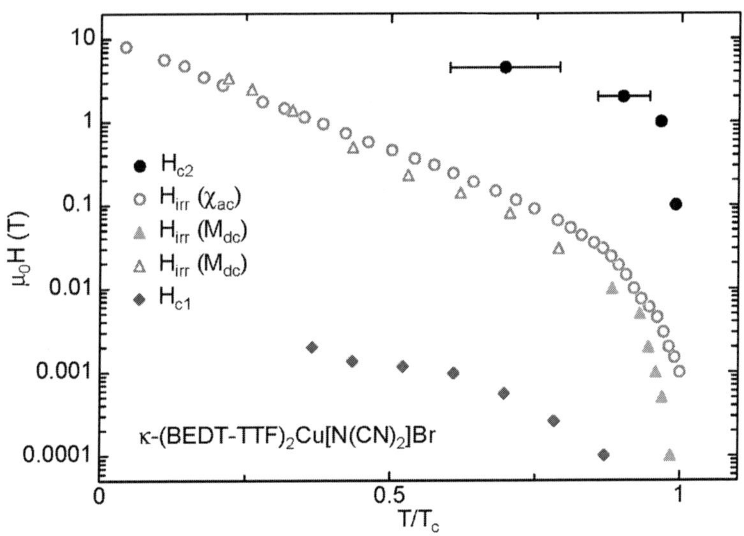


**Figure 3.** Temperature dependences of the lower ($H_{c1}$) and upper ($H_{c2}$) critical fields as well as the irreversibility fields ($H_{irr}$) determined by different techniques for fields applied perpendicular to the ET layers. The data for $H_{c2}$ (closed circles) and $H_{irr}$ (open triangles) are taken from [28].

For fields parallel to the ET layers, $\mu_0 H_{c1}$ is considerably smaller than the earth's field and a reliable determination of the lower critical fields is challenging [31]. From the initial slope of the in-plane upper-critical-field line close to $T_c$, an orbital critical field far beyond 100 T may be estimated (see below). This results in an interplane coherence length much smaller than the layer distance of about 1.6 nm [7]. Therefore, in order to establish a three-dimensional phase coherence, the superconducting layers are Josephson coupled. That indeed has been verified directly by measuring Josephson plasma resonances in magneto-optical experiments [32]. Superconductivity in the "insulating" or badly-conducting anion layers is induced by phase-coherent Cooper pairs in the ET layers. Above a threshold field, $\mu_0 H_{irr}$, for this proximity-induced superconductivity, the layers are effectively decoupled, and the vortices are weakly pinned 2D objects (pancake vortices). This can be realized from the temperature dependence of the irreversibility field, $\mu_0 H_{irr}$. A distinct crossover from 3D vortex fluctuations close to $T_c$ to 2D fluctuations at lower temperatures is observed (Figure 3) [29].

As mentioned, for the 2D organic metals, the destruction of superconductivity due to orbital effects appears at extraordinarily high in-plane magnetic fields. In that case, however, spin effects become relevant. For singlet superconductors, when the Zeeman energy becomes larger than the condensation energy, the Cooper pairs finally break up. This occurs at the Pauli-paramagnetic limit, $\mu_0 H_P = \Delta_0/(\sqrt{2}\mu_B)$, with $\mu_B$ the Bohr magneton and $\Delta_0$ the superconducting energy gap at $T = 0$ [27]. However, even beyond this limit superconductivity can survive in special states where the superconducting order parameter is spatially modulated. The existence of such states were predicted independently by Fulde and Ferrell (FF) [33] as well as Larkin and Ovchinnikov (LO) in 1964 [34].

In the FFLO state, Cooper pairs can only exist with a finite center-of-mass momentum due to the Zeeman-split Fermi surfaces. This results in an oscillating part of the order parameter in real space with a wavelength of the order of the coherence length. Two conditions are necessary for the FFLO state to occur. First, the so-called Maki parameter, $\alpha = \sqrt{2}H_{orb}/H_P$ [35], should be larger than 1.8 [36] and, second, the superconductor needs to be in the clean limit with a mean-free path much larger than the coherence length.

These conditions are favorably fulfilled in many 2D organic superconductors. Mostly, these materials are clean-limit superconductors and, when the magnetic field is oriented exactly parallel to the ET layers, the orbital critical field becomes much larger than the Pauli limit, as mentioned above. Indeed, in a number of works, indications for the existence of the FFLO state have been reported for some of the 2D organic metals [37]. The experimental evidence, however, was mostly based on transport measurements. Only recently, thermodynamic data allowed to conclude the existence of the FFLO state [38].

For $\kappa$-(BEDT-TTF)$_2$Cu(NCS)$_2$, specific-heat measurements in magnetic fields aligned parallel to the 2D layers showed spike-like anomalies for fields above 21 T and temperatures below 3 K. This can be seen clearly in Figure 4, where the specific-heat difference, $\Delta C$, divided by temperature as a function of temperature is plotted. For that, the normal-state specific heat, measured at 14 T applied perpendicular to the layers, has been subtracted for each curve [38]. At about 21 T, the specific-heat anomaly at the superconducting transition sharpens and at 21.5 T a spike in $\Delta C$ occurs. This is evidence for a first-order nature of the phase transition. For higher fields, a second very sharp anomaly appears within the superconducting state. This proves the existence of an additional thermodynamic phase. The latter anomaly shows a well-resolvable hysteresis, while the superconducting transition at higher temperatures has a smaller hysteresis (see [38] for details).

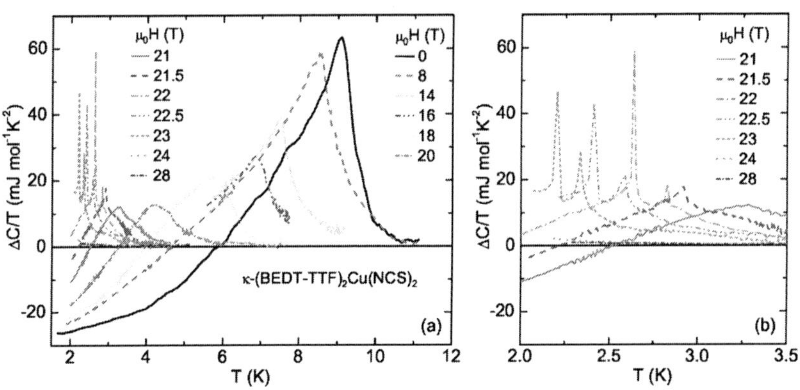


**Figure 4.** (**a**) Temperature dependence of the specific-heat difference between the superconducting and normal state, $\Delta C/T$, of $\kappa$-(BEDT-TTF)$_2$Cu(NCS)$_2$ in magnetic fields applied parallel to the superconducting layers; (**b**) High-field data in an expanded scale.

Further evidence for the existence of the FFLO state has been gathered from magnetic-torque studies (Figure 5a) [39]. Upon lowering the temperature, the magnetic torque, which is proportional to the magnetization, shows an increasingly sharp transition at $\mu_0 H_{c2}$. This indicates that the Pauli limit, $\mu_0 H_P$, is reached. Anyhow, for even lower temperatures superconductivity survives even beyond this field. In addition, a second dip-like feature appears in the torque signal below $\mu_0 H_{c2}$ (vertical arrows in Figure 5), signaling the emergence of the FFLO phase. These results nicely confirm the specific-heat data for the $\mu_0 H_{c2}$ line. There is, however, a discrepancy in the extension of the FFLO phase. Whereas for the specific heat the second anomaly suggests a very narrow region for the additional thermodynamic phase, the dip-like feature in the torque data stays almost constant at about the Pauli limit. This can be explained by the much better in-plane alignment of the magnetic field in the torque measurements made possible by an *in situ* rotation of the torque platform. In the specific-heat experiment for κ-(BEDT-TTF)$_2$Cu(NCS)$_2$ no rotation of the sample platform was possible. Indeed, recent angular-dependent specific-heat measurements of the organic superconductor $\beta''$-(BEDT-TTF)$_2$SF$_5$CH$_2$CF$_2$SO$_3$ supported this notion. These latter data showed that the FFLO state disappears for out-of-plane orientations of the magnetic field as small as 0.5°. A second specific-heat anomaly appeared only for field angles between about 0.2° and 0.3°.

A schematic magnetic phase diagram constructed from above data is shown in Figure 5b. From the very steep initial increase of $\mu_0 H_{c2}$ an orbital critical field of roughly 130 T can be estimated for κ-(BEDT-TTF)$_2$Cu(NCS)$_2$. This large value is caused by the strongly reduced orbital currents for this in-plane field orientation, as mentioned above. The steep initial slope quickly levels off at higher fields upon reaching the Pauli limit. Using $\Delta_0/k_B T_c$ = 2:4,*i.e.*, the normalized superconducting gap as determined in earlier specific-heat studies [40], leads to $\mu_0 H_P$ = 23 T. That fits nicely with the field of about 21 T where the sharp first-order phase transition emerges (Figure 4). Above this field, a strong upturn of the $\mu_0 H_{c2}$ line occurs. Together with this increased slope, the second phase transition appears. This is favorably in line with the expectations for the existence of an FFLO state. Indeed, the 2D organic superconductors are some of the very few materials for which this fascinating spatially inhomogeneous superconducting state can be studied.

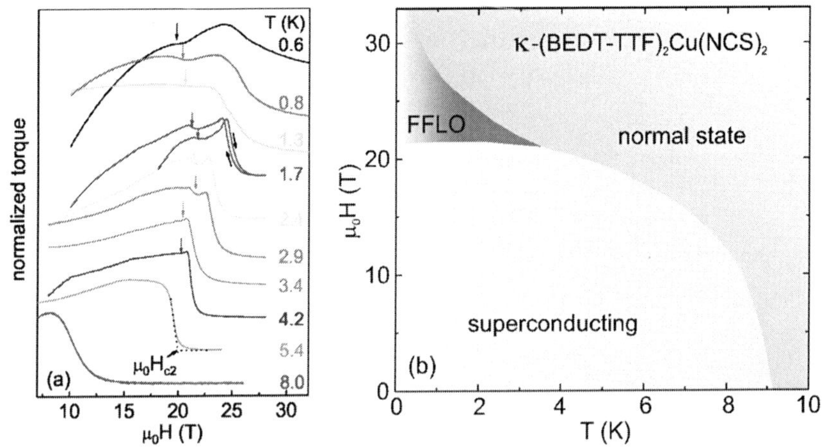


**Figure 5.** (**a**) Magnetic-torque data of κ-(BEDT-TTF)$_2$Cu(NCS)$_2$ at various temperatures for in-plane magnetic fields aligned along the *b* direction. For *T*= 1:7 K, data for both up and down field sweeps, otherwise only down sweeps are shown. The vertical arrows mark the small dip-like features, which are associated with the transition into the FFLO state; (**b**) Schematic phase diagram deduced from the data shown in (**a**) [39].

## 3.2. Nature of the Superconducting State

For the 2D organic superconductors, the controversial experimental situation does not allow a definite conclusion as to whether conventional BCS or unconventional superconductivity exists. It is, however, well proven that the Cooper pairs are in a singlet spin state as evidenced, e.g., by NMR experiments [41,42] and, as discussed above, by the Pauli limitation of the upper critical field and the existence of FFLO states.

One motivation for suggesting an unconventional order parameter of singlet *d*-wave symmetry is based on the schematic *P* – *T* phase diagram of the κ-phase superconductors shown in the left panel of Figure 6 [43]. Next to the superconducting state an antiferromagnetic insulating state exists, resembling phase diagrams observed, e.g., for heavy-fermion and many high-$T_c$ materials. The one κ-phase material known that shows an antiferromagnetic insulating state at ambient pressure is κ-(BEDT-TTF)$_2$Cu[N(CN)$_2$]Cl. A moderate pressure of 300 bar is sufficient to induce superconductivity at a record $T_c$ of about 12.8 K (Figure 6, left)

[44]. A further increase in pressure rapidly suppresses superconductivity. In fact, $\kappa$-(BEDT-TTF)$_2$Cu(NCS)$_2$ displays the largest pressure dependence of $T_c$ yet observed for any superconductor, $\Delta T_c/\Delta p \approx -3$ K/kbar [45].

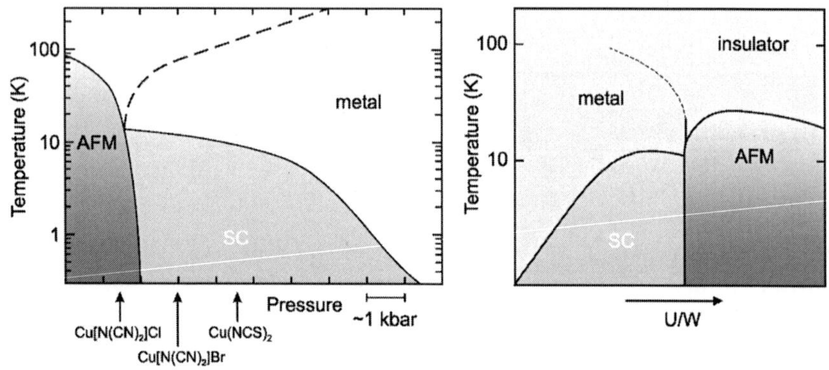


**Figure 6.** (left) Schematic $P - T$ phase diagram for $\kappa$-(BEDT-TTF)$_2X$ salts depicting regions of antiferromagnetic-insulator (AFM) behavior and superconductivity (SC). The arrows indicate starting points for different materials (the anion $X$ is stated) at ambient pressure. The right conceptual phase diagram after [46] for (BEDT-TTF)$_2X$ salts relates the observed phases to a Mott transition. $U$ is the on-site Coulomb repulsion and $W$ denotes the electronic bandwidth.

In contrast to the cuprate high-$T_c$ materials, the crucial parameter for organic superconductors is pressure or chemical substitution, not the charge-carrier concentration. Growing pressure increases the bandwidth $W$ of the electronic bands. Therefore, the phase diagram of the $\kappa$-phase salts might be explained by a bandwidth-controlled Mott transition as visualized by the conceptual phase diagram in the right panel of Figure 6, first proposed by Kanoda [46]. Thereby, pressure controls $U/W$, where $U$ is the on-site Coulomb repulsion. The dashed lines in the$P - T$ phase diagrams delineate different electronic-transport regimes. At high temperatures the resistivity shows a semiconducting behavior, whereas after a maximum in resistance a quadratic temperature dependence of the resistivity appears at low temperatures above $T_c$.

Largely unsettled is the question on the exchange interaction mediating Cooper-pair coupling. Although the proximity of the antiferromagnetic state in the $P - T$ phase diagram (Figure 6) may suggest spin fluctuations responsible for superconductivity, there are a number of experiments showing the existence of a considerable electron-phonon coupling in organic superconductors. Usually, one of the first things investigated to prove electron-phonon interaction as the pairing mechanism is the isotope effect. In the most simple BCS picture, $T_c \propto m^{-1/2}$ is expected, with $m$ being the isotope mass. Other mass dependences, however, are still explicable within BCS theory with a more elaborate electron-phonon coupling theory. For the 2D organic superconductors the experimental situation appears somewhat confusing at first sight.

As shown in Figure 7, a nonuniform $T_c$ shift is found for $\kappa$-phase superconductors when all eight protons in BEDT-TTF are substituted by deuterium [11]. For $\kappa$-(BEDT-TTF)$_2$Cu(NCS)$_2$ (Figure 7a), a large inverse isotope effect appears, whereas for $\kappa$-(BEDT-TTF)$_2$Cu[N(CN)$_2$]Br (Figure 7b) an equally large "normal" isotope effect is found. The observed shifts of 4%–5% in $T_c$ are much larger than could possibly be explained by a simple molecular-mass change. The $T_c$ shift in $\kappa$-(BEDT-TTF)$_2$Cu(NCS)$_2$ can be rationalized by a "geometrical" effect, *i.e.*, by internal uniaxial pressure perpendicular to the planes. Thermal-expansion measurements have shown that uniaxial pressure reduces $T_c$ by about 6 K/kbar in $\kappa$-(BEDT-TTF)$_2$Cu(NCS)$_2$ [47]. Consequently, the inverse isotope effect might be understood by the smaller zero-point spacing of the carbon-deuterium bonds compared to the lighter carbon-hydrogen bonds. This causes a reduced internal pressure and leads to an increased $T_c$ for the deuterated samples [48].

Such a scenario, however, cannot explain the $T_c$ reduction observed for deuterated $\kappa$-(BEDT-TTF)$_2$Cu[N(CN)$_2$]Br samples (Figure 7b). But, as mentioned above, this material is situated close to an insulating state (Figure 6). Indeed, fast-cooled deuterated $\kappa$-(BEDT-TTF)$_2$Cu[N(CN)$_2$]Br samples reflect only partial superconductivity [49]. As visible in Figure 7b, the deuterated sample shows a much higher sensitivity to applied magnetic fields. Already in 1 Oe, the transition curves becomes broad and no diamagnetic saturation is achieved. Anyway, at low enough external field (0.1 Oe) bulk superconductivity for the deuterated sample is found.

Investigations of the isotope effect in 2D organic superconductors were further pursued by substituting four $^{12}$C atoms by $^{13}$C and all eight $^{32}$S atoms by $^{34}$S in the ET molecules [50]. By that, any changes of the anion-hydrogen interaction were avoided. The found $T_c$ reduction of about 1% agrees well with the BCS expectation and provides strong support for a Cooper-pair coupling which is at least partially mediated by phonons.

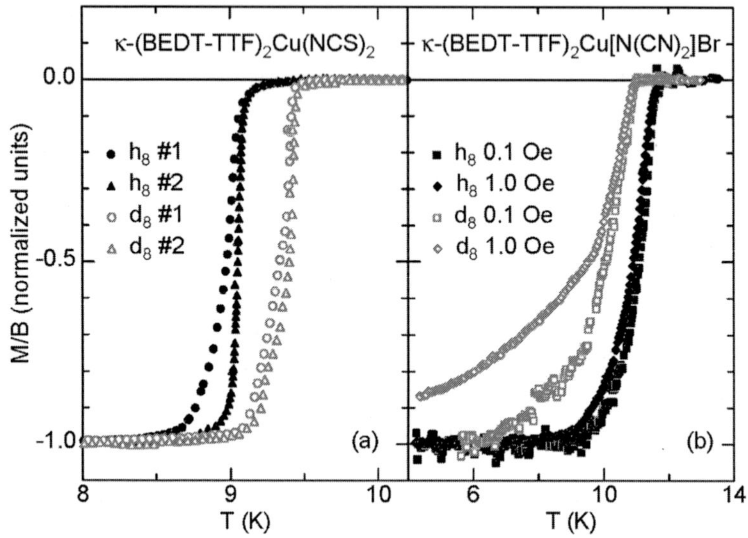


**Figure 7.** Magnetization curves for (**a**) two protonated ($h_8$) and two deuterated ($d_8$) $\kappa$-(BEDT-TTF)$_2$Cu(NCS)$_2$ samples and for (**b**) a protonated and deuterated $\kappa$-(BEDT-TTF)$_2$Cu[N(CN)$_2$]Br sample in two different fields (0.1 Oe and 1 Oe) [11].

Additional support for electron-phonon coupling comes from Raman data which showed a clear hardening of the Raman modes below $T_c$ [51] and from inelastic neutron-scattering experiments [52]. In the latter measurement a significant shift of a phonon branch was observed below $T_c$, an effect hardly seen this clear in any other superconductor. Summarizing, there is numerous experimental evidence for electron-phonon-mediated superconductivity, whereas the argument for a pairing mechanism originating in antiferromagnetic fluctuations is largely based on the $P - T$ phase diagram (Figure 6).

Apart from the coupling mechanism the symmetry of the order parameter is a highly controversial issue. The experimental situation still is rather inconclusive and only some aspects shall be presented here. The superconducting gap structure has been examined, e.g., by scanning tunneling spectroscopy hinting at a dx2−y2dx2−y2symmetry [53]. Thermal-conductivity data [54] also point to *d*-wave symmetry; this time, however, of $d_{xy}$-type symmetry. On the other hand, there exist numerous studies pointing towards a complete *s*-wave-like superconducting gap. These issues are discussed in the excellent review by Lang and Müller [5]. For further discussion of this topic see References [4,7,12–15,30].

In NMR measurements of the spin-lattice relaxation rate, $1/T_1$, three groups almost simultaneously reported comparable results [41,42,55], two of which are shown schematically in Figure 8. Compared to the behavior found for a BCS superconductor such as aluminum (shown as the solid line in Figure 8) [56] several marked differences are observed. First, the Hebel–Slichter coherence peak, *i.e.*, the increase of $1/T_1$ just below $T_c$, is absent. One should note here, that although the occurrence of this peak is an important proof for BCS superconductivity, its absence cannot disprove conventional superconductivity as, e.g., evidenced by the missing Hebel–Slichter peak for the BCS superconductors Nb and V [56]. Second, $1/T_1$ vanishes towards lower temperatures with a $T^3$dependence hinting at a node-like structure of the energy gap. For a complete energy gap over the whole Fermi surface, an exponential temperature dependence, such as for aluminum, is expected. Third, the rapid decrease of $1/T_1$ just below $T_c$ suggests a very fast opening of the gap, that is much faster than usually realized. One should note, however, that all NMR experiments were performed in applied magnetic fields, that might influence the gap structure.

Another powerful measure for deciding whether the superconducting energy gap is nodeless is the specific heat. If the electronic part of the specific heat, $C_e$, vanishes exponentially in the superconducting state, nodes in the energy gap can be ruled out unambiguously. A power-law in the *T* dependence of $C_e$, however, indicates that line or point nodes may be present. In particular, for a *d*-wave symmetry of the gap, a $T^2$ behavior is expected.

Recent unpublished specific-heat data for $\beta''$-(BEDT-TTF)$_2$SF$_5$CH$_2$CF$_2$SO$_3$ in zero field and with 10 T applied perpendicular to the ET planes are shown in Figure 9a on a double-logarithmic scale. These data favorably agree with results of an earlier experiment [57]. Due to the strong phonon contribution to the specific heat the anomaly at $T_c$ = 4:3 K (arrow) is hardly visible in Figure 9a. In the magnetic field of 10 T applied along the *c* direction, the sample is in the normal state.

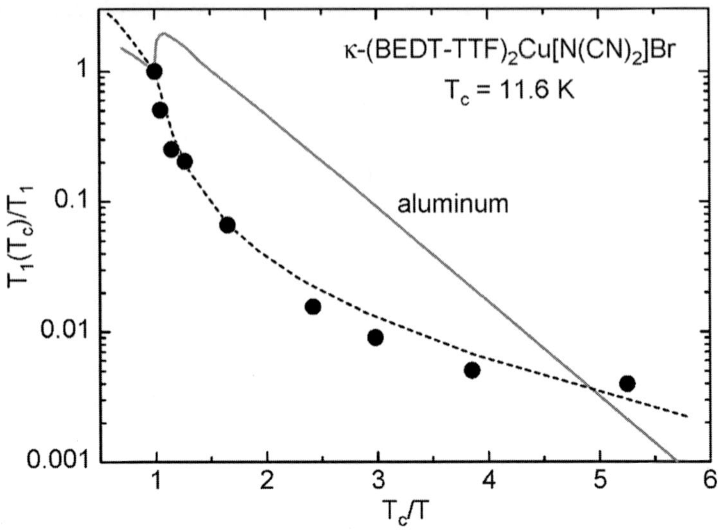


**Figure 8.** Temperature dependence of the normalized spin-lattice relaxation rate. The solid line depicts the behavior for the BCS superconductor aluminum [56]. Data from $^{13}C$ NMR for $\kappa$-(BEDT-TTF)$_2$Cu[N(CN)$_2$]Br in fields parallel to the conducting ET planes are shown as closed circles [41] and dashed line [42]. There is no Hebel–Slichter peak just below $T_c$ and the relaxation rate decreases as $T^3$ rather then exponentially.

The specific-heat difference, $\Delta C$, between the superconducting (0 T) and normal (10 T) state is shown in Figure 9b. In this plot, the anomaly at $T_c$ now is well resolved. It is, however, rather broad due to fluctuation effects. The idealized specific-heat jump at $T_c$ and the temperature dependence of $\Delta C(T)$ does not follow the weak-coupling BCS behavior (dashed line in Figure 9b) [58]. However, the data can be well described when assuming a moderately strong coupling (solid line). Thereby, a BCS-like temperature dependence of the energy gap $\Delta C(T)$ scaled by an appropriate parameter has been assumed [59]. A reasonable description of the data is achieved for the gap ratio $\alpha = \Delta(0)/k_B T_c = 2{:}18$, where $\alpha = 1{:}76$ in the weak-coupling limit.

The inset of Figure 9 shows the exponential vanishing of $C_e \propto \exp(-1{:}9T_c/T\ )$ without any indication of a residual electronic contribution in the superconducting state. Such an exponential dependence has as well been found for three other organic superconductors [40,60,61] in line with reports by other groups

[62,63]. These specific-heat results, therefore, prove the existence of a complete superconducting gap and are strictly against any unconventional nodal order parameter. It is, however, fair to say, that these measurements are not sensitive to a particular direction in *k* space and, therefore, cannot resolve any possible anisotropy of the superconducting gap.

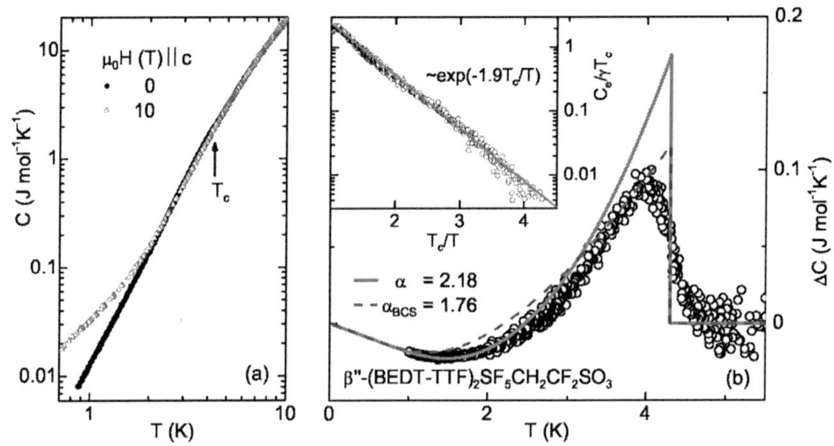


**Figure 9.** (**a**) Specific heat of $\beta$-(BEDT-TTF)$_2$SF$_5$CH$_2$CF$_2$SO$_3$ as a function of temperature in a double-logarithmic scale. Data in the superconducting state in zero field and in the normal state at 10 T applied perpendicular to the BEDT-TTF planes are shown; (**b**) Temperature dependence of the specific-heat difference between the superconducting and normal state. The lines show the BCS behavior for weak coupling (dashed) [58] and strong coupling (solid) [59]. The inset shows the electronic part of the specific heat, $C_e$, divided by $\gamma T_c$ as function of $T_c/T$. The solid line shows the exponential vanishing of $C_e$ towards low T.

Recently, high-resolution specific-heat measurements for two $\kappa$-phase superconductors were reported that indicated $T^2$ dependences of the electronic contributions at low temperatures [64]. Such a dependence would be in favor for a *d*-wave gap scenario. However, in their analysis the authors allowed for different Sommerfeld coefficients, $\gamma$, in the superconducting state than extracted from the normal-state data measured in 14 T [64]. In particular, for the *s*-wave description of the data for $\kappa$-(BEDT-TTF)$_2$ Cu[N(CN)$_2$]Br, $\gamma$ = 20:0 mJ/(mol K$^2$) was used that is 40% smaller than the value in the normal state. This discrepancy is even larger (about 53%) for κ-$(BEDT\text{-}TTF)_2Cu(NCS)_2$. Also for the *d*-wave description, $\gamma$ values smaller than in the normal state were used. Consequently, the analysis in Reference [64] is highly questionable and cannot be taken as evidence for unconventional pairing.

# 4. CONCLUSIONS

With this review, I introduced some selected topics of the rich physics occurring in the quasi-two-dimensional organic metals. Thereby, I laid emphasis on the superconducting properties of these highly anisotropic materials. With their simple electronic structure and the easily accessible complete superconducting phase diagram these charge-transfer salts serve as model systems for 2D superconductors. The observation of the long-time predicted FFLO state allows for a better understanding of superconductivity in general, not only at high magnetic fields. The question on the nature of the superconducting state is still unsettled for the 2D organic metals. Controversial experimental results prevent an unequivocal assignment of the order-parameter symmetry and the coupling mechanism of the Cooper pairs. Some data suggest rather an unconventional *d*-wave-like behavior, whereas other experiments—especially specific-heat results—explicitly prove the existence of a nodeless superconducting energy gap. Further studies, particular experiments that are able to directly probe phase differences of the superconducting wave function, are highly desirable.

# ACKNOWLEDGMENTS

I would like to thank all my coworkers with which I had the luck and honor to cooperate on organic superconductivity over the years, especially the collaborators of the recent years: B. Bergk, R. Beyer, R. Lortz, J.A. Schlueter, and G. Zwicknagl.

# REFERENCES


1. Jérome, D.; Mazaud, A.; Ribault, M.; Bechgaard, K. Superconductivity in a synthetic organic conductor $(TMTSF)_2PF_6$. *J. Phys. Lett.* **1980**, *41*, L95–L98.

2. Ishiguro, T.; Yamaji, K.; Saito, G. *Organic Superconductors*; Springer: Berlin, Heidelberg/Germany, 1998.

3. Jérome, D. Organic conductors: From charge density wave TTF-TCNQ to superconducting $(TMTSF)_2$ $PF_6$. *Chem. Rev.* **2004**, *104*, 5565–5591.

4. Maple, M.B.; Bauer, E.D.; Zapf, V.S.; Wosnitza, J. Unconventional Superconductivity in Novel Materials. In*Superconductivity*; Bennemann, K.H., Ketterson, J.B., Eds.; Springer: Berlin/Heidelberg, Germany, 2008; Volume I, pp. 639–762.

5. Lang, M.; Müller, J. Organic Superconductors. In *Superconductivity*; Bennemann, K.H., Ketterson, J.B., Eds.; Springer: Berlin/Heidelberg, Germany, 2008; Volume II, pp. 1155–1223.

6. Kobayashi, H.; Cui, H.; Kobayashi, A. Organic metals and superconductors based on BETS (BETS =*Bis*(ethylenedithio) tetraselenafulvalene). *Chem. Rev.* **2004**, *104*, 5265–5288.

7. Lang, M. Quasi-two-dimensional organic superconductors. *Supercond. Rev.* **1996**, *2*, 1–115.

8. McKenzie, R.H. Similarities between organic and cuprate superconductors. *Science* **1997**, *278*, 820–821.

9. McKenzie, R.H. A strongly correlated electron model for the layered organic superconductors $\kappa$-$(BEDT\text{-}TTF)_2X$. *Comments Cond. Mat. Phys.* **1998**, *18*, 309–337.

10. Wosnitza, J. Superconducting properties of quasi-two-dimensional organic metals. *Physica C* **1999**, *317–318*, 98–107.

11. Wosnitza, J. Perspectives on the nature of superconductivity in organic metals. *J. Low Temp. Phys.* **1999**, *117*, 1701–1710.

12. Wosnitza, J. Superconductivity in Low-Dimensional Organic Metals. In *Studies of High-Temperature Superconductivity*; Narlikar, A.V., Ed.; Nova Science Publishers: Huntington, VA, USA, 2000; Volume 34, pp. 97–131.

13. Wosnitza, J. Recent aspects of organic metals and superconductors. *Curr. Opin. Solid State Mater. Sci.* **2001**,*5*, 131–141.

14. Singleton, J.; Mielke, C. Quasi-two-dimensional organic superconductors: A review. *Contemp. Phys.* **2002**,*43*, 63–96.

15. Wosnitza, J. Quasi-two-dimensional organic superconductors. *J. Low Temp. Phys.* **2007**, *146*, 641–667.

16. Williams, J.M.; Ferraro, J.R.; Thorn, R.J.; Carlson, K.D.; Geiser, U.; Wang, H.H.; Kini, A.M.; Whangbo, M.H.*Organic Superconductors: Synthesis, Structure, Properties, and Theory*; Prentice Hall: Englewood Cliffs, NJ, USA, 1992.

17. Wosnitza, J. *Fermi Surfaces of Low-Dimensional Organic Metals and Superconductors*; Springer: Berlin/Heidelberg, Germany, 1996.

18. Mori, T.; Kobayashi, A.; Sasaki, Y.; Kobayashi, H.; Saito, G.; Inokuchi, H. Band structures of two types of $(BEDT\text{-}TTF)_2I_3$. *Chem. Lett.* **1984**, *6*, 957–960.

19. Kartsovnik, M.V. High magnetic fields: A tool for studying electronic properties of layered organic metals.*Chem. Rev.* **2004**, *104*, 5737–5782.

20. Kartsovnik, M.V.; Laukhin, V.N.; Pesotskii, S.I.; Schegolev, I.F.; Yakovenko, V.M. Angular magnetoresistance oscillations and the shape of the Fermi surface in $\beta$-$(ET)_2IBr_2$. *J. Phys. I France* **1992**, *2*, 89–99.

21. Kobayashi, A.; Kato, R.; Kobayashi, H.; Moriyama, S.; Nishio, Y.; Kajita, K.; Sasaki, W. Crystal and electronic structures of a new

molecular superconductor, κ-(BEDT-TTF)$_2$I$_3$. *Chem. Lett.* **1987**, *6*, 459–462.

22. Oshima, K.; Mori, T.; Inokuchi, H.; Urayama, H.; Yamochi, H.; Saito, G. Shubnikov-de Haas effect and the Fermi surface in an ambient-pressure organic superconductor [*bis*(ethylenedithiolo) tetrathiafulvalene]$_2$Cu(NCS)$_2$. *Phys. Rev. B* **1988**, *38*, 938–941.

23. Mori, H.; Tanaka, S.; Oshima, M.; Saito, G.; Mori, T.; Murayama, Y.; Inokuchi, H. Crystal and electronic structures of (BEDT–TTF)$_2$[MHg(SCN)$_4$] (M = K and NH$_4$). *Bull. Chem. Soc. Jpn.* **1990**, *63*, 2183–2190.

24. Kartsovnik, M.V.; Andres, D.; Biberacher, W. Hierarchy of the density-wave states and superconductivity in the organic conductor α-(BEDT-TTF)$_2$KHg(SCN)$_4$. *C. R. Chim.* **2007**, *10*, 6–14.

25. Beckmann, D.; Wanka, S.; Wosnitza, J.; Schlueter, J.A.; Williams, J.M.; Nixon, P.G.; Winter, R.W.; Gard, G.L.; Ren, J.; Whangbo, M.-H. Characterization of the Fermi surface of the organic superconductor β″-(ET)$_2$SF$_5$CH$_2$CF$_2$SO$_3$ by measurements of Shubnikov-de Haas and angle-dependent magnetoresistance oscillations and by electronic band-structure calculations. *Eur. Phys. J. B* **1998**, *1*, 295–300.

26. Chandrasekhar, B.S. A note on the maximum critical field of high-field superconductors. *Appl. Phys. Lett.* **1962**, *1*, 7–8.

27. Clogston, A.M. Upper limit for the critical field in hard superconductors. *Phys. Rev. Lett.* **1962**, *9*, 266–267.

28. Lang, M.; Steglich, F.; Toyota, N.; Sasaki, T. Fluctuation effects and mixed-state properties of the layered organic superconductors κ-(BEDT-TTF)$_2$Cu(NCS)$_2$ and κ-(BEDT-TTF)$_2$Cu[N(CN)$_2$]Br. *Phys. Rev. B* **1994**, *49*, 15227–15234.

29. Hagel, J.; Wanka, S.; Beckmann, D.; Wosnitza, J.; Schweitzer, D.; Strunz, W.; Thumfart, M. Vortex dynamics of the organic superconductor κ-(ET)$_2$Cu[N(CN)$_2$]Br. *Physica C* **1997**, *291*, 213–222.

30. Wosnitza, J. Superconducting properties of quasi-two-dimensional organic metals. *Physica C* **1999**, *317–318*, 98–107.

31. Wanka, S.; Beckmann, D.; Wosnitza, J.; Balthes, E.; Schweitzer, D.; Strunz, W.; Keller, H.J. Critical fields and mixed-state properties of the layered organic superconductor $\kappa$-(BEDT-TTF)$_2$I$_3$. *Phys. Rev. B* **1996**, *49*, 9301–9309.

32. Müller, P. Intrinsic Josephson Effects in Layered Superconductors. In *Advances in Solid State Physics*; Helbig, R., Ed.; Vieweg: Braunschweig, Germany, 1994; Volume 34, pp. 1–33.

33. Fulde, P.; Ferrell, R.A. Superconductivity in a strong spin exchange field. *Phys. Rev.* **1964**, *135*, A550–A563.

34. Larkin, A.I.; Ovchinnikov, Y.N. Inhomogeneous state of superconductors. *Zh. Eksp. Teor. Fiz.* **1964**, *47*, 1136–1146. *Sov. Phys. JETP.* **1965**, *20*, 762–768.

35. Maki, K.; Tsuneto, T. Pauli paramagnetism and superconducting state. *Prog. Theor. Phys.* **1964**, *31*, 945–956.

36. Gruenberg, L.W.; Gunther, L. Fulde-Ferrell effect in type-II superconductors. *Phys. Rev. Lett.* **1966**, *16*, 996–998.

37. Zwicknagl, G.; Wosnitza, J. Breaking Translational Invariance by Population Imbalance: The Fulde-Ferrell-Larkin-Ovchinnikov States. In *BCS: 50 years*; Cooper, L.N., Feldman, D., Eds.; World Scientific: Singapore, 2010; pp. 337–371.

38. Lortz, R.; Wang, Y.; Demuer, A.; Böttger, P.H.M.; Bergk, B.; Zwicknagl, G.; Nakazawa, Y.; Wosnitza, J. Calorimetric evidence for a Fulde-Ferrell-Larkin-Ovchinnikov superconducting state in the layered organic superconductor $\kappa$-(BEDT-TTF)$_2$Cu(NCS)$_2$. *Phys. Rev. Lett.* **2007**, *99*, 187002:1–187002:4.

39. Bergk, B.; Demuer, A.; Sheikin, I.; Wang, Y.; Wosnitza, J.; Nakazawa, Y.; Lortz, R. Magnetic torque evidence for the Fulde-Ferrell-Larkin-Ovchinnikov state in the layered organic superconductor $\kappa$-(BEDT-TTF)$_2$Cu(NCS)$_2$. *Phys. Rev. B* **2011**, *83*, 064506:1–064506:7.

40. Wosnitza, J.; Wanka, S.; Hagel, J.; Reibelt, M.; Schweitzer, D.; Schlueter, J.A. Thermodynamic properties of quasi-two-dimensional organic superconductors. *Synth. Met.* **2003**, *133–134*, 201–203.

41. Soto, S.M.D.; Slichter, C.P.; Kini, A.M.; Wang, H.H.; Geiser, U.; Williams, J.M. $^{13}C$ NMR studies of the normal and superconducting states of the organic superconductor $\kappa$-$(ET)_2Cu[N(CN)_2]Br$. *Phys. Rev. B* **1995**,*52*, 10364–10368.

42. Mayaffre, H.; Wzietek, P.; Jérome, D.; Lenoir, C.; Batail, P. Superconducting state of $\kappa$-$(ET)_2Cu[N(CN)_2]Br$ studied by $^{13}C$ NMR: Evidence for vortex-core-induced nuclear relaxation and unconventional pairing.*Phys. Rev. Lett.* **1995**, *75*, 4122–4125.

43. Wzietek, P.; Mayaffre, H.; Jérome, D.; Brazovskii, S. NMR in the 2D organic superconductors. *J. Phys. I France* **1996**, *6*, 2011–2041.

44. Williams, J.M.; Kini, A.M.; Wang, H.H.; Carlson, K.D.; Geiser, U.; Montgomery, L.K.; Pyrka, G.J.; Watkins, D.M.; Kommers, J.M.; Boryschuk, S.J.; *et al.* From semiconductor-semiconductor transition (42 K) to the highest-$T_c$ organic superconductor, $\kappa$-(ET)2 Cu[N(CN)2]Cl ($T_c$ = 12: 5 K). *Inorg. Chem.* **1990**, *29*, 3272–3274.

45. Sadewasser, S.; Looney, C.; Schilling, J.S.; Schlueter, J.A.; Williams, J.M.; Nixon, P.G.; Winter, R.W.; Gard, G.L. Dependence of $T_c$ on hydrostatic pressure in $\beta''$-$(ET)_2SF_5CH_2CF_2SO_3$ and $\kappa$-$(ET)_2Cu(NCS)_2$. *Solid State Commun* **1997**, *104*, 571–575.

46. Kanoda, K. Recent progress in NMR studies on organic conductors. *Hyperfine Interact* **1997**, *104*, 235–249.

47. Müller, J.; Lang, M.; Steglich, F.; Schlueter, J.A.; Kini, A.M.; Geiser, U.; Mohtasham, J.; Winter, R.W.; Gard, G.L.; Sasaki, T.; *et al.* Comparative thermal-expansion study of $\beta''$-$(ET)_2SF_5CH_2CF_2SO_3$ and $\kappa$-$(ET)_2Cu(NCS)_2$: Uniaxial pressure coefficients of $T_c$ and upper critical fields. *Phys. Rev. B* **2000**, *61*, 11739–11734.

48. Kini, A.M.; Schlueter, J.A.; Ward, B.H.; Geiser, U.; Wang, H.H. On the "inverse" isotope effect in organic superconductors: New findings and implications. *Synth. Met.* **2001**, *120*, 713–714.

49. Kawamoto, A.; Miyagawa, K.; Kanoda, K. Deuterated κ-(BEDT-TTF)$_2$ Cu[N(CN)$_2$]Br: A system on the border of the superconductor-magnetic-insulator transition. *Phys. Rev. B* **1997**, *55*, 14140–14143.

50. Kini, A.M.; Carlson, K.D.; Wang, H.H.; Schlueter, J.A.; Dudek, J.D.; Sirchio, S.A.; Geiser, U.; Lykke, K.R.; Williams, J.M. Determination of a mass isotope effect on $T_c$ in an electron-donor-based organic superconductor, κ-(ET)$_2$Cu(NCS)$_2$, where ET represents bis(ethylenedithio)tetrathiafulvalene. *Physica C* **1996**, *264*, 81–94.

51. Pedron, D.; Visentini, G.; Bozio, R.; Williams, J.M.; Schlueter, J.A. Phonon dynamics and superconductivity in the organic crystal κ-(BEDT-TTF)$_2$Cu[N(CN)$_2$]Br. *Physica C* **1997**, *276*, 1–8.

52. Pintschovius, L.; Rietschel, H.; Sasaki, T.; Mori, H.; Tanaka, S.; Toyota, N.; Lang, M.; Steglich, F. Observation of superconductivity-induced phonon frequency changes in the organic superconductor κ-(BEDT-TTF)$_2$Cu(NCS)$_2$. *Europhys. Lett.* **1997**, *37*, 627–632.

53. Arai, T.; Ichimura, K.; Nomura, K.; Takasaki, S.; Yamada, J.; Nakatsuji, S.; Anzai, H. Tunneling spectroscopy on the organic superconductor κ-(BEDT-TTF)$_2$Cu(NCS)$_2$ using STM. *Phys. Rev. B* **2001**, *63*, 104518:1–104518:5.

54. Izawa, K.; Yamaguchi, H.; Sasaki, T.; Matsuda, Y. Superconducting gap structure of κ-(BEDT-TTF)$_2$Cu(NCS)$_2$ probed by thermal conductivity tensor. *Phys. Rev. Lett.* **2002**, *88*, 027002:1–027002:4.

55. Kanoda, K.; Miyagawa, K.; Kawamoto, A.; Nakazawa, Y. NMR relaxation rate in the superconducting state of the organic conductor κ-(BEDT-TTF)$_2$Cu[N(CN)$_2$]Br. *Phys. Rev. B* **1996**, *54*, 76–79.

56. Ginsberg, D.M.; Hebel, L.C. The Green's Function Method. In *Superconductivity*; Parks, R., Ed.; Marcel Dekker: New York, NY, USA, 1969; Volume 1, pp. 193–258.

57. Wanka, S.; Hagel, J.; Beckmann, D.; Wosnitza, J.; Schlueter, J.A.; Williams, J.M.; Nixon, P.G.; Winter, R.W.; Gard, G.L. Specific heat and critical fields of the organic superconductor $\beta''$-(BEDT-TTF)$_2$SF$_5$CH$_2$CF$_2$SO$_3$.*Phys. Rev. B* **1998**, *57*, 3084–3088.

58. Mühlschlegel, B. Die thermodynamischen Funktionen des Supraleiters. *Z. Phys.* **1959**, *155*, 313–327.

59. Padamsee, H.; Neighbor, J.E.; Schiffman, C.A. Quasiparticle phenomenology for thermodynamics of strong-coupling superconductors. *J. Low Temp. Phys.* **1973**, *12*, 387–411.

60. Wosnitza, J.; Liu, X.; Schweitzer, D.; Keller, H.J. Specific heat of the organic superconductor $\kappa$-(BEDT-TTF)$_2$I$_3$. *Phys. Rev. B* **1994**, *50*, 12747–12751.

61. Elsinger, H.; Wosnitza, J.; Wanka, S.; Hagel, J.; Schweitzer, D.; Strunz, W. $\kappa$-(BEDT-TTF)$_2$Cu[N(CN)$_2$]Br: A fully gapped strong-coupling superconductor. *Phys. Rev. Lett.* **2000**, *84*, 6098–6101.

62. Kovalev, A.E.; Ishiguro, T.; Yamada, J.; Takasaki, S.; Anzai, H. The anisotropic low-temperature specific heat of organic superconductor $\kappa$-(BEDT-TTF)$_2$Cu(NCS)$_2$ in magnetic field. *J. Exp. Theor. Phys.* **2001**, *92*, 1035–1037.

63. Müller, J.; Lang, M.; Helfrich, R.; Steglich, F.; Sasaki, T. High-resolution ac-calorimetry studies of the quasi-two-dimensional organic superconductor $\kappa$-(BEDT-TTF)$_2$Cu(NCS)$_2$. *Phys. Rev. B* **2002**, *65*, 140509:1–140509:4.

64. Taylor, O.J.; Carrington, A.; Schlueter, J.A. Specific-Heat Measurements of the Gap Structure of the Organic Superconductors $\kappa$-(ET)$_2$Cu[N(CN)$_2$]Br and $\kappa$-(ET)$_2$Cu(NCS)$_2$. *Phys. Rev. Lett.* **2007**, *99*, 057001:1–057001:4.

2

# MODEL OF PREFORMED HOLE-PAIRS IN CUPRATE SUPERCONDUCTORS

**R. J. Singh**

Physics Department, Aligarh Muslim University, Aligarh, India

## ABSTRACT

A model of preformed hole-pairs in cuprate superconductors has been proposed based on some experimental results i.e., 1) electron paramagnetic resonance spectra of quenched superconductors which show very frequently the fragment $(CuO)_4$ broken off from the $CuO_2$ layer in the structure, 2) 41 meV peak observed in neutron diffraction and nuclear magnetic resonance spectra of superconductors, 3) Heisenberg exchange interaction leading to ferromagnetism observed in CuO which is an essential ingredient of all superconductors and some generally accepted conclusions i.e., a) that the order parameter in superconductors has $d_{x^2-y^2}$ symmetry and b) coherence length is of the order of 15 - 20 Angstrom. Heisenberg exchange interaction between two $(CuO_4)$ plaquettes each containing a lattice hole binds the two holes which are the charge carriers in the cuprate superconductors. It is not very clear whether the hole-pair is in the triplet or singlet state, but the triplet state is supported by the experimental observation of ferromagnetism in the parent material CuO. The proposed hole-pair singlet is different from Zhang-Rice singlet.

## KEYWORDS

High Temperature Superconductor, Electron Paramagnetic Resonance, Preformed Hole-Pairs


## 1. INTRODUCTION

Various mechanisms have been proposed for high temperature superconductivity (HTS), but as yet no consensus has been reached. People have started thinking away from Cooper pairs in BCS theory. New thinking is that hole-pairs are formed even before the transition temperature (Tc) and they condense at Tc. The mechanism of formation of hole-pairs is expected to be quite different from that of Cooper pairs. A model of hole-pairs in cuprate superconductors has already appeared in the reference [1]. But the present model is entirely different. In the reference [1], the pairing between two holes has been shown to be due to the existence of medium of negative dielectric constant between the two holes. In the present model, pairing between two holes has been shown to be due to "exchange interaction" between the two holes. Several mechanisms have been proposed for HTS, but there is no consensus on any mechanism. This led the author to propose yet another mechanism of HTS. Out of the many mechanisms proposed, one will stand out and get universal acceptance. The paper in the reference [1] and the present paper are based on the same set of experiments carried over many years. There is bound to be some overlap in the description of the experimental data. But the hearts of the solutions of the problem of HTS are totally different.

The basic idea of this model has been derived from some experimental results as 1) electron paramagnetic resonance (EPR) spectra of quenched superconductors which very frequently show $(CuO)_4$ fragments broken off from the active $CuO_2$ layer in the cuprates, 2) 41 meV peak observed in neutron diffraction and nuclear magnetic resonance (NMR) spectra of superconductors and 3) Heisenberg exchange interaction leading to ferromagnetism (FM) in CuO in nanoforms, sometimes even up to room temperature (RT) and some generally accepted conclusions as a) that the order-parameter of cuprates has $d_{x^2-y^2}$ symmetry and b) that the coherence length is of the order of 15 - 20 Angstrom (A).

HTS are EPR silent because $Cu^{2+}$ ions are coupled antiferromagnetically. When cuprates are quenched or deoxygenated, some oxygen ions are ejected from the crystal or are displaced from their usual positions, producing small fragments magnetically isolated from the bulk, such as Cu-tetramers and also monomers, dimers and octamers which have been detected by EPR. Besides Cu-monomer ($Cu^{2+}$ or CuO), Cutetramer $(CuO)_4$ is the most frequently occurring species which is probably due to the fact that $(CuO)_4$ is the unit cell of the twodimensional $CuO_2$ sheet in the HTS compounds and also in the constituents of superconductors as $BaCuO_2$, $CaCuO_2$, $SrCuO_2$, $Bi_2CuO_4$ etc. When HTS materials are heated up to temperatures a little higher than the temperatures of their preparations and then suddenly quenched, the $CuO_2$ sheet is most likely to be broken into pieces containing unit cells of the sheet. In this process, many $(CuO)_4$ units may be further broken giving Cu-monomers. The frequent occurrence of $(CuO)_4$ spectra seems to have special bearing with the mechanism of superconductivity in cuprates.

In the **Figure 1**(a) is shown a fully connected $(CuO)_4$ plaquettes and in **Figure 1** (1(b)) an isolated $(CuO)_4$ plaquette whose all the oxygen bonds with $Cu^{2+}$ neighbours have been broken . In the reference [2], it has been shown that the Cu-O bond in HTS is nearly 87.5 % ionic and 12.5% covalent. If eight neighbouring bonds are broken, it means loss of one electron ($12.5 \times 8 = 100\%$) or advent of one hole in the remaining $(CuO)_4$ unit because oxygen is more electronegative than copper. It suggests that if a hole enters the $(CuO)_4$ unit of continuous $CuO_2$ sheet, it becomes equivalent to an isolated $(CuO)_4$ entity and should have the same properties as an isolated $(CuO)_4$entity which are obtained on quenching of HTS materials.

We have studied EPR spectra of nearly all cuprate superconductors [3-10] and their constituents as $BaCuO_2$, $CaCuO_2$, $SrCuO_2$, $BiCuO_4$ etc. after quenching and in each case obtained and analyzed the spectra of $(CuO)_4$ and also other fragments. Similar spectra in cuprates have been obtained by other workers [11-16] also though sometimes with different interpretations.

In the EPR spectra of isolated $(CuO)_4$ entities, four fine structure components were observed showing that the total electronic spin = 2, obtained due to FM coupling of the four spins of four $Cu^{2+}$ ions of $(CuO)_4$ (electronic spin of $Cu^{2+}$ = 1/2) and in some cases, each fine structure component showed further splitting into thirteen hyperfine components , showing the FM coupling of nuclear spins of all the four $Cu^{2+}$ ions giving total nuclear spin = 6 (nuclear spin of Cu = 3/2). Fine structure splitting will take place due to crystal field effects of the surrounding ions on the isolated Cu-tetramers. Coupling of nuclear spins will be possible if all the four holes of the four $Cu^{2+}$ ions circulate around the $(CuO)_4$ framework with a frequency larger than $1/\gamma H_{hfs}$ where $\gamma$ = gyromagnetic ratio and $H_{hfs}$= hyperfine splitting. The expected path of circulation of holes has been shown in the **Figure 2**(a). The circulation of the four holes of the four $Cu^{2+}$ ions around the periphery of $(CuO)_4$ framework (**Figure 2**(a)) will be due to the advent of a hole wandering in the $CuO_2$ sheet in a $(CuO)_4$plaquette which will hop among its oxygen sites· Hopping of the hole among the different oxygen sites will force the holes of $Cu^{2+}$ ions also to move and circulate in the $(CuO)_4$ framework due to electric repulsion between the incoming hole and the $Cu^{2+}$ holes. Electronic spins of the four holes of the four $Cu^{2+}$ ions in the isolated $(CuO)_4$ units order themselves in FM arrangement because the incoming hole hopping among oxygen sites places them in identical conditions and identical situation will be obtained when their spins are aligned in the same direction or are FM ordered. FM ordering of electronic spins of all the four $Cu^{2+}$ holes and their identical motion in the $(CuO)_4$ framework will create identical situation for the nuclear spins of the four Cu-nuclei and hence they will also be pointing in the same direction. Thus electronic spins and also nuclear spins of the four $Cu^{2+}$ ions will be FM aligned.

From above we can expect near equivalence between an isolated $(CuO)_4$ and a $(CuO)_4$ unit in a continuous $CuO_2$ layer in which a hole has entered or the following equation may hold good: an isolated $(CuO)_4$ plaquette = a $(CuO)_4$ plaquette of the continuous $CuO_2$ sheet + a hole inside it.

Based on this equivalence and some experimental results and some generally accepted conclusions enumerated earlier, a model of preformed hole-pairs in cuprate superconductors has been proposed.

The main difference in the situations when a hole enters an isolated $(CuO)_4$ unit and a $(CuO)_4$ unit in a continuous $CuO_2$ sheet is the following. In the case of isolated $(CuO)_4$ unit , all kinds of motion of holes and spin alignments are confined to itself because there is practically no connection with the outside world. In the case of a hole entering the $(CuO)_4$ unit of a continuous $CuO_2$ sheet; as soon as a hole arrives in a $(CuO)_4$area, the spins start arranging themselves in FM alignment and at the same time they start a train of AFM coupling with the neighbouring spins because HTS are basically doped antiferromagnets. A train of AFM coupling may give rise to the formation of spin stripes. When hole-pairs are formed, the AFM coupling with the outside spins will not take place. Because the two holes are bound by exchange forces and remain confined in the $d_{x2-y2}$ orbital.

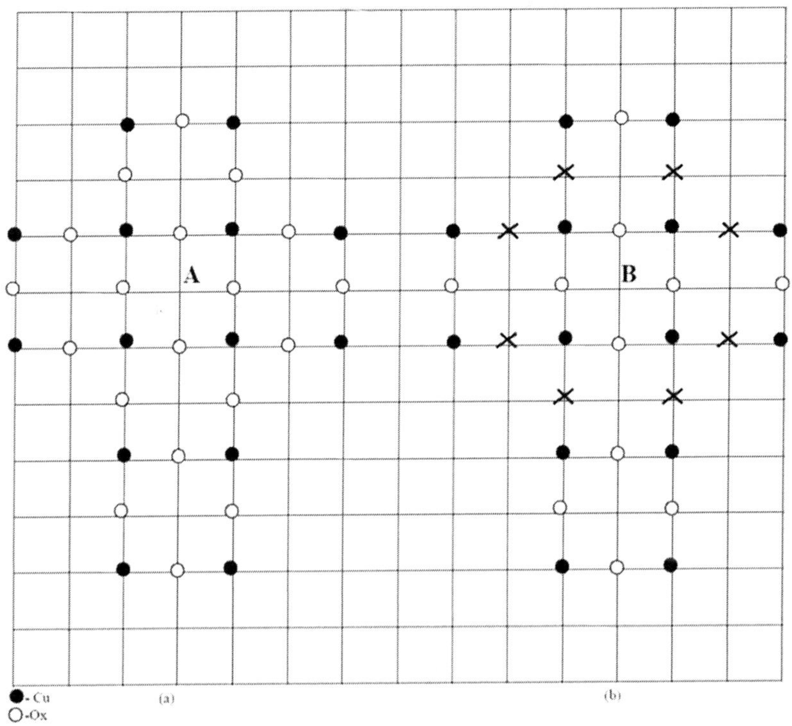


**Figure 1**. (a) An unit cell of $CuO_2$ plane connected on all sides. (b) An unit cell of $CuO_2$ plane whose eight neighbouring bonds have been broken by removal of eight oxygen ions (shown by crosses).

Before proposing the model, let us examine what happens in a $(CuO)_4$ unit in the continuous $CuO_2$sheet when a hole lands into its area. Because of its equivalence with the isolated $(CuO)_4$ unit, as soon as a hole enters a $(CuO)_4$ unit, the spins of the holes of all the four $Cu^{2+}$ ions would be set up in the process of FM alignment. But the question is what will be the direction of spin alignment? The unbiased or preferential direction will be perpendicular to the $CuO_2$ plane. If the alignment is vertically upward of the plane, there is equal probability of vertically downward direction. As a compromise, the aligned spins will oscillate between these two directions. But with what frequency? We associate this frequency with 41 meV peak [17]. 41 meV when expressed in the form of hν, ν comes to be approximately equal to $10^{13}$ hertz or its time-period $\approx 10^{-13}$ sec. So the frequency of oscillation of spin vector $\approx 10^{13}$ hertz. Indirect support for the alignment of spin vector in a direction perpendicular to $CuO_2$ plane comes from the reference [18] in which it has been shown that modest magnetic fields suppress the resonance in the magnetic excitation spectrum of $YBa_2Cu_3O_{6+x}$ significantly when perpendicular to $CuO_2$ plane than for parallel fields. When a hole marches through a column or row of $(CuO)_4$ plaquettes, the process of phase alignment of spin vectors of the four holes of the four $Cu^{2+}$ ions in $(CuO)_4$ units follows it. In the cell which we call first, let us assume that the spins have arranged themselves in the U state(i.e., all spin vectors pointing vertically upward) and as the hole advances, a sequence of well defined spin aligned states U (Ms = 2), Uo (Ms = 0), D (Ms = –2), Do (Ms = 0) is obtained. These symbols U, Uo etc. have been explained in the caption of the **Figure 2**(b). Ms represents the combined projection of the four spins of the four $Cu^{2+}$ ions in a $(CuO)_4$ plaquette on Z-axis which is perpendicular to $CuO_2$plane i.e., XY plane. These well defined states will not be obtained in consecutive $(CuO)_4$ units. This sequence of well defined spin states will be found in $(CuO)_4$ units at fixed distances between them. The journey of the hole through one round of these well defined spin states is completed when it traverses a length of ten $(CuO)_4$ units as shown in the Figure (3). It means that the successive well defined spin states will come after the hole crosses two and a half plaquettes [see Figure (3)]. It also means that a hole will take $10^{-14}$sec. to cross a plaquette because ten $(CuO)_4$units are crossed in $10^{-13}$sec which is the time period of spin oscillation.

Another important thing that will happen to $(CuO)_4$ entity is that all the four $Cu^{2+}$ holes have to circulate around the outer boundary of $(CuO)_4$ [see **Figure 2**(a)]. Circulation of $Cu^{2+}$ holes around the boundary of $(CuO)_4$ plaquette is a necessity if coupling of nuclear spins of four copper ions has to be materialized whose signature is provided by EPR spectra. But here again the question is with what frequency? $Cu^{2+}$ holes will circulate around the boundary of $(CuO)_4$ plaquette? For this let us calculate the thermal velocity of holes (mass of hole = mass of electron = m = $9.11 \times 10^{-28}$ gm.) at the temperature of 92 K = Tc of the workhorse $YBa_2Cu_3O_{7-\delta}$. In the twodimensional plane , thermal energy = $2 \times 1/2\,KT$ = KT, because thermal energy per degree of freedom =1/2 KT. Putting numerical values, thermal energy = KT = $1.38 \times 10^{-16} \times 92 = \frac{1}{2} mv^2$ = (kinetic energy of hole). Thus velocity of hole v = $5.28 \times 10^6$ cm/sec = $5.28 \times 10^{14}$ A/sec. The next question is what will be the perimeter of the path through which $Cu^{2+}$ holes will circulate. It will not be simply 3.84x4 A, as the side of the square $(CuO)_4$ unit = 3.84 A [3.84 A is one side of the unit cell of the $YBa_2Cu_3O_{7-\delta}$; the unit cell has been taken as a square]. If so, holes have to pass through the interior of the nucleus which is certainly not possible. They must maintain the same distance from the Cu nuclei as the outermost electron will maintain. According to Bohr theory (taking the reported value of effective Z of Cu = 13.2), the radius of Cu-3d orbit has been calculated to be 0.3615A. So circulating along the path ABCDA shown in the **Figure 2**(a), the holes will keep away from the Cu-nuclei by this distance and circulate along the outer boundary. So the total path length = 4 [3.84 + 2 (0.3615)] = 18.252 A. and frequency of circulation of holes = $(5.28 \times 10^{14}/18.252) \approx 2.9 \times 10^{13}$hertz or time-period = $0.345 \times 10^{-13}$sec.

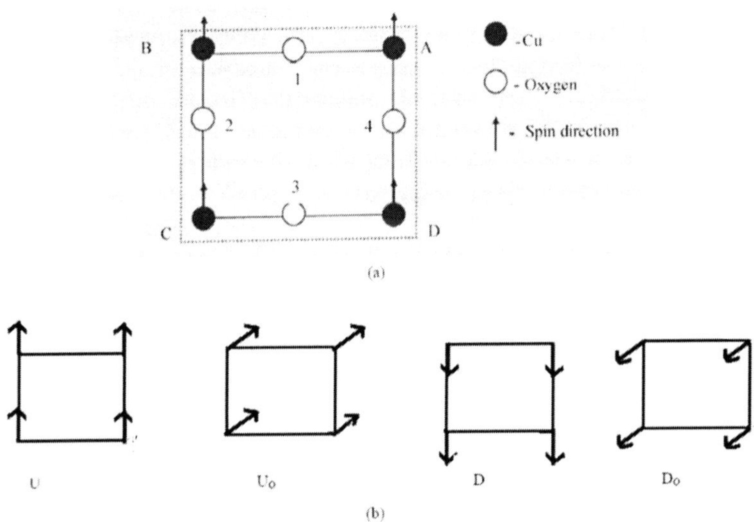


**Figure 2**. (a) Dotted lines show the path of circulation of the four holes of four $Cu^{2+}$ ions of a $(CuO)_4$ plaquette, (b) When all the spins are aligned vertically above the $CuO_2$ plane, then Ms = 2 and the state is denoted by the symbol U; coming back to the plane when spin vectors are in $CuO_2$plane i.e., Ms = 0, the sate is called Uo; again spins aligning vertically below the plane, Ms = –2 and the state is called D; again reversing direction and coming back to the plane,. Ms = 0 and the state is called Do. Resultant spin vector changes angle by 90° in the consecutive states U, Uo, D, Do and then U again.

We are assuming the velocities of the both the holes wandering in the $CuO_2$ plane and the holes of $Cu^{2+}$ to be equal i.e., $5.28 \times 10^{14}$ A/sec. It may not sound very reasonable. One hole is wandering nearly free on the $CuO_2$ sheet and the other is associated with $Cu^{2+}$ ion. But the $Cu^{2+}$ holes while circulating on the outer boundary of $(CuO)_4$ plaquette may be treated as free as the wandering hole. On this argument, the velocities of the both kinds of holes has been taken to be equal.

As we have mentioned earlier, in accordance with the occurrence of the 41 meV peak [17] that the wandering hole spends $10^{-14}$sec in one $(CuO)_4$ plaquette and the $Cu^{2+}$ hole circulating on the boundary of the $(CuO)_4$ unit should also be circulating for the same time when the wandering hole is within the bounds of $(CuO)_4$ unit. We have already assumed that the circulatory motion of $Cu^{2+}$ is induced by the advent of the wandering hole in its area and it persists as long as the wandering hole is within its boundaries. In $10^{-14}$sec, the wandering hole can enter a plaquette, say, at position 1,then jump to position 2 and then exit the plaquette from the position 3 (see the**Figure 2**(a)). In the same time of $10^{-14}$ sec, each $Cu^{2+}$ hole can cover one and a half time the length of a side of the square.

In this kind of circulatory motion, two things are noted, 1) spin magnets are getting closer and farther from the center of the square path ABCDA [see the **Figure 2**(a)] and 2) charged particles are moving in a closed path. None of the $Cu^{2+}$ holes completes one round of the path ABCDA, but if these holes are considered indistinguishable, the closed path has been traversed one and a half time by a hole. Under this condition, magnetic fields should be created by both the causes mentioned just above.

Let us calculate magnetic fields due to both causes. Due to the first cause, the magnetic field produced at the center of the circulatory path = $4\ \mu/r^3$ (where $\mu$ = Bohr magneton = magnetic moment associated with the spin momentum of one electron/hole and r = distance of magnetic moments from the center of the path ABCDA at which magnetic field is being estimated. If all the spin vectors are in phase and pointing vertically upward of the $CuO_2$ plane , each one occupying one corner of the path (see the **Figure 2**(a)), the magnetic field at the center of the path = 1853 gauss. If they are situated at the middle points of the sides of the square path, then the magnetic field = 5241 gauss. But the field direction is changing so fast that (time-period = $10^{-13}$ sec) that the time average value of the magnetic field will be zero. Only on the time-scale of less than $10^{-13}$sec, finite value of magnetic field may be observed. The magnetic field produced by the second cause has been calculated to be 72 gauss (by the standard method that field = $\mu_0/2 \times I/r$, if r is taken as the radius of the circle whose circumference has been taken to be equal to the perimeter of the path ABCDA =18.252 A).

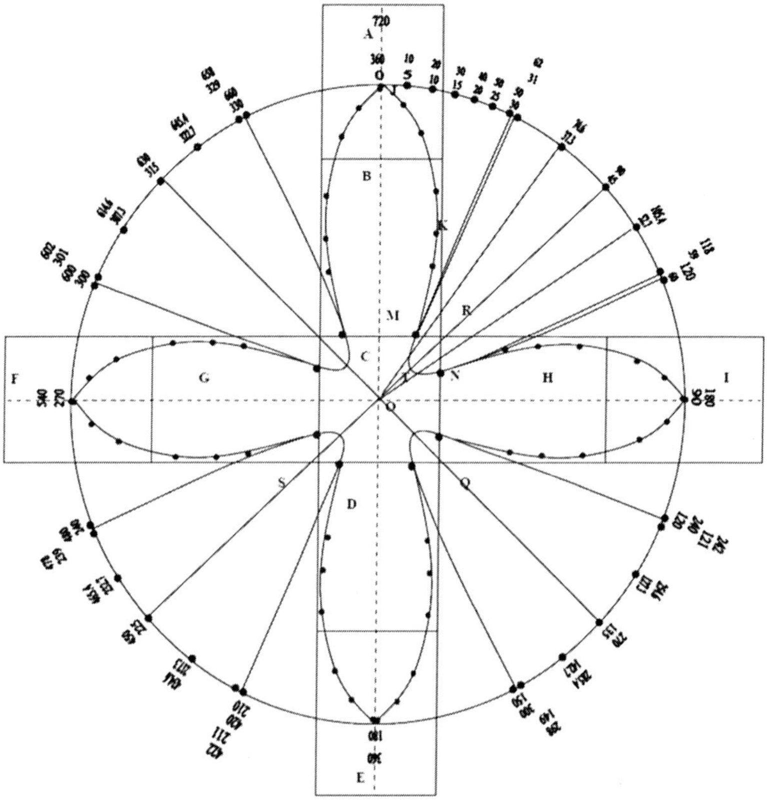


**Figure 3**. Formation of hole-pairs. All the squares A.B.C.D.E.F.G.H.I are the unit cells of the two-dimensional $CuO_2$ plane. Both the holes traverse the same path but at any instant they are on the opposite sides of the central cell.

Hsu et al. [19] argued that due to circulatory current in the $CuO_2$ plane, a real magnetic field should be produced which they tested by muon spin resonance and neutron scattering but were unable to detect any field probably due to fast temporal and spatial change of magnetic field .The magnetic field is of transient nature and its estimation is quite difficult but in the present case where the time scales are of the order of $10^{-13}$ sec, the presence of magnetic field can be experienced by the system.

So far we have shown that when the wandering hole enters a $(CuO)_4$ plaquette it experiences a magnetic field generated by the spin magnetic moments of the holes associated with the $Cu^{2+}$ ions and their circulation around the boundary of $(CuO)_4$ plaquette. We will see further that this magnetic field will guide the motion of the wandering hole in the network of $(CuO)_4$ plaquettes in the $CuO_2$ plane. Next we try to show how the two wandering holes form hole-pairs by virtue of Heisenberg exchange interaction.

## 2. FORMATION OF HOLE-PAIRS

Before discussing formation of hole-pairs, let us be clear about the two kinds of holes of different origins which have entered our discussion. One kind of holes are the set of four holes, each one associated with one of the four $Cu^{2+}$ ($3d^9$) ions which circulate about the periphery of $(CuO)_4$framework as discussed earlier. These holes will be called $Cu^{2+}$ holes. The other kind of holes are those which are always present in the superconductors. In cuprate superconductors, doping is believed to create holes on the oxygen sites in the $CuO_2$ layer. They wander in the $CuO_2$ plane with thermal velocity. They will be called lattice holes.

In analogy with the reference [20], the binding between two holes seems to be through the Heisenberg exchange interaction because conditions of binding in the two cases are quite similar. There are a large number of lattice holes wandering in the two-dimensional $CuO_2$ plane When two holes traveling in the opposite directions in the same column or row .of the $CuO_2$ continuous sheet come at a distance of nearly four $(CuO)_4$ units as shown in the Figure (3) they might feel a mutual force of attraction due to Heisenberg exchange interaction The separation of four $(CuO)_4$ units has been chosen because the coherence length is known to be 15 - 20 A or nearly equal to the length of four $(CuO)_4$ plaquettes kept side by side.

In the reference [20], the strength of Heisenberg exchange interaction for a series of small free radicals in aqueous solution has been measured. In this case, the critical exchange distance lies between one and three times the hard sphere encounter distance. The situation we are imagining in the cuprate superconductors appears to be similar to that of free radicals noted above due to the following similarities; 1) when a lattice hole enters a neutral $(CuO)_4$ unit, it may behave like a free radical., 2) two $(CuO)_4$ plaquettes each with a hole in it are situated three $(CuO)_4$ blocks away [we have taken two $(CuO)_4$ plaquettes each with a hole in it at a distance of three plaquettes in agreement with coherence length], 3) in between the $Cu^{2+}$ plaquettes each containing a hole the space is neutral (plaquettes without holes) as between radicals in aqueous solution. Thus there is enough probability of exchange interaction between the two $(CuO)_4$ plaquettes, each containing one hole. When the two holes proceed further, there will be a separation of only one block, then also the exchange interaction should continue as the radicals in the aqueous solution. Still further when the two holes will enter the same $(CuO)_4$ plaquette, the exchange interaction should not occur as happens in the present case. In the case of free radicals in aqueous solution also, exchange interaction does not take place at a distance less than one hard sphere encounter distance.

Let two lattice holes reach A and E cells three blocks apart simultaneously. At this separation they will experience exchange force. There are chances of two kinds of interactions: a) spins of $Cu^{2+}$holes both in A and E cells are ferromagnetically coupled and b) they are antiferromagnetically coupled. Discussion of both cases (a) and (b) are on similar lines. In the FM coupling, if the $Cu^{2+}$holes in the A cell are in U state, the spins of the $Cu^{2+}$ holes in the cell E will spontaneously attain the U state under the effect of exchange interaction whatever may be their initial configuration. In the case of AFM coupling, the spins of $Cu^{2+}$ holes in A and E cells will be in U and D states respectively under the effect of exchange interaction. It may be mentioned here that the spins of $Cu^{2+}$ holes and spins of lattice holes do not compound together, because there is no evidence of compounding of these two kinds of holes in EPR spectra of HTS or their constituents. So the motion of lattice holes are governed by the magnetic field produced by the $Cu^{2+}$ holes of the cells in which they are situated at a particular instant. For the sake of clarity, only case (a) of FM coupling will be discussed at length and case (b) will be taken at the end and its main result will be discussed.

We first discuss the case (a) of FM coupling. The lattice holes from both the cells A and E will proceed towards each other with thermal velocity. But as mentioned earlier, when they proceed along the cells, the spin configurations of $Cu^{2+}$ holes will go on changing from U state towards $U_0$state and so on Now let us estimate the exchange interaction energy between the $Cu^{2+}$ ions of the plaquettes in A and E.cells. There are several theoretical methods of calculating exchange energy. They are quite involved and do not always give satisfactory results. There are also empirical values of exchange interaction which are obtained by specific heat data. But in the present problem, we are interested in the properties of only the $CuO_2$ plane in the cuprate superconductors and not the whole superconductor for which there may not be any specific data. For calculating Heisenberg exchange integral J, we use the standard formula [21]

$$J = 3KTc/2zs(s+1) \tag{1}$$

assuming that Tc is the transition temperature of the superconductor under consideration at which the ferromagnetic coupling of holes vanishes and consequently superconductivity also vanishes. In (1), k = Boltzmann constat =1.38 × $10^{-16}$ ergs, Tc is the transition temperature of the $YBa_2Cu_3O_{7-\delta}$, = 92 K, z = number of nearest neighbours of $Cu^{2+}$ ion = 4, s = spin = 1/2. Here z = 4 because every $Cu^{2+}$ ion in the cell A has 4 $Cu^{2+}$ nearest neighbours lying in the cell E and vice versa. Putting numerical values in (1), J = 3.9675 meV. The exchange interaction Eex also has been calculated by the standard formula

$$Eex = -2J\textstyle\sum_j S_i \cdot S_j \tag{2}$$

Putting numerical values in (2), Eex = –7.935 meV. This exchange interaction energy Eex is also the exchange interaction energy between two lattice holes, one in the cell A and other in the cell E , because lattice holes and $Cu^{2+}$ holes in the respective cells are inextricably connected. However, exchange interaction energy between two lattice holes in A and E cells calculated by the formulae (1) and (2) gives the same values which are as follows: $J = 3KTc/2zs(s+1)$ =15.87 meV (with z = 1) and $Eex = -2J\sum_j S_i \cdot S_j$ = –7.935 meV(with i = j = 1). Thus the two lattice holes are bound by Heisenberg exchange energy and may form a hole-pair which are charge carrier in the superconductors. This hole-pair is different from Zhang-Rice pair which is a singlet, but the present one is a triplet. Zhang-Rice singlet [22] is formed between a lattice hole and a $Cu^2$ hole. The present hole-pair is formed between two lattice holes.

Better statistical approximations show that for different crystal structures, J can be increased by a factor more than one. For SC, BCC, ECC this factor is 0.5/0.28, 0.5/0.325, 0.5/0.346 respectively. In the present case, the system under consideration is two-dimensional $CuO_2$ crystal and the factor is not known by which J or Eex should be multiplied. If Eex is multiplied by the average of these factors = 1.59, it becomes equal to 12.6 meV. We are not sure about the multiplying factor but it is certain that Eex must be greater than the calculated value of 7.935 meV or anywhere between 7.935 and 12.6 meV which in terms of temperature is equivalent of 95 K to 151 K which covers nearly the whole range of Tc's of cuprates.

In the case of the two lattice holes one in each of A and E cells, the spin directions are vertically upward of the $CuO_2$ plane and the separation is three blocks of $(CuO)_4$. There will be strong exchange interaction. But in a very short time ($\approx 10^{-14}$ sec), the hole from the A-side will cross over to B-cell and that from E-side to D-cell [Figure (3)]. After they have entered the new cells, there will be rephasing of the spins of the $Cu^{2+}$ in the B and D-cells. The spin vectors of $Cu^{2+}$ holes in B and D-cells will be tilted from the vertical direction by an angle $\omega t$ ($\omega$ = angular velocity of spin oscillation and t = time elapsed for the lattice hole to come to the new cells B and D from A and E respectively). Change of angle by $\omega t$ may be understood from the **Figure 2**(b). It means that the vertical component of spin vectors of each of the $Cu^{2+}$ holes has decreased by a factor $\text{Cos}\,\omega t$ and a horizontal component of the spin vectors of each of the $Cu^{2+}$ holes will be equal to the magnitude of the original vertical component multiplied by $\sin\omega t$ The spin vectors of $Cu^{2+}$ ions of B and D cells will remain parallel to each other though inclined to the vertical direction and the separation between the two cells will be at least the length of one $(CuO)_4$ plaquette. In this case also the exchange interaction will be continued but with less vigour than that between A and E cells because of the trend shown by the exchange interaction to reduce when the separation is further reduced. As they proceed further, both of the lattice holes enter the cell C , where the vertical component of spin vectors become negligibly small and finally at or near the center of the cell , they become zero and so also the horizontal component. Moreover, both the lattice holes are enclosed in the same cell and very close to each other and hence there is no probability of exchange interaction taking place. More about the motion and behaviour of the two lattice holes enclosed in the same cell will be discussed in detail in the next section.

## 3. ORDERPARAMETER OF THE HOLE-PAIR

We will show now. how the magnetic field produced by the spin moments of $Cu^{2+}$ holes in different $(CuO)_4$ units control the motion of lattice holes and how $d_{x^2-y^2}$ order parameter is obtained. We describe the motion of the two holes mathematically and show that their path resembles the $d_{x^2-y^2}$ orbitals of electrons in atoms.

In the Figure (3), the squares A,B,C,D,E,F,G,H,I are the unit cells of the two-dimensional $CuO_2$lattice, each cell being a square of side = 3.84 A which is approximately equal to the size of $(CuO)_4$plaquette or unit cell of the twodimensional $CuO_2$ plane in Y-123 superconductor. When two holes enter the **Figure 3** from opposite sides, one from the A side and the other from the E side, they feel an attractive force due to exchange interaction between them and continue moving towards each other with thermal velocity. Because they are charged particles, their motion will be guided by the magnetic field produced by the $Cu^{2+}$ holes of the $(CuO)_4$ plaquettes in which they are situated at any instant. The motion of a hole is confined to XY plane and its velocity can have only two components Vx and Vy. The velocity of a charged particle does not change by magnetic field; only its direction is changed and $V^2$ is always equal to $Vx^2 + Vy^2$. Motion of a charged particle of charge = e in XY plane under the action of the magnetic field Hz (in the Z-direction) is given by the following equations.

$$dVx/dt = (e/mc)\,Hz\,Vy \tag{3}$$

$$dVy/dt = -(e/mc)\,Hz\,Vx \tag{4}$$

$$dVz/dt = 0 \tag{5}$$

where m is the mass of the charged particle. The Equation (5) goes to zero and need not be considered further.

Before proceeding further we will clarify some symbols and their different forms in which they are used here. They are 1) angular velocities ω and ω', 2) Vx and Vy to be used in the form of ($V \sin \omega t$) and ($-V \cos \omega t$) respectively and 3) Hz to be used as $Hz \, \mathrm{Cos}\, \omega\, t$.

1) Under the effect of exchange interaction and magnetic field generated by the $Cu^{2+}$ holes and its own thermal velocity a lattice hole (numbered 1) starting from A-cell goes to C, then to I, back to C, then to E. In the meantime, the spins of the $Cu^{2+}$ holes start from U-state at A goes to Uo at C, then to D at I, then to Do at C and to U again at E. We see from the Figure (3) that the geometrical angle from A to E is 180°, but the spins of the $Cu^{2+}$complete 360° in going from U to U state. To differentiate between the two, the spin angles will be denoted by ω and the geometrical angle by ω'. Obviously ω' = 2 ω. It is to be noted that ω' is the angular frequency of rotation of a charged particle with charge = e and mass = m in XY plane with the magnetic field in the Z-direction and it is equal to eHz/mc (where c = velocity of light).

2) A lattice hole arrives at A with thermal velocity V and moves forward under the effect of magnetic fields generated by $Cu^{2+}$ holes. Velocity V has now two components Vx and Vy. In the Figure (3), X-axis is from the origin O (center of the cell C) towards I and Y-axis is from O towards A. The components of velocity V are

$$Vy = -V \cos\omega t \tag{6}$$

(-ve sign is because the hole is proceeding opposite to the positive Y-axis; $\cos\omega t$ is used because Vx and Vy depends on the inclination of the spin vectors from the vertical direction expressed by the angle ωt) and

$$Vx = V \sin\omega t \tag{7}$$

3) In (3) and (4), Hz represents the vertical magnetic field. If (3) and (4) are to be used to describe the motion of holes, the vertical magnetic field should be expressed as $Hz \cos \omega t$, because $Cu^{2+}$ spin vectors leans away from the vertical direction as they move from A with angular velocity ω and the vertical component of magnetic field decreases as $\cos \omega t$.). The vertical magnetic field both in A or E cells are equal to Hz because, here ωt = 0 in the general expression of magnetic field Hz Cos ωt. In the light of above clarifications, (3) and (4) can be written as

$$\mathrm{d}Vx/\mathrm{d}t = (e/mc)\, Hz \cos(\omega' t/2) \cdot Vy \tag{8}$$

$$\mathrm{d}Vy/\mathrm{d}t = -(e/mc)\, Hz \cos(\omega' t/2) \cdot Vx \tag{9}$$

Vx in (7) and Vy in (6) can be written with the help of (a ) part above as

$$Vx = V \sin \omega' t/2 \tag{10}$$

$$Vy = -V \cos(\omega' t/2) \tag{11}$$

Differentiating (10) and (11) with respect to t, we get

$$\mathrm{d}Vx/\mathrm{d}t = \omega'/2 \cdot V \cdot \cos \omega' t/2 \tag{12}$$

$$\mathrm{d}Vy/\mathrm{d}t = \omega'/2 \cdot V \cdot \sin \omega' t/2 \tag{13}$$

Equating dVx/dt and dYy/dt from (12) and (13) with those of (8) and (9) respectively, we have

$$(e/mc)Hz\cos(\omega't/2)\cdot Vy = \omega'/2\cdot V\cdot\cos\omega't/2 \tag{14}$$

$$(e/mc)Hz\cos(\omega't/2)\cdot Vx = \omega'/2\cdot V\cdot\sin\omega't/2 \tag{15}$$

From (14),

$$Vy = \left[\omega'/2\cdot V\cdot\cos\omega't/2\right]/\left[((e/mc))\,Hz\cos(\omega't/2)\right]$$

$$= \omega'/2\cdot V\cdot mc/eHz \tag{16}$$

And from (15),

$$Vx = -\omega'/2\cdot V\cdot\sin\omega't/2/(e/mc)Hz\cos(\omega't/2)$$

$$\omega'/2\cdot V\cdot(mc/eHz)\tan\omega't/2 \tag{17}$$

$V^2 = Vx^2 + Vy^2$, and substituting Vx and Vy from (16) and (17), we have

$$(\omega'/2\cdot V\cdot mc/eHz)^2\left[\sec^2\omega't/2\right], or\ V = \pm$$
$$V^2 = \left(\omega'/2\cdot V\cdot mc/eHz\right)^2\left[1+\tan^2\omega't/2\right] =$$

$$(\omega'/2\cdot V\cdot mc/eHz)[\sec\omega't/2],$$

$$\text{or } 1 = \pm(\omega'/2\cdot mc/eHz)$$

$$\left[\sec\omega't/2\right], \text{or } \sec\omega't/2 = \pm 2/\omega'\cdot \mathrm{eHz/mc} \tag{18}$$

We have mentioned earlier that eHz /mc = ω', so (18) can be written as

$$\sec\omega't/2 = \pm 2 \tag{19}$$

Further discussion of the problem has to be done in terms of spin angles ω, so (19) in terms of ω becomes

$$\text{Sec}\ \omega t = \pm 2 = \pm(1/\omega)\cdot(2\omega) \tag{20}$$

Until now, all the above eqns are associated with the velocities of particles (holes) and to convert them to actual coordinates, both sides of (20) are integrated, which gives

$$\int \sec \omega t \ \mathrm{d}t = \pm \int (2/\omega)\cdot(\omega)\mathrm{d}t$$

$$\text{or}, 1/\omega \cdot \text{lin}\left|\tan(\omega t/2)+\pi/4\right| = \pm(2/\omega)\cdot(\omega t)$$

Because the integral

$$\int \sec \omega t \ \mathrm{d}t = 1/\omega.\text{lin}\left|\text{Tan}(\omega t/2)+\pi/4\right|$$

$$\text{or}, 1/2 \cdot \text{lin}\left|\tan(\omega t/2)+\pi/4\right| = \pm.\ \omega t$$

$$\text{or}, 1/2 \cdot R \cdot \text{lin}\left|\tan(\omega t/2)+\pi/4\right| = \pm R \cdot \omega t \tag{21}$$

$$\text{or}, R \cdot \omega t = \pm 1/2 \cdot R \cdot \text{lin}\left|\tan(\omega t/2)+\pi/4\right| \tag{21'}$$

where R is some length used for plotting Equation (21').

(21') can be plotted in a special way Left t hand side (LHS) of (21') will describe a circle with radius = R when t is changed from 0 to T to 2T, where T is the time period of the rotation of spin vectors of $Cu^{2+}$ holes. R has been given a nominal value of 7.68 A which is twice the side (3.84 A) of the unit cell of the $CuO_2$ plane of Y-123 superconductor. The right hand side (RHS) including the term R/2 (=3.84 A) will indicate the position of the lattice hole at an angle ωt. The magnitude of RHS of (21') is shown by a straight line from the position of the angle ωt on the circumference of the circle towards the center of the circle. The tip of the line (shown by dots) represents the position of the lattice hole. The values of RHS of (21') have been shown in the **Table 1**. The positions of the lattice holes at different spin angles (ωt) have been shown in the Figure (3). In **Table 1**, in the range of spin angles 180° - 360° and 540° - 720°, the values of RHS of (21') comes to be negative. But they have to be plotted in the same way as those values which are positive. It is due to the fact that at all ωt values, the holes are directed towards the center of the figure by exchange forces.

**Table 1**. ωt = angle which the spin vector of the hole makes with a direction perpendicular to the $CuO_2$ plane . In the rotation of 360° in the Cartesian system, the spin vector completes two oscillations and so ωt varies from 0° - 720°. The expression 3.84. $|\text{lin}|\text{Tan}(\omega t/2+\pi/4)|$ gives the position of the hole at the angle ωt in the Figure (3).

| $\Omega t$ (1) | 3.84. $\lvert\text{lin}\rvert\text{Tan}(\omega t/2+\pi/4)$ (2) | (1) | (2) | (1) | (2) | (1) | (2) |
|---|---|---|---|---|---|---|---|
| 0 | 0.000 | 180 | 0.000 | 360 | 0.000 | 540 | 0.000 |
| 20 | 1.367 | 200 | −1.367 | 380 | 1.367 | 560 | −1.367 |
| 40 | 2.927 | 220 | −2.927 | 400 | 2.927 | 580 | −2.927 |
| 60 | 5.053 | 240 | −5.053 | 420 | 5.053 | 600 | −5.053 |
| 62 | 5.333 | 242 | −5.333 | 422 | 5.333 | 602 | −5.333 |
| 74.6 | 7.680 | 254.6 | −7.680 | 434.6 | 7.680 | 614.6 | −7.680 |
| 80 | 9.354 | 260 | −9.354 | 440 | 9.354 | 620 | −9.354 |
| 89 | 18.207 | 269 | −18.207 | 449 | 18.207 | 629 | −18.207 |
| 90 | Indeterminate | 270 | Indeterminate | 450 | Indeterminate | 630 | Indeterminate |
| 91 | 18.207 | 271 | −18.207 | 451 | 18.207 | 631 | −18.207 |
| 100 | 9.354 | 280 | −9.354 | 460 | 9.354 | 640 | −9.354 |
| 105.4 | 7.680 | 285.4 | −7.680 | 465.4 | 7.680 | 645.4 | −7.680 |
| 118 | 5.333 | 298 | −5.333 | 478 | 5.333 | 658 | −5.333 |
| 120 | 5.053 | 300 | −5.053 | 480 | 5.053 | 660 | −5.053 |
| 140 | 2.927 | 320 | −2.927 | 500 | 2.927 | 680 | −2.927 |
| 160 | 1.367 | 340 | −1.367 | 520 | 1.367 | 700 | −1.367 |
| 180 | 0.000 | 360 | 0.000 | 540 | 0.000 | 720 | 0.000 |

Moreover, the ± sign in (21') can make all the quantities positive as required. So it is better to write (21) or (21') in the following form.

$$1/2.R.\left|\text{lin}\right|\tan\left(\omega t/2+\pi/4\right)\left|\ \right|=R.\ \omega\ t \tag{22}$$

Till now, we have mainly discussed the motion of the hole1. The conditions governing the motion of the hole 2 is identical to that of the hole1 and they execute the same kind of motion, only difference being that the hole 1 starts from the A-cell and the hole 2 starts from the E-cell at the same time, Both the holes repeatedly traverse the four lobes in the **Figure 3** in the same time but are always situated diametrically opposite to each other. In the **Figure 3** we see that the path of the pair of lattice holes (1 and 2) has four lobes around the center which is like the orbitals of electrons in $d_{x^2-y^2}$ configuration. We started with the pre-knowledge that the symmetry of the hole-pair is $d_{x^2-y^2}$ type and developed a model which corroborates it, demonstrates the correctness of our logic.

On examining the **Figure 3**, two important questions arise, 1) why the hole 1 coming from A on reaching C is deflected towards I-cell and the hole 2 towards the F-cell and 2) why the two holes on reaching extreme ends of the figure bounce back. To answer these questions, we will examine the**Figure 3** and the **Table 1** in detail. At 62°, the position of the hole 1 is just at the upper boundary of the cell C. From 62° to 74.6°, the position of this hole is inside the cell. At 74.6° its position is just at the center of the cell C or at the middle of the diagonal PQ in the cell C. Beyond 74.6° it crosses the diagonal PQ through the center. Between 74.6° and 105.4° the position of the hole is beyond the diagonal PQ. Near 90°, the positions of the hole is beyond the premises of the cell. Specially at 90°, the hole seems to go to infinity (see the table 1). At 118°, the position of the hole is at the RHS boundary of the cell C. At 180° the position of the hole is at the extreme right at the center of the cell I. On reaching 180°, it turns back and the path between 180° and 360⁰ is of similar nature to that between 0° and 180° but in different quadrant of the circle (see **Figure 3**).

Now we consider the motion of the hole 2 which starts its journey from 360° where as the hole 1 had started its journey from 0° The position of the hole 2 at an angle (360 + θ)° is equivalent to the position of the hole 1 at an angle θ, but always in opposite quadrants. For example, the position of the hole 2 from 360° to 540° lying in the third quadrant is just opposite to that of the hole 1from 0° to 180° lying in the first quadrant of the circle. Near 450° the position of the hole 2 is beyond the diagonal PQ towards the first quadrant. At 450° it seems to go to infinity. At 540° the hole 2 is at the left extreme in the cell F. On reaching 540° the hole 2 turns back and the path between 540° and 720° is of similar nature to that between 360° to 540° but in the fourth quadrant. As time advances, both the holes navigate the full circle but always remaining opposite to each other. Both of them bounce back when they reach the four extreme positions shown in the **Figure 3**.

When the holes 1 and 2 enter simultaneously the cell C there is practically no magnetic field generated by the $Cu^{2+}$holes to guide their motion because in the cell C, the spins of the $Cu^{2+}$ holes are in the $U_0$ state or very near it where no magnetic field is generated by the $Cu^2$ holes (see **Figure 2**(b)). At such close separation there is no probability of exchange interaction to exist. When exchange interaction takes place between two electrons or two holes, the dominant interaction between them is attractive interaction and the usual electric and magnetic effects become negligible. When two holes or electrons come very close to each other in the absence of exchange interaction, the electric and magnetic effects become prominent but the former substantially outweighs the latter. So in the absence of exchange interaction, the trajectory will be decided by the electric repulsion between the two holes particularly when both lie in the cell C.

The entry points for the holes1 and 2 in the cell C are at the angles 62° and 422° respectively. As the angles ωt exceeds 62° and 422° for holes 1 and 2 respectively, they come closer inside the cell C where there is no exchange interaction effect, they will repel and push back each other from their calculated positions. As the angle ωt increases from 62° to 90° for the hole 1 and from 422° to 450° for the hole 2, the separation becomes less and less, repulsion stronger and stronger and mutual pushing back more and more.

At 90° for the hole1and 450° for the hole 2, they experience maximum repulsion and so maximum push back from their expected positions which are at the center of the cell C. Beyond 90° for the hole 1 and beyond 450° for the hole 2 ,repulsion gradually decreases and the push back is less. From 90° to 118°, the hole 1 experiences the same kind of repulsion and push back as from 90° to 62°. The hole 1 takes the path MTN [see **Figure 3**] and at 118° enters the cell H on the right hand side. Similarly the hole 2 takes the path KLJ from 422° to 478° and enters the cell G on the left hand side. Thus both the holes change their routes after entering the cell C The hole 1 will cross the cell H with its thermal velocity modified by the magnetic field produced by $Cu^{2+}$ holes in that cell. After the cell H, the hole 1 will enter the area of the cell I. At the same time and by similar process the hole 2 will enter the area of the cell F. There is stronger strength of exchange interaction when one hole is in the F-cell and the other in the I-cell than that when they were in the G and H-cells. After reaching the extreme ends, both the holes will bounce back and return to the C-cell again and deflected left and right as they had on entering this cell earlier. Thus they will continue traversing the orbital of $d_{x^2-y^2}$ symmetry.

But the question is why the holes 1 and 2 bounce back from their extreme positions in I and F-cells respectively and why not continue their forward journey beyond I and F. As calculated by (2), the exchange interaction energy is greater than 7.935 meV and may go up to 12.6 meV.

The maximum kinetic energy a hole can have = $1/2mv^2$

where m = mass of the hole taken to be the mass of an electron and v = velocity of hole = $5.28 \times 10^6$ cm./sec. Thus kinetic energy = 7.937 meV. The holes on reaching the extreme points, each one has kinetic energy = 7.937 meV and the attractive exchange energy between them lying somewhere between 7.935 and 12.6 meV. Thus the attractive energy exceeds the kinetic energy and will pull the holes back from the extreme positions i.e., there will be a mirror action at the extremities. Thus the two holes will remain confined to the $d_{x^2-y^2}$ shaped space. It is also obvious that the two holes spend more time in the cell C than an equal path in any other region because of the reduction of their velocities due to repulsion between them. In all the quadrants they traverse a parabolic path like MTN with reduced speed and spend much time around the center. It indicates that in $d_{x^2-y^2}$ symmetry of the order parameter, there may be slight admixture of s-wave character also as suggested by many workers.

Exchange energy has been assumed and its strength calculated in the $CuO_2$ layer of Y-123 superconductor (using the data of transition temperature and unit cell dimensions). $CuO_2$ is the active layer and common to all the cuprate superconductors. So the mechanism proposed here may be applied to all the cuprates. There is also experimental evidence of the existence of Heisenberg exchange interaction in the $CuO_2$ layer of CuO which is the parent compound of all the cuprates. Ferromagnetism [23, 24] in CuO has been observed and explained on the basis of exchange interaction.

Now we discuss the case (b). When two holes proceeding from opposite sides come at a distance of three $(CuO)_4$ plaquettes in between them, there is equal probability that they will couple in FM or AFM order. In FM coupling, $Cu^{2+}$ spin directions both in A and E cells initially will be along (+z or upward of the $CuO_2$ plane) direction, but in AFM coupling, initially the spin direction in A will be along (+z) direction and in E along the (−z) direction. But the trajectory of the two holes in both the cases will remain the same.as shown in the **Figure 3**. This is due to the direction of force exerted by magnetic field on a conductor carrying current due to hole (+ charge). But in the case of AFM coupling, the two holes will be paired in singlet state But this singlet state will be different from that of Zhang-Rice [22]. Zhang-Rice singlet is due to the AFM combination of a $Cu^{2+}$ hole and a lattice hole. The present singlet is due to the AFM combination of two lattice holes. In FM coupling 92 K (Tc of 123-superconductor) has been taken as Curie temperature. In AFM coupling, 92 K has been considered as Neel temperature In FM coupling, triplet will be the ground state (state of lowest energy) , but in AFM coupling, singlet will be ground state. It is difficult to choose which one will be the ground state, triplet or singlet. Going with the experimental evidence [23,24] that ferromagnetism is observed in CuO, there seems to be more likelihood of the coupling of two lattice holes in the triplet state.

When we say that preformed hole-pairs may provide a valid mechanism of HTS, a question arises as to what causes difference in the Tc's of different superconductors. Probably it is the influence mostly of nearest layers of the $CuO_2$ layer e.g., BaO in Y-123, CaO in Bi-, BaO in Hg-based superconductors etc that brings about the difference. Different neighbouring layers may change the chemical properties of $CuO_2$ layer which probably decides the strength of hole-pair binding.

We have calculated the Heisenberg exchange interaction energy between hole spins, but there is also additional hyperfine interaction between hole spins and nuclear spins of $Cu^{2+}$ ions as evidenced by hyperfine splitting of spectral lines in EPR spectra as mentioned earlier. Its effect on exchange interaction must be taken into account. Hole-nuclear spin interaction is of type AI.S (A = hyperfine splitting constant, I = nuclear spin, S = hole spin). The correction due to this effect on exchange interaction has been calculated by Cottam et al. [25]. who found it to be quite small and proportional to $AT^{1/2}$. Taking average value of A = 100 gauss (from EPR spectra) and T = 92 K, the correction is ≈0.011 meV which can be neglected.

In the reference (1) we have shown that there is possibility of hole-pair formation at any temperature and their number increases with the lowering of temperature and they undergo Bose-Einstein condensation at Tc. There are many researchers who have explained different phenomena connected with superconductors with the help of preformed hole-pairs. Stankowski et al. [26] concluded from the pressure dependence of the phase diagram of HTS materials that Fermions start to bind in local pairs at temperatures much higher than Tc. Demser et al. [27] in their time-resolved measurement of quasiparticle relaxation times as a function of temperature and doping level in Y-123 superconductor noted that there is no evidence of any change in the low energy gap structure at Tc, which implies the existence of preformed hole-pairs up to much higher temperature than Tc. Researchers [28] at Princeton University have photographed preformed hole-pairs at least 50°C above the transition temperature and noted that as the temperature is lowered , they multiply and eventually at Tc they undergo B.E condensation. Researchers [29] at Brookhaven National Laboratory have reported imaging of preformed hole-pairs much before Tc and have noted that the pairing occurs only along the direction in which copper atoms are bonded with oxygen atoms. In the present model, it is quite obvious that the pairing occurs only along the direction of CuO bonds. In forming pairs between the two holes, they proceed head on towards each other along the CuO direction in the $d_{x^2-y^2}$ type orbital and also while receeding from each other, they follow the CuO direction.

# 4. CONCLUSIONS

It has been observed in EPR experiments that when cuprate superconductors are quenched from temperatures a little higher than the temperature of their preparations, there are produced $(CuO)_4$ fragments magnetically isolated from the bulk. In isolated $(CuO)_4$ plaquettes, spins of $Cu^{2+}$ holes have been found to be FM aligned and they produce magnetic field. Isolated $(CuO)_4$ plaquette have been shown to be one short of electron or have one hole in comparison to a connected $(CuO)_4$ plaquette. It is argued that when a hole wandering in the continuous $CuO_2$ sheet enters a particular $(CuO)_4$ plaquette, it should behave as an isolated plaquette having all four $Cu^{2+}$ holes FM aligned and producing magnetic field that will guide the motion of the incoming hole. If two holes proceeding from opposite sides come at a distance of 4 - 5 plaquettes in the $CuO_2$ sheet, they are paired by

Heisenberg exchange energy. The two holes move together in $d_{x^2-y^2}$ configuration. The complicated motion in $d_{x^2-y^2}$ configuration has been explained. The proposed hole-pair may exist either in triplet or in singlet state but the former has some experimental support. The proposed singlet is different from Zhang-Rice singlet. Many references experimentally confirming formation of holepairs have been cited.

# REFERENCES

1. R. J. Singh, "Preformed Hole Pairs in Cuprate Superconductors," International Journal of Modern Physics B, Vol. 23, No. 1, 2009, pp. 53-76.

2. Q. B. Meng, Z. J. Wu and S. Y. Zhang, "Evaluation of the Energy Barrier Distribution in Many-Particle Systems Using the Path Integral Approach," Journal of Physics: Condensed Matter, Vol. 10, No. 5, 1998, pp. L85-L88.

3. Alex Punnoose, B. P. Maurya, Jilson Mathew, M. Umar, M. I. Haque and R. J. Singh, "EPR Observation of $Cu^{2+}$–$Cu^{2+}$ Pairs in Cupric Oxide Powders," Solid State Communications, Vol. 88, No. 3, 1993, pp. 195-198.

4. R. J. Singh, Alex Punnoose, J. Mathew, B. P. Maurya, et al., "S=1 and S=2 EPR Signals in Modified CuO and $BaCuO_2$," Physical Review B, Vol. 49, No. 2, pp. 1346- 1349.

5. R. J. Singh, M. Ikram, A. Punnoose, B. P. Maurya and Shakeel Khan, "Copper Tetramers in High-Temperature Superconductors," Physics Letters A, Vol. 208, No. 4-6, 1995, pp. 369-374.

6. A. Punnoose and R. J. Singh, "EPR Studies of High-Tc Superconductors and Related Systems," International Journal of Modern Physics B, Vol. 9, No. 10, 1995, pp. 1123-1157.

7. S. Khan, M. Ikram, A. Singh and R. J. Singh, "EPR Study of Deoxygenated $La_2CuO_4$," Physica C, Vol. 281, No. 2-3, 1997, pp. 143-148.

8. Shakeel Khan, Arti Singh and R. J. Singh, "EPR Study of $La_{2-x}Sr_xCuO_4$ [M=Ba,Sr]," Solid State Communications, Vol. 106, No. 9, 1998, pp. 621-626.

9. S. Khan, A. Singh and R. J. Singh, "EPR Study of $La_{1.854}Sr_{0.146}CuO_4$," Physica C, Vol. 325, No. 3-4, 1999, pp. 165- 172.

10. R. J. Singh, P. K. Sharma, A. Singh and S. Khan, "EPR Spectra of Deoxygenated High Temperature Superconductors," Physica C, Vol. 356, No. 4, 2001, pp. 285-296.

11. N. Guskos, et al., "EPR Measurements on the $Cu^{2+}$ Ion in the High-Tc Superconductors $MBa_2Cu_3O_7-\delta$," Physica Status Solidi, Vol. 165, No. 1, 1991, pp. 249-253.

12. S. K. Mishra and L. E. Misiak, "EPR Study of High-Tc Superconductor $Y_{0.9}Ho_{0.1}Ba_2Cu_3O_{7-\delta}$," Solid State Communications, Vol. 72, 1989, pp. 351-357.

13. J. T. Lue and P. T. Wu, "Observation of Spin-Waves and Chemical Shifts in the High-Temperature Superconducting $YBa_2Cu_3O_7-\delta$ Systems," Solid State Communications, Vol. 66, No. 1, 1988, pp. 55-58.

14. S. Tyagi, M. Barsoum and K. V. Rao, "Electron Spin Resonance in $Y_1Ba_2Cu_3O_y$," Physics Letters A, Vol. 128, No. 3-4, 1988, pp. 225-227.

15. Y. Ishikara and K. Shige Matsu, "EPR Observation of $Cu_2$+---$Cu_2$+ Pairs In Cupric Oxide Powders," Journal of the Physical Society of Japan, Vol. 61, 1992, pp. 3067- 3069.

16. K. Kindo, M. Honda, T. Kohashi and M. Date, "Electron Spin Resonance in Cupric Oxide," Journal of the Physical Society of Japan, Vol. 59, 1990, pp. 2332-2335.

17. D. N. Basov and T. Timusk, "Electrodynamics of HighTc Superconductors," Reviews of Modern Physics, Vol. 77, No. 2, 2005, pp. 721-779.

18. P. C. Dai, H. A. Mook, G. Aeppli, S. M. Hayden and F. Dogan, "Resonance as a Measure of Pairing Correlations in the High-Tc Superconductor $YBa_2Cu_3O_{6.6}$," Nature, Vol. 406, 2000, pp. 965-968.

19. T. C. Hsu, J. B. Marston and I. Affleck, "Effective Hamiltonian for the Superconducting Cu Oxides," Two Observable Features of the Staggered-Flux Phase at Nonzero Doping," Physical Review B, Vol. 43, No. 4, 1991, pp. 2866- 2877.

20. D. M. Bartels, A. D. Trifunac and R. G. Lawler," Observations of Heisenberg Spin Exchange between Reactive Free Radicals," Chemical Physics Letters, Vol. 152, No. 1, 1988, pp. 109-115.

21. C. Kittel, "Introduction to Solid State Physics," 7th Edition, JohnWiley & Sons, Singapore, 1995, p. 446.

22. F. C. Zhang and T. M. Rice, "Effective Hamiltonian for the Superconducting Cu Oxides," Physical Review B, Vol. 37, No. 7, 1988, pp. 3759-3761.

23. H.-M. Xiao, L.-P. Zhu, X.-M. Liu and S.-Y. Fu, "Anomalous Ferromagnetic Behavior of CuO Nanorods Synthesized via Hydrothermal Method," Solid State Communications, Vol. 141, No. 8, 2007, pp. 431-435.

24. H. W. Qin, Z. L. Zhang, X. Liu, Y. J. Zhang and J. F. Hu, "Room-Temperature Ferromagnetism in CuO Sol–Gel Powders and Films," Journal of Magnetism and Magnetic Materials, Vol. 322, No. 14, 2010, pp. 1994-1998.

25. M. G. Cottom and M. J. Jones, "Theory of Nuclear Spin Interactions in Ferromagnetic Insulators. I. The Thermodynamic Properties," Journal of Physics C, Vol. 6, 1973, pp. 1020-1036.

26. J. Stankowski, et al., "Pressure Crossing Point of Tc(0 ) and Tc(P). Parabolic Dependence Tc(c) and Tc(P). Parabolic Dependence Tc(c) in Composite YBCO~PSTI_ ~ Studied by MMMA," Materials Science (Poland), Vol. 22, No. 3, 2004, p. 175.

27. J. Demsar, B. Podobnik, J. E. Evetts, G. A. Wagner and D. Mihailovic, "Evidence for Crossover from a Bose-Einstein Condensate to a BCS-Like Superconductor with Doping in $YBa_2Cu_3O_{7-\delta}$ from Quasiparticle Relaxation Dynamics Experiments," Europhysics Letters, Vol. 45, No. 3, 1999, pp. 381-386.

28. K. K. Gomes, A. N. Pasupathy, A. Pushp, S. Ono, Y. Ando and A. Yazdani," Visualizing Pair Formation on the Atomic Scale in the High-Tc Superconductor $Bi_2Sr_2CaCu_2O_{8+\delta}$," Nature, Vol. 447, 2007, pp. 569-572.

29. http:// www.physorg.com/news14510552.html

3

# X-RAY SPECTROSCOPY STUDIES OF IRON CHALCOGENIDES

**Chi Liang Chen and Chung-Li Dong**

[1] Institute of Physics, Academia Sinica, Nankang, Taipei, National Synchrotron Radiation Research Center (NSRRC), Hsinchu,, Taiwan

## 1. INTRODUCTION

A recent study that identified high temperature superconductivity in Fe-based quatenary oxypnictides has generated a considerable amount of activity closely resembling the cuprate superconductivity discovered in the 1980s (Kamihare et al., 2008; Takahashi et al., 2008; Ren et al., 2008). This system is the first in which Fe plays an essential role in the occurrence of superconductivity. Fe generally has magnetic moments, tending to form an ordered magnetic state. Neutron-scattering experiments have demonstrated that mediated superconducting pairing may originate from magnetic fluctuations, similar to our understanding of that in high-*Tc* cuprates (de la Cruz et al., 2008; Xu et al., 2008). Binary superconductor $FeSe_x$ is another example of a Fe-based superconductor with a less toxic property, leading to the discovery of several superconducting compounds (Hsu et al., 2008). The *Tc* value of FeSe is ~8 K in bulk form and exhibits a compositional dependence such that *Tc* decreases for over-doping or under-doping of compounds (McQueen et al., 2009; Wu et al., 2009), as does that of high-*Tc*cuprates. FeSe has received a significant amount of attention owing to its simple tetragonal symmetry P4/nmm crystalline structure, comprising a stack of layers of edge-sharing $FeSe_4$ tetrahedron. The phase of FeSe heavily depends on Se deficiency and annealing temperature. While 400 °C annealing reduces the non-superconducting NiAs-type hexagonal phase and increases the PbO-type tetragonal superconducting phase (Hsu et al., McQueen et al., 2009; Wu et al., 2009; Mok et al., 2009), the role of Se deficiency remains unclear. Notably,this binary system is isostructural with the FeAs layer in quaternary iron arsenide. Also, band-structure calculations indicate that FeSe- and FeAs-based compounds have similar Fermi-surface structures (Ma et al., 2009), implying that this simple binary compound may significantly contribute to efforts to elucidate the origin of high-temperature superconductivity in these emerging Fe-based compounds. Therefore, although the electronic structure is of great importance in this respect, spectroscopic measurements are still limited.

According to investigations on how fluorine doping (Kamihara et al., 2008; Dong et al., 2008) and rare earth substitutions (Yang et al., 2009) influence the superconductivity in $LaO_{1-x}F_xFeAs$ compounds, x-ray absorption spectroscopy (Kroll et al., 2008), x-ray photoemission spectroscopy (Malaeb et al., 2008) and resonant x-ray inelastic scattering (Yang et al., 2009) results, Fe 3*d* states hybridize with the As 4*p*states, leading to a situation in which itinerant charge carriers (electrons) are responsible for superconductivity. Most of these studies suggest moderate to weak correlate correlations in this system. Photoemission spectroscopy (PES) measurements (A. Yamasaki et al., 2010) support the density of state (DOS) calculations on the FeSe*x* system. These results indicate the Fe-Se hybridization and itinerancy with weak to moderate electronic correlations (Yoshida et al., 2009), while recent theoretical calculations have suggested strong correlations (Aichhorn et al., 2010; Pourret et al., 2011). While fluorine substitution leads to electron doping in the $LaO_{1-x}F_xFeAs$ system, exactly how Se deficiency may bring in the mobile carriers in the $FeSe_x$ system to ultimately lead to superconductivity remains unclear. Therefore, this study elucidates the electronic structure of $FeSe_x$ ($x$=1~0.8) crystals by using XAS Fe and Se K-edge spectra. Powder x-ray diffraction (XRD) measurements confirm the lattice distortion. Analytical results further demonstrate a lattice distortion and Fe-Se hybridization, which are responsible for producing itinerant charge carriers in this system.

As mentioned earlier, although band-structure calculations indicate that FeSe and FeAs-based compounds have similar Fermi-surface structures, the poor quality of crystals arising from serious oxidization at their surfaces inhibit spectral measurements on pure (stoichiometric) FeSe. Also, FeSe exhibits an unstable crystalline structure. Therefore, investigating the effect of chemical substitution, at either the Se or Fe site, is a promising means of maintaining or improving the superconducting behavior on one hand and stabilizing the crystal structure on the other. Te doping of the layered FeSe with the PbO structure modifies its superconducting behavior, with a maximum *Tc* of

~ 15 K when Te replaces half of the Se. The improvement of *Tc*, which is correlated with the structural distortion that originates from Te substitution, is owing to the combined effect of lattice disorder, arising from the substitution of larger ions, and electronic interaction. Since layered $FeSe_{1-y}Te_y$ crystals are readily cleaved and highly crystalline, x-ray spectra of layered $FeSe_{1-y}Te_y$ crystals can provide clearer information about the electronic structure than those of FeSe crystals. Therefore, this study investigates the electronic properties of $FeSe_{1-y}Te_y$ ($y$=0~1) single crystals by using XAS and RIXS. XAS is a highly effective means of probing the crystal field and electronic interactions. The excitation-induced energy-loss features in RIXS can reflect the strength of the electron correlation. During their experimental and theoretical work on Fe-pnictides, Yang et al. (2009) addressed the issues regarding the Fe-based quaternary oxypnictides. However, few Fe-Se samples of this class have been investigated from a spectroscopic perspective. Angle-resolved photoemission (ARPES) combined with DFT band structure calculation on "11" Fe-based superconductor $FeSe_{0.42}Te_{0.58}$ reveals effective carrier mass enhancement, which is characteristic of a strongly electronic correlation (Tamai et al., 2010). This finding is supported by a large Sommerfeld coefficient, ~ 39 mJ/mol K (de la Cruz et al., 2008; Sales et al., 2009) from specific heat measurement. This phenomenon reveals that the FeSe "11" system is regarded as a strongly correlated system. Moreover, its electronic correlation differs markedly from that of "1111" and "122" compounds, perhaps due to the subtle differences between the *p-d* hybridizations in the Fe-pnictides and the FeSe "11" system. This postulation corresponds to the observation of *p-d*hybridization, as discussed later. This postulation is also supported by recent DMFT calculations, which demonstrate that correlations are overestimated largely owing to an incomplete understanding of the hybridization between the Fe *d* and pnictogen *p* states (Aichhorn et al., 2009). Nakayama et al. (2010) discussed the pairing mechanism based on interband scattering, which has a signature of Fermi surface nesting in ARPES. Based on the SC gap value, their estimations suggest that the system is highly correlated (Nakayama et al., 2010). Moreover, a combined electron paramagnetic resonance (EPR) and NMR study of $FeSe_{0.42}Te_{0.58}$ superconductor has indicated the coexistence of electronic itinerant and localized states (Arčon et al., 2010). The coupling of the intrinsic state with localized character to itinerant electrons exhibits similarities with the Kondo effect, which is regarded as a typical interaction of a strongly correlated electron system. The localized state is characteristic of strong electron correlations and makes the FeSe "11" family a close relative of high-*Tc*superconductors. Comparing the XAS and RIXS spectra reveals that $FeSe_{1-y}Te_y$ is unlikely a weakly correlated system, thus differing from other Fe-based quaternary oxypnictides.

The charge transfer between Se-Te and the Se 4*p* hole state induced by the substitution is strongly correlated with the superconducting behavior. Above results suggest strong electronic correlations in the FeSe "11" system, as discussed later in detail.

## 2. EXPERIMENTS

FeSe*x* crystals were grown by a high temperature solution method described elsewhere (Wu et al., 2009; Mok et al., 2009). Crystals measuring 5 mm x 5 mm x 0.2 mm with (101) plate like habit could be obtained by this method. Three compositions results of FeSe*x* crystals ($x$=0.91, 0.88 and 0.85) are presented here for comparison and clarity. Additionally, large layered single crystals of high-quality $FeSe_{1-y}Te_y$ were grown using an optical zone-melting growth method. Elemental powders of $FeSe_{1-y}Te_y$ were loaded into a double quartz ampoule, which was evacuated and sealed. The ampoule was loaded into an optical floating-zone furnace, in which 2 x 1500 W halogen lamps were installed as infrared radiation sources. The ampoule moved at a rate of 1.5 mm/h. As-grown crystals were subsequently homogenized by annealing at 700 ~800 °C for 48 hours, and at 420 °C and for another 30 hours. Chemical compositions of $FeSe_{1-y}Te_y$ single crystals were determined by a Joel scanning electron microscope (SEM) coupled with an energy dispersive x-ray spectrometer (EDS) (Yeh et al., 2009; Yeh et al., 2008). In the Te substitution series, the composition of nominal $y$=0.3 was $FeSe_{0.56}Te_{0.41}$; that of nominal y=0.5 was $FeSe_{0.39}Te_{0.57}$; that of nominal $y$=0.7 was $FeSe_{0.25}Te_{0.72}$, and that of nominal $y$=1.0 was $FeTe_{0.91}$. The grown crystals were characterized by a Philips Xpert XRD system; $T_c$ was confirmed by both transport and magnetic measurements (Wu et al., 2009; Mok et al., 2009).

X-ray absorption spectroscopy (XAS) provides insight into the symmetry of the unoccupied electronic states. The measurements at the Fe K-edge of chalcogenides were carried out at the 17C1 and 01C Wiggler beamlines at the National Synchrotron Radiation Reach Center (NSRRC) in Taiwan, operated at 1.5GeV with a current of 360mA. Monochromators with Si (111) crystals were used on both the beam lines with an energy resolution $\Delta E/E$ higher than $2 \times 10^{-4}$. Absorption spectra were recorded by the fluorescence yield (FY) mode at room temperature by using a Lytel detector (Lytle et al., 1984). All spectra were normalized to a unity step height in the absorption coefficient from well below to well above the edges, subsequently yielding information of the unoccupied states with p character. Standard Fe and Se metal foils and oxide powders, $SeO_2$, FeO, $Fe_2O_3$ and $Fe_3O_4$ were used not only for energy calibration, but also for comparing different electronic valence states. Since surface oxidation was assumed to interfere with the interpretation of the spectra, the FeSe*x* crystals were cleaved *in situ* in a vacuum before recording the spectra.

The unoccupied partial density of states in the conduction band was probed using XAS, while information complementing that obtained by XAS was obtained using XES. Those results reveal the occupied partial density of states associated with the valence band. Detailed x-ray absorption and emission studies were conducted. Next, tuning the incident x-ray photon energies at resonance in XAS yields the RIXS spectrum, which is used primarily to probe the low-excited energy-loss feature which is symptomatic of the electron correlation. XAS and XES measurements of the Fe $L_{2,3}$-edges were taken at beamlines 7.0.1 and 8.0 at the Advanced Light Source (ALS) at Lawrence Berkeley National Laboratory (LBNL). In the Fe *L*-edge x-ray absorption process, the electron in the Fe 2*p* core level was excited to the empty 3*d* and 4*s* states and, then, the XES spectra were recorded as the signal associated partial densities of states with Fe 4*s* as well as Fe 3*d* character. The RIXS spectra were obtained by properly selecting various excitation energies to record the XES spectra, based on the x-ray absorption spectral profile. The XAS spectra were obtained with an energy resolution of 0.2 eV by recording the sample current. Additionally, the x-ray emission spectra were recorded using a high-resolution grazing-incidence grating spectrometer with a two-dimensional multi-channel plate detector with the resolution set to 0.6 eV (Norgdren et al., 1989). Surface oxidization is of priority concern in Fe-based superconductors; to prevent oxidation of the surface, all data were gathered on a surface of the sample that was cleaved *in situ* in a vacuum with a base pressure of $2.7 \times 10^{-9}$ torr.

# 3. RESULTS AND DISCUSSION

## 3.1. Microstructure Of Fese And Fete

Figure 1 (a) shows the tetragonal crystal structure of FeSe and its building blocks, i.e. Se-Fe tetrahedra and Fe-Se pyramidal sheets. Figure 1 (b) shows the electronic energy level of the individual constituent elements and FeSe*x* (indicating the hybridization band). Figure 2 shows the XRD patterns of the FeSe*x*(*x*=0.9, 0.88 and 0.85) crystals, which represent the superconducting FeSe phase. The patterns are correlated with the P4/nmm and indexed in the figure. Weak hexagonal reflections are also observed. According to this figure, the main diffraction peak (101) shown expanded in the inset shifts to a lower 2 as x decreases, indicating an increase in the lattice parameters. The lattice parameters calculated from these patterns are *a*=*b*=3.771Å, *c*=5.528Å for *x*=0.9, *a*=*b*=3.775Å, *c*=5.528Å for *x*=0.88 and*a*=*b*=3.777Å, *c*=5.529 Å for *x*=0.85. Experimental results indicate that the *a*=*b* parameter increases incrementally as x decreases. Simultaneously, a markedly smaller change occurs in the *c*-parameter. Thus, the *a*-*b* plane variation is surpasses that of the c axis. These lattice parameters are very close to those described in the literature for Se deficient powders (Hsu et al., 2008).

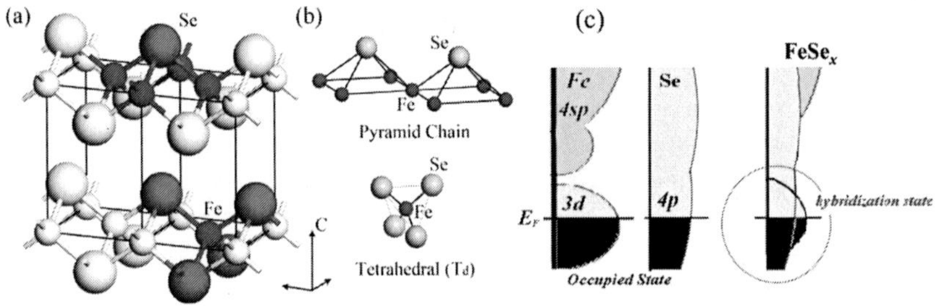


**Figure 1.** (a) Illustration of the crystal structure of tetragonal FeSe*x*, where the blue (small balls) and red (large balls) denote Fe and Se, respectively; the pyramid chain and tetrahedral arrangements are marked in color in the unit cell. Hybridization is shown in two color bonds; (b) the local symmetry of Se atom (in pyramid chain) and Se atom (in tetrahedral geometry) shown separately; (c) energy level diagram of the FeSe*x* system along with individual elements. The hybridization and unoccupied states in the FeSe are highlighted by a circle.

Conversely, FeTe with the same tetragonal crystal structure is stable up to a significantly higher temperature, ~1200 K. As is expected, replacing Se atoms within FeSe with Te stabilizes the tetragonal phase at a synthetic temperature close to or above 731 K. This observation correlates well with our X-ray diffraction analysis. This phenomenon is likely owing to that Te, which has a larger atomic size than Se, inhibits interatomic diffusion in the FeSe lattice. In contrast, Se atoms move easily in the larger FeTe lattice. The lattice parameters calculated from $FeSe_{1-y}Te_y$ patterns are increased with a increasing *y* (Yeh et al., 2009; Yeh et al., 2008). Analysis results indicate that *Tc* and gamma angle of the distorted lattice both reach a maximum value at $FeSe_{0.56}Te_{0.41}$ (~50% Te substituted). This correlation between the gamma angle and *Tc* indicates that *Tc* heavily depends on the level of lattice distortion and the distance of the Fe-Fe bond in the Fe-plane. Results of above studies correspond to the electron-orbital symmetry based on the XAS measurements, as discussed below.

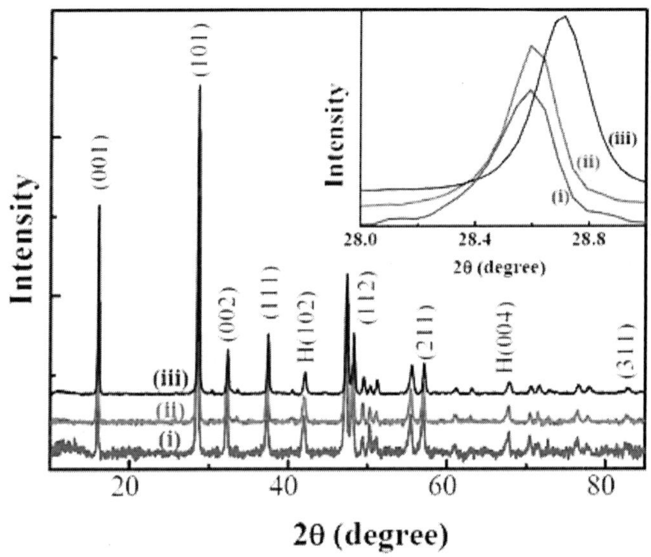


**Figure 2.** Powder XRD patterns of FeSe*x* crystals with *x*= (i) 0.85, (ii) 0.88 and (iii) 0.9. The patterns are fitted to the P4/nmm space group and indexed. Hexagonal phase reflections are denoted by a prefix H.

## 3.2. Electronic Structure Results Based On X-Ray Spectroscopy

### 3.2.1. Fesex Crystals

XAS spectra of the transition metal Fe K-edge ($1s \rightarrow 4p$) in Fig. 3(a) are mostly related to the partial density of $4p$ states of the iron site (Fig. 1(c)). Unoccupied states in the $3d$ (due to quadruple transition) and 4sp bands are sensitive to the local structure and the type of the nearest neighbors (de Groot et al., 2009; Chang et al., 2001; Longa et al., 1999). Above spectra can therefore obtain information about the changes in the electronic states possibly originating from changes in the environment of the Fe ions such as Se vacancies in the present case. Figure 3 shows the Fe K-edge absorption spectra of FeSe*x*crystals along with the standards Fe, FeO, $Fe_2O_3$ and $Fe_3O_4$. All spectra are normalized at the photon energy ~100 eV above the absorption edge at $E_0$ = 7,112 eV (the pure Fe K absorption edge energy). The spectra of the crystals appear to be close to the Fe metal foil, indicating that the crystals are free from oxidation. The recent Fe *L*-edge ($2p_{2/3} \rightarrow 3d$ transition) spectra measured before and after cleaving the samples in UHV further confirm this observation, as shown in the inset of Fig 3(a). The oxidation peak is observed in the crystal before cleaving but it does not appear after cleaving, indicating the possible formation of a thin oxide layer on the surface during handling. Such a thin layer may negligibly impact the deeper penetrating K-edge measurements. Since our measurements are made after cleaving the crystals under vacuum this possibility of surface oxidation is even eliminated. The following section describes in more detail the FeSe*x* electronic structure of the 3*d* states obtained by XAS measurements and resonant inelastic x-ray scattering (RIXS) at the Fe $L_{2,3}$ edges. The observations agree with those of Yang *et al.* (2009) on Fe-pnictides 1111 and 122 systems, as well as those of Lee *et al.* (2008) when using first principles methods to study FeSe*x* system. Therefore, our results on the oxygen free FeSe*x* crystals eliminate the possibility of oxygen in superconductivity, as is case in the $LaO_{1\text{-}x}F_xFeAs$ system.

These spectra reveal three prominent features, $A_1$, $A_2$ and $A_3$ (Fig. 3(a)), of which, $A_1$ could be assigned to the 3*d* unoccupied states originating from the Fe-Fe bonds in metallic iron. The features $A_2$and $A_3$ refer to the unoccupied Fe 4*sp* states. The rising portion of the broad feature $A_2$ (~7,118.8 eV) appears as a broad peak labeled e1 at ~7,116.8 eV, which appears to be well separated in the first derivative plots of the spectra shown in Fig. 3(b). While not observed in the spectra of the reference Fe metal foil or oxides, this broad peak is a part of the Fe 4*sp* band. The $e_1$ feature appears at an energy between those of the Fe metal and FeO and, therefore, originates from a different interaction, as discussed later. Based on these first derivative plots, this study also evaluates a formal charge of Fe, in conjunction with the three standards FeO ($Fe^{2+}$), $Fe_2O_3$ ($Fe^{3+}$) and $Fe_3O_4$ ($Fe^{2.66+}$). Extrapolating the energy of FeSe0.88 with those of the standards allows us to obtain a formal charge of ~1.8+ for Fe in these crystals, thus establishing the electronic charge of Fe in the covalency (2+). Our results further indicate that the peak (energy) position does not increase in energy with a decreasing *x*, implying that the effective charge (valence) of Fe does not change with *x*. This finding is consistent with the above Fe *L*-edge spectra as well as RIXS analysis (Tamai et al., 2010). Thus, the possible electronic configurations of Fe in the ground state can be written as $3d^{6.2}$ or $3d^{6}4s^{0.2}$, indicating a mixture of monovalency ($3d^{6}4s^{1}$ or $3d^{7}$) and divalency ($3d^{6}$).

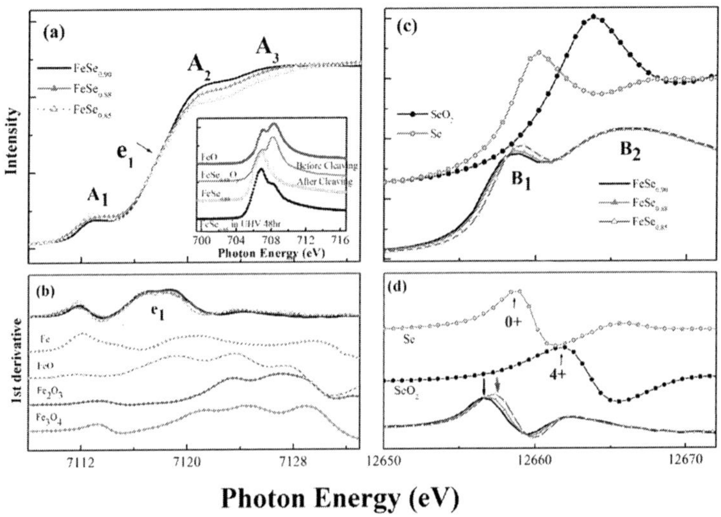


**Figure 3.** (a) Fe K-edge ($1s \rightarrow 4p$) absorption spectra of $FeSe_x$ with different Se contents, and the inset shows the XAS Fe $L_3$-edge spectra of $FeSe_{0.88}$ crystal before and after cleaving *in situ* in a vacuum (b) the first derivative plots of the same spectra. The spectra of the standards - Fe-metal foil, FeO, $Fe_2O_3$and $Fe_3O_4$ are also given alongside the sample spectra. (c)The Se K-edge ($1s \rightarrow 4p$) absorption spectra FeSe*x* with different Se contents, along with the standards Se metal and $SeO_2$ and (d) the first derivative plots of the same spectra.

Figure 3(c) shows the Se K-edge spectra of the FeSe*x* crystals and Se and $SeO_2$ standards, while Fig. 3(d) displays the corresponding first derivative plots to highlight energy changes in the spectra. The spectra represent mainly Se character without any trace of $SeO_2$, indicating the absence of oxidation, even in the deeper layers of the FeSe*x* crystals. The spectra exhibit two peaks $B_1$ and $B_2$. The $B_1$feature at photon energy around 12,658 eV, formally assigned to the transitions $1s \rightarrow 4p$, increases slightly in intensity as well as shifts to a higher energy as x is decreased. This suggests an increase in the Se 4*p* unoccupied states i. e. in the upper Hubbard band (UHB). According to the first derivative plots, a formal charge of ~2.2- is obtained for Se in the $x$=0.88 crystal by interpolation with the energies of the standards Se and $SeO_2$ (as in the case of Fe). This finding agrees with a total charge of 0 when the formal charges of Fe and Se are added ($Fe^{1.8+}$ $Se_{0.88}{}^{2.2-}$), thus establishing the role of the electronic charge of Se in the covalency (2-).

The excess negative charge of -0.2 found in Se can be explained as follows. The electronic charge of Fe in the covalent FeSe*x* should be 2+. However, in this work, a formal charge of 1.8+ is obtained, implying that some electronic charge is returned to Fe due to the Se deficiency. However, the covalent charge of Se should be 2- yet 2.2- is obtained here, which is beyond the 6 electron occupancy of the 4*p*state orbitals. The Se K-edge spectra reveal a hole increase with a decreasing $x$; however, no change in the Fe K-edge suggests a change in valence. According to the soft x-ray photoemission spectroscopy (XPS) measurements of Yamasaki et al. (2010), the close distance between Se and Fe may increase the covalence of the Fe-Se bond due to the hybridization of Se 4*p* and Fe 3*d* states. Yoshida et al. (2009) also observed an adequate correlation between their DOS calculations and the XPS spectra, which show a feature corresponding to the Fe 3*d*-Se 4*p* hybridization. Theoretical calculations of Subede *et al.* (2008) also suggest this Fe-Se hybridization. From the above discussion, we can infer that the $B_1$ feature (Fig. 3(c)) represents the Fe 3*d*-Se 4*p* hybridization band (Fig. 1(c)). Additionally, the increased electronic charge on Fe mentioned above is due to itinerant electrons in the Fe-Se hybridization bond and appears as a hole increase in the Se K-edge spectra. Correlating this finding with the decreasing transition width in the resistance measurements (Mok et al., 2009) obviously reveals that the charge carriers responsible for superconductivity are itinerant electrons in a manner similar to the itinerant holes in the case of cuprates. Oxygen annealing in case of $YBa_2Cu_3O_{6+}$, oxidizes $Cu^{2+}$ to $Cu^{3+}$ through the hybridization of Cu 3*d*-O 2*p* states. A previous study (Grioni et al., 1989; Merz et al., 1998) assigned the $Cu^{3+}$ state to the empty state in the Cu-O bonds, which is also referred to as the $3d^9$L ligand state, where a hole is located in the oxygen ions surrounding a 'Cu site' (L, a ligand hole, tentatively label this as $3d^8$-like). These itinerant holes are responsible for superconductivity. Similarly, in the case of FeSe*x*, the Se deficiency is bringing about Fe 3*d*-Se 4*p*hybridization leading to itinerant electrons. According to Mizuguchi *et al.* (2008), changes in bond lengths likely result in a reduction in the width of the resistive transition due to external pressure.

Experimental results indicate that the intensity of the $A_2$ feature diminishes as $x$ is decreased, indicating a lattice distortion that increases with a decreasing $x$. Additionally, the change in the $A_2$feature is larger than that of the $A_3$ feature. Notably, the $A_2$ feature could be associated with $p_{xy}$ ($\sigma$) and A3 to $p_z$ ($\pi$) orientations since multiple scattering in the XAS from $p$ orbitals could reveal different orbital orientations, and also owing to the nature of the p orbitals, i.e. $p_{xy}$ and $p_z$. We thus speculate that a larger distortion occurs in the a-b plane (Fe-Fe distance) than in the c axis. This is also seen from the XRD measurements (Fig. 2) where the change in the $a$=$b$ parameter is more profound than that of the $c$ parameter (Fe-Se distance). Therefore, the Fe orbital structure changes from 4$p$ to a varying (modulating) coordination with a decreasing $x$. The broad feature $B_2$, at ~ 20 eV above the Se K-edge appears at the same energy in the entire FeSe$x$ spectra, is not affected by the Se deficiency and is assigned to the multiple scattering from the symmetrical Se 4$p$ states in the coordination sphere that are correlated to the local structure of the Se ions (de Groot et al., 2009). This finding correlates with the XRD results where the c parameter remains nearly unchanged. This occurrence becomes obvious when examining Fig. 1, where Se is found at the tip of the Fe-Se pyramid.

Since Se is located at the apex of the tetrahedral pyramid chain in the FeSe$x$ lattice (Fig. 1), removing a Se ion with a formal negative charge (-2) from the lattice would result in a Se vacancy with an effective positive charge and also repulse the surrounding positively charged Fe atoms. This finding corresponds to the distortion in the ab plane, as discussed above. Additionally, according to Lee et al. (2008), the Fe atoms around the vacancy may function similar to magnetic clusters.

### 3.2.2. Electron Correlations Of Fese1-Ytey

As discussed above, due to the unstable phase of stoichiometry FeSe and efforts to more thoroughly understand the origin of superconductivity in this class of materials, of worthwhile interest is to investigate the effect of chemical substitution on FeSe in order to maintain or improve the superconducting property and stabilize the crystal structure. Figure 4(a) compares Fe $L_{2,3}$-edges XAS spectra of $FeSe_{1-y}Te_y$ before and after cleavage *in situ*, in which an oxidized iron foil is used as a reference sample. Of the two XAS lines, $A_1$ and $B_1$, at the $L_3$-edge of the oxidized iron foil, $B_1$ are prominent from the uncleaved samples, indicating that the sample surface is seriously contaminated by oxygen. The shoulder-like line $B_1$, which originates from Fe-O bonding in cleaved samples, is smeared. As is well known, iron oxide yields $B_1$. However, this line is missing from the spectra of cleaved $FeSe_{1-y}Te_y$ samples, indicating that iron oxide is only a very minor constituent. This finding reflects the importance of cleavage *in situ* for making spectral measurements of samples in this class.Figure 4(b) further illustrates how sensitive the FeSe (without Te doping) is, in which it oxidizes very easily, even in an ultra-high vacuum chamber with a base pressure of ~2.7x10$^9$ torr. The study of $FeSe_{0.88}$ crystal in an earlier section observed this phenomenon. In this work, XAS of pure FeSe is obtained, and the pure compound oxidizes within 12 hours, as indicated by a comparison with a doped sample ($FeSe_{1-y}Te_y$, y=0.5). The high-energy shoulder structure, as denoted by a dash line, is associated with the Fe-O bond; it is absent right after *in-situ* cleavage of either sample. This structure appears again in the FeSe sample when placed in a vacuum for 12 hr, yet does not reappear in the doped samples. This finding demonstrates that pure FeSe suffers from serious surface oxidization. Capable of enhancing the Tc value, Te doping can also stabilize the crystal structure. Figure 4(c)presents Fe $L_{2,3}$-edges XAS spectra of $FeSe_{1-y}Te_y$, by using Fe metal, oxidized foil and $Fe_2O_3$ as reference samples. Unoccupied Fe 3*d* orbitals are probed using XAS. The spectra reveal two major transitions. Governed by the dipole selection rule, the transition is mainly due to $2p^63d^6$-$2p^53d^7$transition, in which an Fe 2*p* electron is excited into an empty 3*d* state. Spin-orbit splitting separates the 2*p* state into $2p_{3/2}$ and $2p_{1/2}$ states, yielding two prominent absorption features around 707 eV ($L_3$) and 720 eV ($L_2$). XAS at the transition-metal *L*-edge probed unoccupied 3*d* states, which are sensitive to the chemical environment, valence state, crystal field and 3*d* electronic interactions. The ratio of intensities of the two major absorption features is largely determined by the high- or low-spin ground states through the crystal-field effect. XAS of $Fe_2O_3$ reveals a strongly split structure at both $L_2$ and $L_3$absorption edges, which is formed by the interplay of crystal field and electronic interactions. Fe-O does not contribute to the $FeSe_{1-y}Te_y$ spectra, and no spectral profile resembles that of $Fe_2O_3$ or oxidized foil. As is generally assumed, the shoulder at the high-energy tail (dotted portion) of the Fe $L_3$line in the $FeSe_{1-y}Te_y$ samples is associated with covalent $sp^3$ bonds between Fe 3*d* and Se 4*p*/Te 5*p*states (Bondino et al., 2008). $FeSe_{1-y}Te_y$ yields no observable line splitting or change in intensity ratio, implying a weak crystal-field effect and also that the Fe ion favors a high-spin ground state. Line shapes in the spectra of these cleaved crystals in the UHV chamber resemble those of iron metal, indicating metallic nature and a localized 3*d* band (Yang et al., 2009). As is anticipated, the variation in the full width at half maximum (FWHM) of the Fe $L_3$ peak in the XAS and the width of the Fe 3*d*band with localized character is smaller than those of Fe because the Fe-Fe interaction is stronger (Nekrasov et al., 2008).

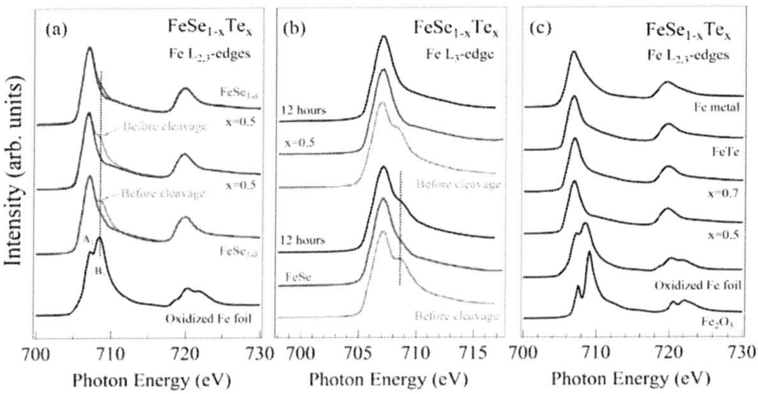


**Figure 4.** (a) Comparison of Fe $L_{2,3}$-edge XAS of $FeSe_{1-y}Te_y$ before and after cleavage *in situ*, using oxidized iron foil as a reference. (b) FeSe is highly sensitive, even in the vacuum; in addition, it oxidizes within 12 hours. (c) Fe $L_{2,3}$-edge XAS of $FeSe_{1-y}Te_y$ cleaved *in situ*. Samples of Fe metal, oxidized foil and $Fe_2O_3$ serve as references.

Figure 5(a) describes the RIXS (lower part) obtained at selected energies (letters a-g), as denoted by arrows in XAS (upper part). The upper RIXS spectrum is obtained at an excitation photon energy of 735 eV, far above the Fe $L_3$-edge absorption threshold. This spectrum is commonly referred to as the so-called non-resonant normal emission spectrum. This non-resonant spectrum reveals two main fluorescent features around 704 eV and 717 eV, resulting from de-excitation transitions from occupied 3*d* states to the $2p_{3/2}$ and $2p_{1/2}$ holes, respectively, which are thereby refilled. The RIXS recorded with various excitation energies, a-f, include the strong 704 eV peak, as observed in the non-resonant spectrum. At an excitation energy close to that of the $L_2$ absorption feature, the line at 717 eV emerges and remains at the same energy as the excitation energy increases beyond the $L_2$ edge. As indicated by arrows in spectra d and e, small bumps appear at energies of approximately 710 and 716 eV. These energies track well the excitation energies and, therefore, are caused by elastic scattering. As magnified in the inset, the fluorescent feature of both samples does not follow the excitation energy and well overlaps the line at 704 eV in the non-resonant spectrum. Notably, RIXS includes no energy loss feature, which is generally associated with electron correlation and excitations.

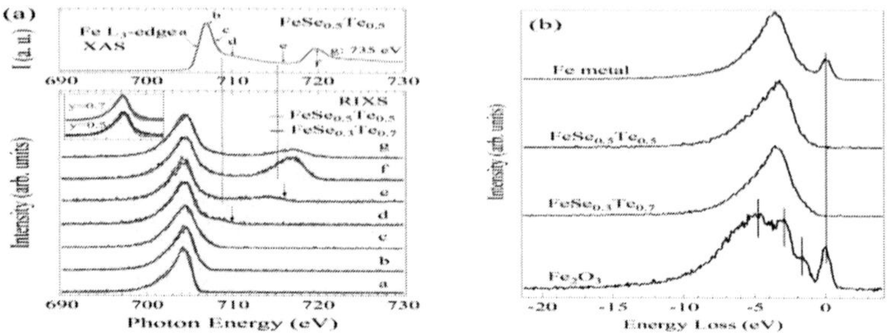

**Figure 5.** (a) RIXS (lower part) obtained at selected energies (a-g), as denoted by arrows in XAS (upper part). Inset shows the fluorescent feature of both samples, which overlaps well with the line at 704 eV in the non-resonant spectrum. (b) Comparison of RIXS resonantly excited at 708 eV for samples with*y*=0.5 and 0.7 and reference samples $Fe_2O_3$ and pure iron metal.

Figure 5(b) compares RIXS obtained with resonant excitation at 708 eV from samples with *y*=0.5 and 0.7 and from reference samples $Fe_2O_3$ and pure iron metal. They are displayed with an energy loss scale; elastic scattering produces the feature at an energy loss of zero. Notably, $Fe_2O_3$ RIXS exhibit more complex energy-loss features, which are marked by short vertical lines. These features are enhanced at particular excitation energies because of inter-electronic transitions, and are identified as*d-d* excitations (Duda et al., 2000). Fe metal yields one main fluorescent line and a more symmetric profile. Spectra of $FeSe_{1-y}Te_y$ are also obviously dominated by only a single line and resemble that of Fe metal, implying metallic character. Analysis results of the Coulombic interaction and Fe 3*d*bandwidth indicate the lack of excitation-induced energy-loss features in RIXS of iron-pnictide samples, implying a weak correlation in the iron-pnictide system (Yang et al., 2009). Therefore, known metallic iron and insulating iron oxide must be compared further to reveal the importance of electron correlation and metallicity.

Just as XAS demonstrates the density of unoccupied states (DOS) in *d* orbitals, non-resonant XES are interpreted simply as revealing the occupied DOS of d states, which is dominated by Fe 3*d* orbitals near the

Fermi level. Figure 6 (a) shows the non-resonant Fe $L_{2,3}$-XES of $Fe_2O_3$, $FeSe_{1-y}Te_y$ and Fe metal. The spectra of $FeSe_{1-y}Te_y$ and Fe are nearly identical in both shape and energy, indicating that the experimental determination of the contribution of occupied Fe 3*d* states in $FeSe_{1-y}Te_y$ closely resembles that of Fe metal. The ratio of intensities of $L_2$ and $L_3$ XES should equal 1/2 for free atoms, reflecting the statistical populations of the $2p_{1/2}$ and $2p_{3/2}$ energy levels. However, in metals, Coster-Kronig transitions markedly reduce this ratio, revealing the correlations and metallicity of a system (Raghu et al., 2008). Figure 3(b) presents ratio $I_2/I_3$, i.e. the ratio of integrated intensities of the XES $L_2$and $L_3$ lines. $FeSe_{1-y}Te_y$ samples with $y$=0.5 and $y$=0.7 have almost identical $I_2/I_3$ intensity ratios. These ratios are closer to that of correlated $Fe_2O_3$ than to that of the metal, indicating that the character of $FeSe_{1-y}Te_y$ more closely resembles that of correlated $Fe_2O_3$. This finding corresponds to that of the Fe 3*d* states in $FeSe_{1-y}Te_y$ systems, which exhibit more localized than itinerant character, as denoted by a smaller FWMH in XAS than that of Fe-pnictides.

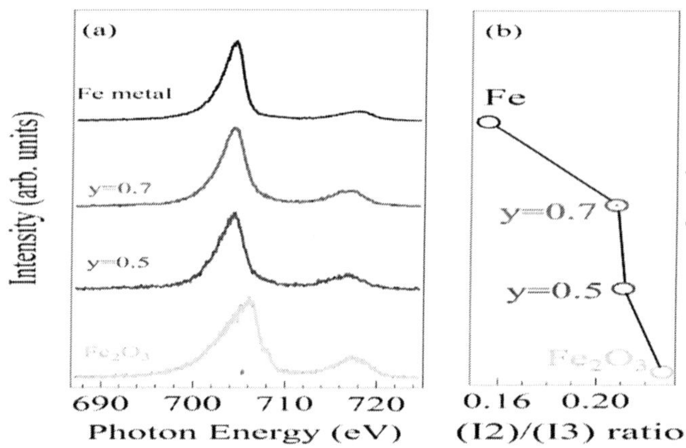


**Figure 6.** (a) Comparison of non-resonant Fe $L_{2,3}$-XES of $Fe_2O_3$, $FeSe_{1-y}Te_y$ and Fe metal. (b) Ratio of integrated intensities ($I_2/I_3$) of XES $L_2$ and $L_3$ lines.

The similarity between XAS and XES of $FeSe_{1-y}Te_y$ and those of Fe metal signifies the importance of metallicity in this system. This metallicity is further and directly ascertained from absorption-emission spectra, which, with a carefully calibrated energy scale, can reveal both the occupied and the unoccupied electronic densities of states around the Fermi level. Figure 7(a) displays the absorption-emission spectra, revealing the DOS of Fe 3*d* states across the Fermi level. An arrow indicates the intersection of XAS and XES, indicating that $FeSe_{1-y}Te_y$ can be regarded as metallic in nature. Figure 7(b) shows the overlain spectra of $FeSe_{1-y}Te_y$ and Fe. The Fe spectrum has a single line, which is narrower than the corresponding line of $FeSe_{1-y}Te_y$. This dominant line in the spectrum of $FeSe_{1-y}Te_y$is slightly broader than that of Fe, perhaps owing to Se hybridization state in the lower-energy region (Yamasaki et al., 2010; Subede et al., 2008).

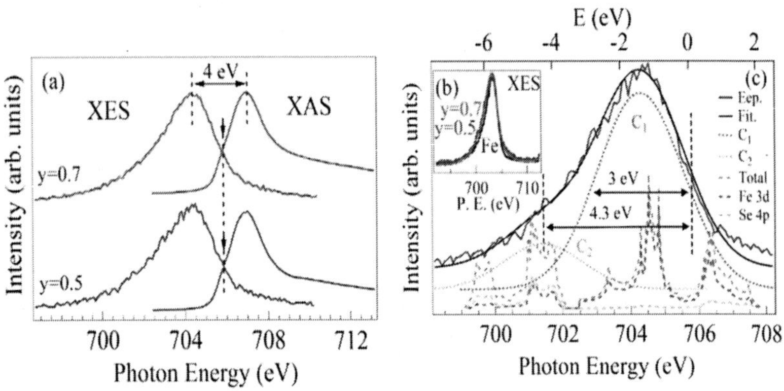


**Figure 7.** (a) Absorption-emission spectrum of $FeSe_{1-y}Te_y$ ($y$=0.5 and 0.7). Intersection yields denotes $E_F$. (b) Spectra of $FeSe_{1-y}Te_y$ ($y$=0.5 and 0.7) and Fe metal. A deviation originates from Fe-Se hybridized states. (c) Fe $L_3$-XES line is fitted by two components (dotted lines). The dashed lines refer to calculated DOS (from Subede et al., 2008).

Iron pnictides are weakly correlated systems, unlike high-$T_c$ cuprates. However, a theoretical work has established that Fe-Se may not be correlated as weakly as Fe pnictides (Aichhorn et al., 2010). An important question that arises concerns whether compounds in the FeSe family exhibit a weak correlation, like that of iron pnictides, or a strong correlation, similar to that in high-$T_c$ cuprates. Therefore, information concerning bandwidth of the Fe 3*d* states must be obtained from spectral results. An estimate of the FWHM in $FeSe_{1-y}Te_y$ XES yields a width of 4.1 eV. This value reflects the fact that core-hole lifetime and multiplet broadening, and are not associated with the Fe 3*d* bandwidth.Figure 7 (c) displays a deconvoluted spectrum that includes a single dominant line ($C_1$) and a low-energy contribution ($C_2$), which is consistent with published x-ray data (Freelon et al., 2010). The main line is associated primarily with Fe 3*d* bands; in addition, the contribution in the low-energy region is interpreted as originating from the hybridization of Fe 3*d* and Se 4*p* states (Kurmaev et al., 2009). The low-energy shoulder originates from hybridization of Fe 3*d* and As 4*p* states in the FeAsFO "1111" system. Analytical results indicate that the FeSe "11" system is simpler than the "1111" system: the low-energy contribution therefore originates from hybridization of Fe 3*d*-Se 4*p* without calculation forecasts. This hybridization of Fe 3*d*-Se 4*p* is also consistent with recent density functional calculations (Subede et al., 2008). According to the projected density of states (adopted from Subede et al, (2008)) in the bottom of Fig. 7(c), peak $C_1$ is mainly due to the Fe 3*d* states. Peak $C_2$ lies far below the Fermi level, and can be ascribed to the hybridized Fe 3*d*-Se 4*p* states. Width of the dominant peak in Fe $L_3$ XES is ~3 eV. Given the instrumentation resolution, this value is taken as a guide for an upper limit of the Fe 3*d* bandwidth. Estimation results of ~3 eV for the Fe 3*d* bandwidth correlate with the on-off-resonance photoemission results, revealing that Fe dominates the binding energy range of 0-3 eV (Yamasaki et al., 2010). The low-energy contribution at 4.3 eV below $E_F$ is attributed to Se 4*p* states, which corresponds to the on-off-resonance photoemission results (Yamasaki et al., 2010) and DOS calculation (Subede et al., 2008). The degree of electron correlation implies the relative magnitudes of the Coulomb interaction $U$ and bandwidth $W$. Additionally, $U$ is simply taken as the energy difference between the occupied and unoccupied states near $E_F$, and is estimated to be ~4 eV. This value matches values presented in another work, in which $U$ of the FeSe system has been predicted to be ~4 eV (Aichhorn et al., 2010). The values imply that the magnitude of $U$ is larger than that of the upper limit of the estimated Fe 3*d* bandwidth. The result $U/W>1$ in this work reflects a strong electronic correlation due to competition between the localized effect of $U$ and the itinerant character of $W$(Zaanen et al., 1985).

Various experiments and theoretical calculations have been undertaken, especially for the "1111" and "122" systems, to elucidate electron correlation in these Fe-based compounds. Now, the results herein concerning the "11" system are compared with those obtained elsewhere for "1111" and "122" compounds. Comparing the photoemission spectrum (PES) of Fe with the Cu 2*p* core-level spectra of $CeFeAsO_{0.89}F_{0.11}$ (1111) (Bondino et al., 2008) and high-$Tc$ cuprates ($La_{2-x}Sr_xCuO_4$ and $Nd_{2-x}Ce_xCuO_{4-\delta}$) (Steeneken et al., 2003) reveals the absence of satellite structures with higher binding energies in the Fe 2*p* spectra. This finding implies the presence of strongly itinerant Fe 3*d* electrons, which contrasts with the observation from the Cu 2*p* spectra in high-*Tc* cuprate. Further comparing the Fe $L_3$ XAS spectra reveals the lack of a well-defined multiplet structure, which is indicative of the delocalization of 3*d* band states, and subsequently providing complementary evidence of the itinerant character of Fe 3*d* electrons. Strongly itinerant 3*d* electrons are thus an important characteristic of the Fe-based superconductor, implying that it does not exhibit as strong correlations as a high-*Tc* cuprate. A previous study has examined the extent of electron correlations in PrFeAsO by using coupled x-ray absorption and emission spectroscopy (Freelon et al., 2010). From the spectroscopic results herein, the bandwidth of the Fe 3*d* states is estimated to be ~ 2eV, which is similar to or larger than the theoretical Coulomb parameter U2 eV. The relative magnitudes of U and W thus simply suggest that the "1111" system exhibits weak to intermediate electron correlations (Freelon et al., 2010). A detailed study involving the theoretical calculations for both "1111" and "122" Fe-pnictides was performed, with its results correlating well with experimental x-ray spectroscopic results (Yang et al., 2009). A cluster calculation was performed to highlight the role of strong Coulomb correlations; spectral features associated with strong correlations are shown in the theoretical results, yet are suppressed in the experimental data. Based on a direct comparison of the energy position of this feature in the experimental data and that in the cluster simulation, an upper limit of the Hubbard U of 2 eV is determined; this value is markedly smaller than the Fe 3*d* bandwidth. This finding suggests that Fe-pnictides should be viewed as a weakly correlated system. Although the XAS results herein resemble those in previous studies for Fe-pnictides and do not show the well-defined multiplet structure that is exhibited by Fe ionic oxides, two important differences are observed upon closer inspection: (a) the high-energy shoulder intensity of $FeSe_{1-y}Te_y$ is lower than that of Fe-pnictides, and (b) the linewidth of the main peak in the XAS spectrum of Fe metal is smaller than that of Fe-pnictides (Yang et al., 2009). The weak and broad shoulder is owing to hybridization of Fe 3*d*- Z n*p* states (Z: *sp* element). Thus, this reduction in intensity implies less hybridization than exhibited by other Fe-based "1111" and "122" systems. This finding is consistent with the observation that the linewidth of the Fe 3*d* main peak from FeSe is narrower than those from the "1111" and "122" systems. The presence of localized Fe 3*d*electrons in the FeSe system is also confirmed by XES, which can extract information about the bandwidth of the Fe 3*d* states. As is estimated, the bandwidth in this work is ~ 3 eV, although this value is larger than that, ~ 2eV, found in a previous study of the "1111" system (Bondino et al., 2008). However, according to our spectroscopic results, the Coulomb parameter U is estimated to be ~ 4eV, which exceeds values reported elsewhere, which range from 0.8 eV to 2 eV (Yang et al., 2009;Bondino et al., 2008; Nekrasov et al., 2008). Recent calculation studies also support the Coulomb parameter U~ 4eV in this work. Above differences suggest that the FeSe system differs from "1111" and "122" compounds, and the FeSe "11" system is probably not a weakly correlated system.

The electronic properties discussed above concern Fe 3*d* orbitals. The correlation between the superconducting behavior and the band structures, as well as interaction of the *d*-*d* and *d*-*p* states, may also play a vital role in these systems, as revealed by the XANES tests on FeSe*x* single crystals at section 3.3.1. As mentioned earlier, although doping of Te (y=0.5) enhances the Tc value, under- or over-doping decreases this value. Origin of the enhancement must be clarified. The Fe K-edge spectral profiles of the $FeSe_{1-y}Te_y$ are identical and appear to remain unaffected by Te doping (Fig. 8), which is consistent with observations of Fe *L*-edge XAS and RIXS. This finding suggests that the valence state and the electronic structure around the iron sites resemble each other. Additionally, the electronic configuration depends heavily on the hybridization between the orbitals of the Fe (3*d*) – Se(4*p*)/Te(5*p*) states. Accordingly, the electronic properties of 4*p*(5*p*)-character, which are identified from the Se (Te) K-edge XAS spectra, must be further examined. Figures 9 (a) and 9 (b) display the Se and Te K-edge spectra, respectively. The insets highlight the energy shifts in these spectra. Se K-edge spectra have a spectral line shape resembling that in FeSe*x* series. The spectra are explained by a projection to the local electronic transition from the Se inner 1*s* to the outer 4*p* state. The spectra reveal two peaks, $A_2$and $B_2$. Peak $A_2$ at a photon energy of around 12,658 eV originates from Fe-Se hybridized states. The peak of the doped sample has an increased intensity and shifts to a higher energy. According to related investigations (Joseph et al, 2010), the increase in peak $A_2$ intensity is owing to an increase in the strength of Fe 3*d*-Se 4*p* hybridization. The upward shift in energy is the result of an increase in the valence states. The measured energies of the absorption edge follow the order of increasing photon energy $y$=0.3, 0.7 and 0.5, as displayed in the inset. Figure 10 plots those values. Notably, the position of the absorption edge is related to the valence state; the valence state is highest at $y$=0.5, implying that the carriers between Fe and Se are itinerant electrons in the Fe-Se hybridization band, causing a valence change upon Te substitution. This increase in the valence is an indication of the change in coordination geometry and the increase in the number of 4*p* holes. The spectra thus provide evidence that the 4*p* holes are increased the most upon Te substitution when $y$=0.5. The change in the number of holes that is caused by the variation in the strength of the Fe 3*d*-Se 4*p* hybridization band is also confirmed in the x-ray absorption study of the Se deficiencies in the FeSe*x* system.

Additionally, the number of holes changes with the superconductivity (Fig. 10) in a manner resembling that in cuprate, whose itinerant holes, via the hybridization of Cu 3*d*-O 2*p* states, are responsible for superconductivity. Consequently, the change in valence implies a change in the number of holes by Fe 3*d*-Se 4*p*hybridization, subsequently forming itinerant holes, as occurs in Se-deficient FeSe. Notably, the broad feature $B_2$ appears around 20 eV above the absorption threshold in all spectra. According to the earlier discussion of section 3.2.1 and Joseph et al., (2010) this feature is attributed to multiple scatterings from symmetrical Se 4*p* states in the coordination sphere, which is related to the local structure of the Se site. This broad feature appears at the same energy in all of the FeSe*x* spectra and remains unaffected by the Se deficiency. This finding implies similar multiple scattering from Se 4*p* states and the same local structure around Se ions.

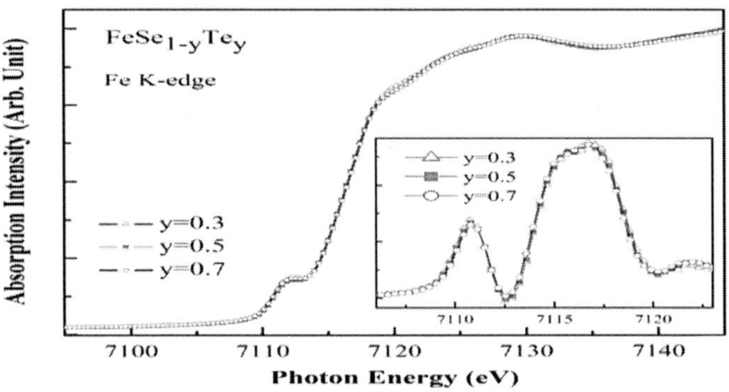


**Figure 8.** Fe K-edge XAS spectral profiles of $FeSe_{1-y}Te_y$ are identical and appear to remain unaffected by Te doping. Insets show details of 1st differential spectra, which suggest a similar electronic state around Fe site.

However, this broad feature does not appear at the energy at which it occurred in the work of Joseph et al., (2010) because of the change in the local geometry around the Se site upon Te substitution. This finding suggests local inhomogeneity, which correlates with the local inhomogeneity that was evident in XANES and EXAFS studies (Joseph et al., 2010). EXAFS analysis results indicate that the Fe-Se and Fe-Te bond lengths in $FeSe_{0.5}Te_{0.5}$ differ from each other, revealing a distinct site occupation and local in-homogeneity. A detailed polarization study of the Se K-edge demonstrates changes in the $A_2$and $B_2$ peaks with the xy and z characteristic of the *p* states (Joseph et al., 2010). Owing to the natural characteristic of the *p* orbitals, the multiple scattering in the XAS results that are associated with *p*orbital symmetry reveals their different orbital orientations, thus reflecting $p_{xy}$ and *pz* character. This feature shifts to a lower energy owing to local inhomogeneity caused by doping. This finding is supported by the work of Joseph et al., (2010) and implies distortion of the tetrahedral symmetry at the Se sites. Te K-edge spectra in Fig. 9 (b) include one edge feature that is associated with 1*s* to 5*p*transitions in the coordination sphere, which reflect the local envelopment of the Te ions. The inset inFig. 9 (b) describes details of the absorption edge. The photon energy associated with the chemical shifts increases in the order $y$=0.5, 0.7 and 0.3, i.e. a trend which contradicts the Se K-edge observations. The $y$=0.5 substitution exhibits the lowest valence state, a finding which contrasts with the Se K-edge results, possibly owing to that the charge is gained in the 5*p* orbital at the Te site. The critical and corresponding energy shift in the Se and Te K-edge features upon Te substitution is consequently indicative of an increase in the 4*p* holes and a decrease in the 5*p* holes at $y$=0.5. The tetragonal phase of FeSe has a planar sub-lattice layered structure with Se ions at the tips of the pyramid chain and an Fe plane between Se ions. The substituted Te has an ionic radius which exceeds that of Se, subsequently increasing the hybridization of Fe 3*d*-Se 4*p*/Te 5*p*. Comparing Se and Te K-edges reveals an expected charge transfer between Se and Te: as Te is doped into tetragonal FeSe crystals, the number of 4*p* holes is increased by Fe 3*d*–Se 4p/Te 5*p* hybridization. These results are consistent with an earlier study of the structural distortion that is associated with variation in the angle and electron-transport properties (Yeh et al., 2008). Importantly, the change in the number of *p* holes between Fe-Se and Fe-Te may determine superconducting behavior.

Te doping causes structural distortions in FeSe, as revealed by detailed x-ray refinement (Yeh et al., 2008). Doping expands the lattice because the ionic radius of Te exceeds that of Se. As the doping concentration increases, angle varies, subsequently increasing the bond length along the *c*-axis and altering the density of states at the Fermi level (Yeh et al., 2008), which corresponds to density-functional calculations (Subede et al., 2008). $T_c$ and angle display a similar trend: both reach their maxima at $y$=0.5. Figure 6 plots $T_c$ against Te doping ($T_c$ is denoted by a red star); a simple sketch of the varying angle is also shown. Since the electronic structure around the Fe site in $FeSe_{1-y}Te_y$ does not significantly change, exactly how Se 4*p* holes affect superconductivity should be examined. Either the energy shift or the area under the absorption feature of XAS yields the hole concentration. Therefore, in this study, the energy shift with respect to pure FeSe is determined from the Se K-edge of $FeSe_{1-y}Te_y$ and is presented in Fig. 10 as a black circle. Evolution of the edge shift reasonably estimates the hole concentration. Variation in the number of Se 4*p* holes is closely related to the change in transition temperature. The correlation between the Se 4*p* hole concentration and $T_c$ suggests that $T_c$ depends more heavily on the variation in the number of 4*p* holes than on the Fe-Fe interaction in the Fe plane, a claim which is consistent with the absence of a significant change in the Fe XAS and RIXS. This finding is also consistent with the asymmetric expansion of the tetragonal lattice in relation to the structure of $FeSe_{0.5}Te_{0.5}$ – by 8 % along the *c*-axis, but only about 0.6% along the *a*-axis (Yeh et al., 2008); the Fe-Fe interaction in the Fe plane has negligible impact herein. Conversely, the role of the ligand holes and the subsequent effect of charge transfer are important. These effects may originate from Fe acting as a superexchange medium, which does not observably change from the perspective of the Fe site, a finding which is consistent with previous studies (Fang et al., 2008;Yildrim et al., 2008; Si et al., 2008).This finding may provide an important foundation for understanding the origin of superconductivity in the family of compounds considered in this study.

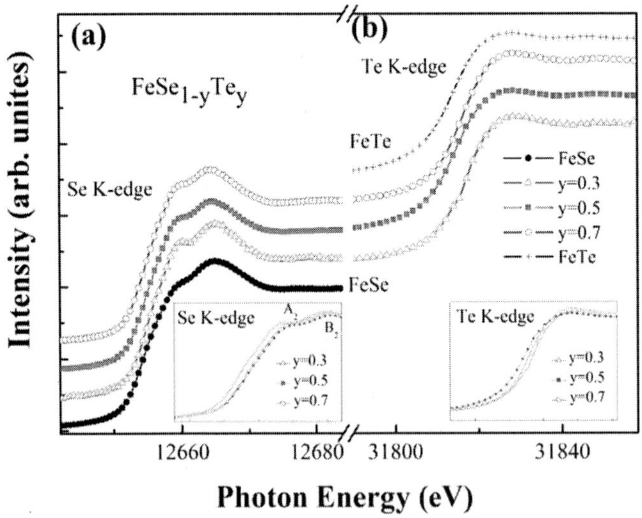


**Figure 9.** XAS of (a) Se K-edge and (b) Te K-edge. Insets show details of energy shifts of (a) Se K and (b) Te K-edge upon Te doping.

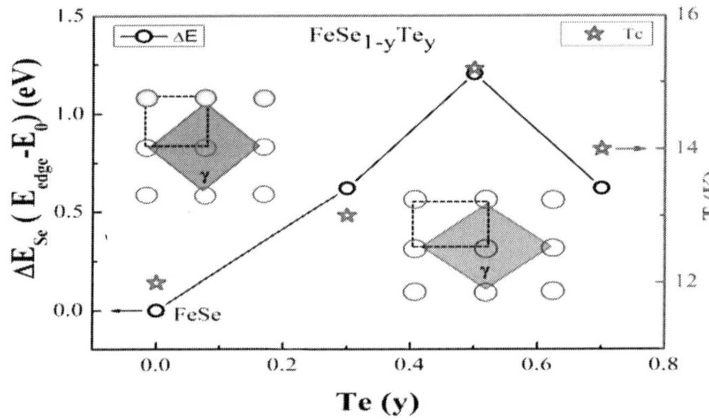


**Figure 10.** Energy shift of Se K-edge (black circle) and *Tc* as functions of Te doping. Sketch of variation of angle also shown.

# 4. CONCLUSION

This study elucidates the electronic properties related to the electron correlation and superconductivity of $FeSe_x$ and $FeSe_{1-y}Te_y$, with reference to measurements of XAS and RIXS. Spectroscopic data exhibit the signature of Fe 3d localization and different hybridization effects from those of "1111" and "122" systems. The charge balance considerations from p-hole also result in itinerant electrons. Fluctuation in the number of ligand 4p holes may arise from the charge transfer between Se and Te in the $FeSe_{1-y}Te_y$ crystals. Analysis results indicate that the superconductivity in Fe-based compounds of this class is strongly associated with the ligand 4p hole state. Additionally, the variation of Tc correlates well with the structural deformation and the change in the Se 4p holes. Moreover, the symmetry of Fe in the ab plane changes from the 4p orbital to modulating (varying) coordination geometry. XRD measurements indicate that this lattice distortion that increases with Se deficiency and the Te doped. Tetragonal FeSe with a PbO structure not only has the same planar sub-lattice as layered Fe-based quaternary oxypnictides, but also exhibits a structural stability upon Te substitution; it is a promising candidate for determining the origin of Tc in Fe-based superconductors. A fundamental question concerning the role of Fe magnetism in these superconductors is yet to be answered. The importance of charge transfer and the ligand 4p hole state should be considered as well.

## ACKNOWLEDGEMENT

The authors would like to thank the National Science Council of the Republic of China, Taiwan (Contract Nos. NSC-98-2112-M-213-006-MY3 and NSC-099-2112-M-001-036-MY3) for financially supporting this research. M. K. Wu, Y. Y. Chen, S. M. D. Rao, and K. W. Yeh at Academia Sinica are appreciated for providing study samples and their valuable discussions. J.-F. Lee, T. S. Chan, C. W. Pao, J. M. Chen, and J. M. Lee of NSRRC are commended for their valuable discussions and experimental support. J.-H. Guo and W. L. Yang of Advanced Light Source are gratefully acknowledged for their experimental support.


## REFERENCES


1. M. Aichhorn, S. Biermann, T. Miyake, A. Georges, M. Imada, 2010Theoretical evidence for strong correlations and incoherent metallic state in FeSe. Physical Review B, 82: 064504.

2. M. Aichhorn, L. Pourovskii, V. Vildosola, M. Ferrero, O. Parcollet, T. Miyake, A. Georges, S. Biermann, 2009Dynamical mean-field theory within an augmented plane-wave framework: Assessing electronic correlations in the iron pnictide LaFeAsO. Physical Review B, 80:085101.

3. D. Arčon, P. Jeglič, A. Zorko, A. Potočnik, A. Y. Ganin, Y. Takabayashi, M. J. Rosseinsky, K. Prassides, 2010Coexistence of localized and itinerant electronic states in the multiband iron-based superconductor FeSe0.42Te0.58., Physical Review B, 82:140508(R)

4. F. Bondino, E. Magnano, M. Malvestuto, F. Parmigiani, M. A. Mc Guire, A. S. Sefat, B. C. Sales, R. Jin, D. Mandrus, E. W. Plummer, D. J. Singh, N. Mannella, 2008Evidence for strong itinerant spin fluctuations in the normal state of CeFeAsO0.89F0.11 iron-oxypnictide superconductors, Physical Review Letters, 10126 7001267004

5. C. L. Chang, C. L. Chen, C. L. Dong, G. Chern, J. Lee, F. , L. Y. Jang, 2001X-ray absorption near edge structure studies of Fe1xNixOy thin films Journal of Electron Spectroscopy and Related Phenomena, 114-116:545-548.

6. C. Cruz, Q. de la Huang, J. W. Lynn, Ratcliff. I. I. J. W. Li, J. L. Zarestky, H. A. Mook, G. F. Chen, J. L. Luo, N. L. Wang, P. Dai, 2008Magnetic order close to superconductivity in the iron-based layered LaO1-xFxFeAs systems. Nature (London), 453899902

7. F. de Groot, G. Vankó, P. Glatzel, 2009The 1s x-ray absorption pre-edge structures in transition metal oxides. Journal of physics: Condensed Matter, 21: 104207.

8. J. Dong, H. J. Zhang, G. Xu, Z. Li, G. Li, W. Z. Hu, D. Wu, G. F. Chen, X. Dai, J. L. Luo, Z. Fang, N. L. Wang, 2008Competing orders and spin-density-wave instability in La(O1−xFx)FeAs. Europhysics Letters, 83:27006.

9. L. C. Duda, J. Nordgren, G. Dräger, S. Bocharov, T. Kirchner, 2000Minimal two-band model of the superconducting iron oxypnictides, Journal of Electron Spectrosocpy and Related Phenomena, 110275285

10. M. H. Fang, H. M. Pham, B. Qian, T. J. Liu, E. K. Vehstedt, L. Spinu, Z. Q. Mao, 2008Superconductivity close to magnetic instability in Fe(Se1-xTex)0.82, Physical Review B, 78224503224507

11. B. Freelon, Y. S. Liu, C. R. Rotundu, S. D. Wilson, J. Guo, J. L. Chen, W. L. Yang, C. L. Chang, P. A. Glans, P. Shirage, A. Iyo, R. J. Brigeneau, 2010X-ray absorption and emission spectroscopy study of the effect of doping on the low energy electronic structure of PrFeAsO1-, Journal of the Physical Society of Japan, 79: 074716-1-074716-7.

12. M. Grioni, J. B. Goedkoop, R. Schoorl, F. M. F. de Groot, J. C. Fuggle, F. Schäfers, E. E. Koch, G. Rossi, J. Esteva, M. , R. C. Karnatak, 1989Studies of copper valence states with Cu L3 x-ray-absorption spectroscopy, Physical Review B, 3915411545

13. F. C. Hsu, J. Y. Luo, K. W. Yeh, T. K. Chen, T. W. Huang, P. M. Wu, Y. C. Lee, Y. L. Huang, Y. Y. Chu, D. C. Yan, M. K. Wu, 2008Superconductivity in the PbO-type structure -FeSe. Proceedings of the National Academy of Sciences (PNAS), 1051426214264

14. B. Joseph, A. Iadecola, A. Puri, L. Simonelli, Y. Mizuguchi, Y. Takano, N. L. Saini, 2010Evidence of local structural inhomogeneity in FeSe1xTex from extend x-ray absorption fine structure, Physical Review B, 82: 020502- 020505(R).

15. B. Joseph, A. Iadecola, L. Simonelli, Y. Mizuguchi, Y. Takano, T. Mizokawa, T. , N. L. Saini, 2010A study of the electronic structure of FeSe1xTex chalcogenides by Fe and Se K-edge x-ray absorption near edge structure measurements, Journal of Physics: Condensed Mater, 22: 485702485706.

16. Y. Kamihara, T. Watanabe, M. Hirano, H. Hosono, 2008Iron-Based Layered Superconductor La[O1-xFx]FeAs (x=0.05–0.12) with Tc = 26 K. Journal of The American Chemical Society, 13032963297

17. T. Kroll, S. Bonhommeau, T. Kachel, H. A. Dürr, J. Werner, G. Behr, A. Koitzsch, R. Hübel, S. Leger, R. Schönfelder, A. K. Ariffin, R. Manzke, F. M. F. de Groot, J. Fink, H. Eschrig, B. Büchner, M. Knupfer, 2008Electronic structure of LaFeAsO1−xFx from x-ray absorption spectroscopy. Physical Review B, 78:220502.

18. E. V. Kurmaev, J. A. Mc Leod, A. Buling, N. A. Skorikov, A. Moewes, M. Neumann, M. A. Korotin, Y. A. Izyumov, N. Ni, P. C. Candield, 2009Contribution of Fe 3d states to the Fermi level of CaFe2As2, Physical Review B 80

19. K. Lee, W. , V. Pardo, W. E. Pickett, 2008Magnetism driven by anion vacancies in superconducting -FeSe1-x, Physical Review B, 78174502174506

20. S. D. Longa, A. Arcovito, B. Vallone, A. C. Castellano, R. Kahn, J. Vicat, Y. Soldo, J. L. Hazemann, 1999Polarized X-ray absorption spectroscopy of the low-temperature photoproduct of carbonmonoxy-myoglobin, Journal of Synchrotron Radiation, 611381147

21. F. W. Lytle, R. B. Greegor, D. R. Sandstrom, E. C. Marques, J. Wong, C. L. Spiro, G. P. Huffman, F. E. Huggins, 1984Measurement of soft x-ray absorption spectra with a fluorescent ion chamber detector. Nuclear Instruments and Methods in Physics Research Section A: Accelerators, Spectrometers, Detectors and Associated Equipment, 226542548

22. F. , W. Ji, J. Hu, Z. Y. Lu, T. Xiang, 2009First-Principles Calculations of the Electronic Structure of Tetragonal -FeTe and -FeSe Crystals: Evidence for a Bicollinear Antiferromagnetic Order. Physical Review Letters, 102:177003

23. W. Malaeb, T. Yoshida, T. Kataoka, Fujimori. A. Atsushi, M. Kubota, K. Ono, . H. Usui, K. Kuroki, R. Arita, H. Hosono, 2008Electronic Structure and Electron Correlation in LaFeAsO1xFx and LaFePO1-xFx. Journal of the Physical Society of Japan, 77:093714.

24. T. M. Mc Queen, Q. Huang, V. Ksenofontov, C. Felser, Q. Xu, H. Zandbergen, . Y. S. Hor, J. Allred, A. J. Williams, D. Qu, J. Checkelsky, Ong, N. P. , R. J. Cava, 2009Extreme sensitivity of superconductivity to stoichiometry in Fe1+Se. Physical Review B, 79:014522.

25. M. Merz, N. Nücker, P. Schweiss, S. Schuppler, C. T. Chen, V. Chakarian, . J. Freeland, Y. U. Idzerda, M. Kläser, G. Müller-Vogt, Th. Wolf, 1998Site-specific x-ray absorption spectroscopy of Y1-xCaxBa2Cu3O7-y: Overdoping and role of apical oxygen for high temperature superconductivity, Physical Review Letters, 8051925195

26. Y. Mizuguchi, F. Tomioka, S. Tsuda, T. Yamaguchi, Y. Takano, 2008Superconductivity at 27 K in tetragonal FeSe under high pressure, Applied Physics Letter, 93152505152507

27. B. H. Mok, S. M. Rao, M. C. Ling, K. J. Wang, C. T. Ke, P. M. Wu, C. L. Chen, F. C. Hsu, T. W. Huang, J. Y. Luo, D. C. Yan, K. W. Yeh, T. B. Wu, A. M. Chang, M. K. Wu, 2009Growth and Investigation of Crystals of the New Superconductor -FeSe from KCl Solutions. Crystal Growth and Design, 9732603264

28. K. Nakayama, T. Sato, P. Richard, T. Kawahara, Y. Sekiba, T. Qian, G. F. Chen, J. L. Luo, N. L. Wang, H. Ding, T. Takahashi, 2010Angle-Resolved Photoemission Spectroscopy of the Iron-Chalcogenide Superconductor Fe1.03Te0.7Se0.3: Strong Coupling Behavior and the Universality of Interband Scattering. Physical Review Letters, 105:197001.

29. I. A. Nekrasov, Z. V. Pchelkina, M. V. Sadovskii, 2008Electronic structure of prototype AFe2As2 and ReOFeAs high-temperature superconductors: A comparison, JETP Letters, 88144149

30. J. Norgdren, G. Bray, S. Cramm, R. Nyholm, J. E. Rubensson, N. Wassdahl, 1989Soft x-ray emission spectroscopy using monochromatized synchrotron radiation (invited). Review of Scientific Instruments. 60:1690.

31. A. Pourret, L. Malone, A. B. Antunes, C. S. Yadav, P. L. Paulose, B. Fauque, K. Behnia, 2010Strong correlation and low carrier density in Fe1+yTe0.6Se0.4 as seen from its thermoelectric response. Physical Review B, 83:020504.

32. S. Raghu, X. L. Qi, C. X. Liu, D. J. Scalapino, S. C. Zhang, 2008Minimal two-band model of the superconducting iron oxypnictides, Physical Review B, 77: 220503-220506 (R).

33. Z. Ren, A. , W. Lu, J. Yang, W. Yi, X. Shen, L. , Z. Li, C. , G. , C. , X. Dong, L. , L. Sun, L. , F. Zhou, Z. Zhao, X. , 2008Superconductivity at 55 K in iron-based F-doped layered quaternary compound Sm[O1-xFx]FeAs. Chinese Physics Letters, 2522152216

34. B. C. Sales, A. S. Sefat, M. A. Mc Guire, R. Y. Jin, D. Mandrus, Y. Mozharivskyj, 2009Bulk superconductivity at 14 K in single crystals of Fe1+yTexSe1−x. Physical Review B, 79, 094521.

35. Q. Si, E. Abrahams, 2008Strong correlations and magnetic frustration in the high Tc iron pnictides, Physical Review Letters, 101076401076404

36. P. G. Steeneken, L. H. Tjeng, G. A. Sawatzky, A. Tanaka, O. Tjernberg, G. Ghiringhelli, N. N. Brookes, A. A. Nugroho, A. A. Menovsky, (2003, 2003Crossing the gap from p- to n-type doping: nature of the states near the chemical potential in La2-xSrxCuO4 and Nd2-xCexCuO4-, Physical Review Letters 90

37. A. Subedi, L. Zhang, D. J. Singh, M. H. Du, 2008Density functional study of FeS, FeSe, and FeTe: Electronic structure, magnetism, phonons, and superconductivity, Physical Review B 78

38. H. Takahashi, K. Igawa, K. Arii, Y. Kamihara, M. Hirano, H. Hosono, 2008Superconductivity at 43 K in an iron-based layered compound LaO1-xFxFeAs. Nature (London, United Kingdom), 453376378

39. A. Tamai, A. Y. Ganin, Bacsa. E. J. Rozbicki, W. Meevasana, P. D. C. King, M. Caffio, R. Schaub, S. Margadonna, K. Prassides, M. J. Rosseinsky, F. Baumberger, 2010Strong Electron Correlations in the Normal State of the Iron-Based FeSe0.42Te0.58 Superconductor Observed by Angle-Resolved Photoemission Spectroscopy. Physical Review Letters, 104:097002

40. M. K. Wu, F. C. Hsu, K. W. Yeh, T. W. Huang, J. Y. Luo, M. J. Wang, H. H. Chang, T. K. Chen, S. M. Rao, B. H. Mok, C. L. Chen, Y. L. Huang, C. T. Ke, P. M. Wu, A. M. Chang, C. T. Wu, T. P. Perng, 2009The development of the superconducting PbO-type -FeSe and related compounds. Physica C, 469 (9-12):340-349.

41. G. Xu, W. M. Ming, Y. G. Yao, X. Dai, S. C. Zhang, Z. Fang, 2008Doping-dependent phase diagram of LaOMAs (M=V-Cu) and electron-type superconductivity near ferromagnetic instability. Europhysics Letters, 82:67002.

42. A. Yamasaki, Y. Matsui, S. Imada, K. Takase, H. Azuma, T. Muro, Y. Kato, A. Higashiya, A. Sekiyama, S. Suga, M. Yabashi, K. Tamasaku, T. Ishikawa, K. Terashima, H. Kobori, A. Sugimura, N. Umeyama, H. Sato, Y. Hara, N. Miyagawa, I. Ikeda, 2010Electron correlation in the FeSe superconductor studied by bulk-sensitive photoemission spectroscopy. Physical Review B, 82: 184511.

43. W. L. Yang, A. P. Sorini, C. Chen, C. , B. Moritz, W. Lee, S. , F. Vernay, P. Olalde-Velasco, J. D. Denlinger, B. Delley, J. Chu, H. , J. G. Analytis, I. R. Fisher, Z. A. Ren, J. Yang, W. Lu, Z. X. Zhao, Brink. J. van den, Z. Hussain, Z. Shen, Devereaux, T.2009Evidence for weak electronic correlations in iron pnictides. Physical Review B, 80:014508.

44. K. W. Yeh, Z. W. Huang, Y. L. Huang, T. K. Chen, F. C. Hsu, P. M. Wu, Y. C. Lee, Y. Y. Chu, C. L. Chen, J. Y. Luo, D. C. Yan, M. K. Wu, 2008Tellurium substitution effect on superconductivity of the α-phase iron selenide. Europhysics Letters, 84:37002.

45. K. W. Yeh, C. T. Ke, T. W. Huang, T. K. Chen, Y. L. Huang, P. M. Wu, M. K. Wu, 2009Superconducting FeSe1−xTex Single Crystals Grown by Optical Zone-Melting Technique. Crystal Growth Design, 9:4847.

46. T. Yildrim, 2008Origin of the 150-K anomaly in LaFeAsO: competing antiferromagnetic interactions, frustration, and a structural phase transition, Physical Review Letters, 101057010057013

47. R. Yoshida, T. Wakita, H. Okazaki, Y. Mizuguchi, S. Tsuda, Y. Takano, H. Takeya, K. Hirata, T. Muro, M. Okawa, K. Ishizaka, S. Shin, H. Harima, M. Hirai, Y. Muraoka, T. Yokoya, 2009Electronic Structure of Superconducting FeSe Studied by High-Resolution Photoemission Spectroscopy. The Physical Society of Japan, 78: 034708.

48. J. Zaanen, G. A. Sawatzky, J. W. Allen, 1985Band gaps and electronic structure of transition-metal compounds, Physical Review Letters, 55418421

4

# SYNTHESIS AND CHARACTERIZATION OF SUPERCONDUCTING $CA_{1-X}NA_XFFEAS$

**Klaus K. Wolff [1], Larysa Shlyk [1], Markus Bischoff [1], Eva Rose [2], Rainer Niewa [1] and Thomas Schleid [1,*]**

[1] Institut für Anorganische Chemie, Universität Stuttgart, Pfaffenwaldring 55, 70569 Stuttgart, Germany

[2] 1. Physikalisches Institut, Universität Stuttgart, Pfaffenwaldring 57, 70550 Stuttgart, Germany

## ABSTRACT

A representative of the fluoride-containing iron pnictide high-temperature superconductors, namely CaFFeAs, was doped with sodium up to the composition $Ca_{0.86}Na_{0.14}FFeAs$ for the first time. Single crystals with an edge length in the range of 0.1 – 2.0 mm were obtained via solid-state and flux syntheses, respectively. The composition of the crystals was verified by means of single crystal X-ray diffractometry and energy dispersive X-ray spectroscopy (EDX). Measurements of the electrical resistivity, as well as the magnetization on a crystal of $Ca_{0.89}Na_{0.11}FFeAs$ both show a transition to the superconducting state on cooling to 34.5 K. Investigations of the upper critical fields reveal an anisotropy ratio of about five. The lattice parameters and molar volumes increase with rising sodium content. This effect is clearly observable for the *c*-axis and the volume, whereas the increase of the *a*-axis is rather minor.

## KEYWORDS

high-$T_c$; iron arsenides; superconductors

## 1. INTRODUCTION

With the detection of superconductivity in iron pnictides in 2008 [1,2], a new class of high-temperature superconductors [3] was found 22 years after the discovery of this phenomenon in oxocuprates [4]. The so-called 1111 phases can be derived from LaOFeAs [5,6] with a tetragonal ZrSiCuAs-type structure, in which isotypical cationic $[LaO]^+$ and anionic $[FeAs]^-$ layers alternate. An isotypic class is the series, *Ae*FFeAs (*Ae* = Ca, Sr, Eu) [7–9], which results from the exchange of oxygen by fluorine and lanthanum by divalent alkaline-earth metals or europium, respectively. Meanwhile, several structurally related types of such iron pnictides became known [10]. Analogous to the oxocuprates, the superconductivity occurs in two-dimensional antiferromagnetically ordered layers, based on iron-centered edge-linked pnictogen tetrahedra. Within the 1111 compounds, superconductivity is only observed if a partial reduction (electron doping) or oxidation (hole doping) in the $[FeAs]^-$ layers occurs [11]. Critical temperatures between $T_c$ = 26 and 43 K for $LaO_{1-x}F_xFeAs$ [2,12] and 55 to 58 K for $SmO_{1-x}F_xFeAs$ [13,14] were achievable for fluoride-doped samples, whereas the strontium substituted oxide $La_{1-x}Sr_xOFeAs$ [15] exhibits a transition temperature of $T_c$ = 25 K.

Up to now, only electron-doped derivatives of the fluorine containing the 1111 compound CaFFeAs, like $CaFFe_{1-x}Co_xAs$ [7] and $Ca_{1-x}Ln_xFFeAs$ (*Ln* = Pr, Nd) [16], are reported to superconduct. After the discovery of superconductivity in 1111 phases, hole doping on LaOFeAs with $Ca^{2+}$ was already investigated without the detection of any superconducting phases [2]; the reason why electron doping was claimed as a critical factor for the generation of superconductivity in this system at that time. However, this view was disproven by hole-doped $La_{1-x}Sr_xOFeAs$ with $T_c$ = 25 K for $x$ = 0.13 three months later [15].

# 2. RESULTS AND DISCUSSION

In the present work, we show that it is possible to attain high-temperature superconductivity on alkaline metal-doping of a fluorine containing the 1111 phase for the first time. CaFFeAs [7] was doped with sodium, resulting in the solid-solution series, $Ca_{1-x}Na_xFFeAs$ ($x = 0 - 0.14$), prepared via a solid-state reaction method. $Ca_{1-x}Na_xFFeAs$ consists of small platelet-like crystal aggregates (Figure 1) sufficient in size and quality for single crystal X-ray analysis (about $0.1 \times 0.1$ mm$^2$). Larger single crystals with a composition of $Ca_{0.89}Na_{0.11}FFeAs$ and a size up to $2 \times 2$ mm$^2$ were grown in a flux of sodium chloride for physical measurements.

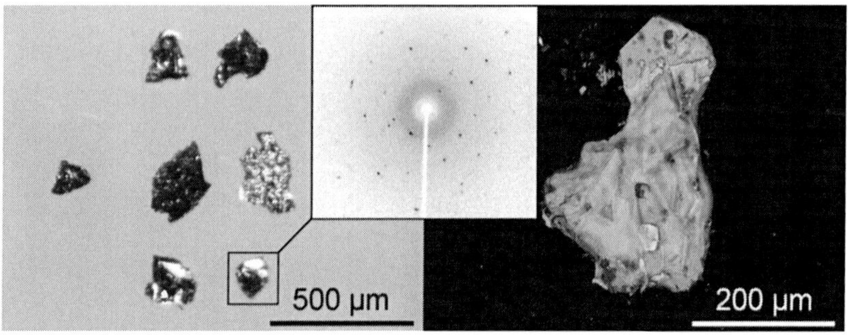


**Figure 1.** Crystal aggregates of $Ca_{1-x}Na_xFFeAs$ viewed via an optical (**left**) and an electron microscope (**right**). The inset shows a Laue-image of a representative single crystal.

## 2.1. Crystal Structure and Crystallographic Data of $Ca_{1-x}Na_xFFeAs$

The 1111 phase CaFFeAs [7] crystallizes in the tetragonal ZrCuSiAs-type structure, representing a filled PbFCl variant, in the space group *P*4/*nmm* with $Z = 2$. The structure consists of alternating isotypic cationic $[FCa_{4/4}]^+$ and anionic $[FeAs_{4/4}]^-$ layers of edge-connected tetrahedra (Figure 2).

Sodium partly substitutes calcium in $Ca_{1-x}Na_xFFeAs$ on the 2*c* position. From the four crystallographically independent atoms, iron and fluorine are both tetrahedrally coordinated, whereas arsenic and calcium are surrounded in the form of square antiprisms built from four counter ions of each kind (Figure 3).

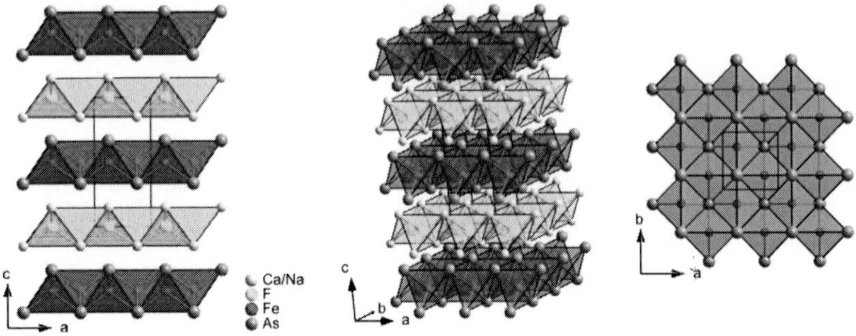


**Figure 2.** Crystal structure of $Ca_{1-x}Na_xFFeAs$. (**a**) The view along [010]; (**b**) a three-dimensional picture and (**c**) top view of a $[FeAs_{4/4}]^-$ layer along [001].

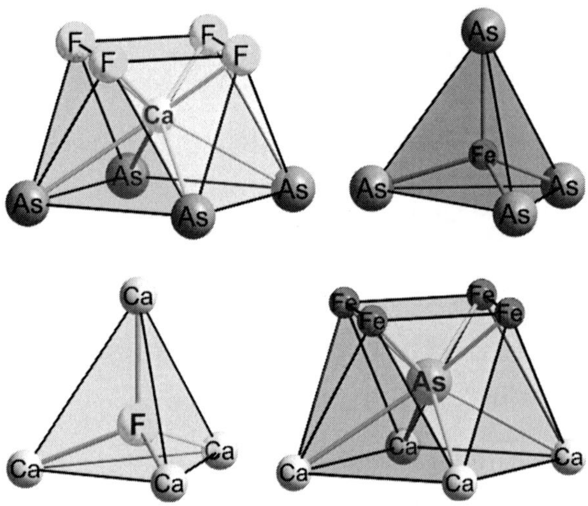


**Figure 3.** Coordination polyhedra around the four crystallographically independent atoms in $Ca_{1-x}Na_xFFeAs$.

Crystallographic data for selected representatives of the substitution series $Ca_{1-x}Na_xFFeAs$ are listed in Table 1. More detailed crystallographic information can be found in the supplementary part (Tables S1–S4).

**Table 1.** Crystallographic data for selected crystals of $Ca_{1-x}Na_xFFeAs$.

| **$x$ in $Ca_{1-x}Na_xFFeAs$** | | **0** | **0.03** | **0.08** | **0.14** |
|---|---|---|---|---|---|
| Crystal system | | tetragonal | | | |
| Space group | | $P4/nmm$ ($Z$ = 2) | | | |
| Lattice parameters | $a$/pm | 387.57(9) | 387.57(2) | 387.65(2) | 387.68(3) |
| | $c$/pm | 858.4(2) | 859.48(5) | 859.89(6) | 859.87(9) |
| | $c/a$ | 2.215 | 2.218 | 2.218 | 2.218 |
| $V_m/cm^3 \cdot mol^{-1}$ | | 38.83 | 38.87 | 38.91 | 38.91 |
| $D_x/g \cdot cm^{-3}$ | | 4.89 | 4.87 | 4.84 | 4.82 |
| Diffractometer | | Nonius KappaCCD (Bruker AXS, Karlsruhe, Germany) | | | |
| Radiation | | Mo-$K\alpha$ ($\lambda$ = 71.07 pm) | | | |
| Collected reflections | | 1589 | 2278 | 3232 | 1622 |
| Unique reflections | | 118 | 145 | 211 | 118 |
| $R_{int}$, $R_\sigma$ | | 0.083, 0.028 | 0.060, 0.018 | 0.068, 0.021 | 0.104, 0.039 |
| $R_1$, $wR_2$ | | 0.076, 0.157 | 0.018, 0.044 | 0.022, 0.053 | 0.026, 0.058 |
| $GooF$ | | 1.178 | 1.168 | 1.166 | 1.165 |
| Solution and refinement | | Program package *SHELX-97* [17,18] | | | |

## 2.2. Phase Analysis of $Ca_{1-x}Na_xFFeAs$ and Discussion

Powder X-ray diffractometry verifies the presence of the target compound in each sample of the series. For $x \geq 0.1$, however, an increasingly larger portion of secondary phases arises (Figure 4).

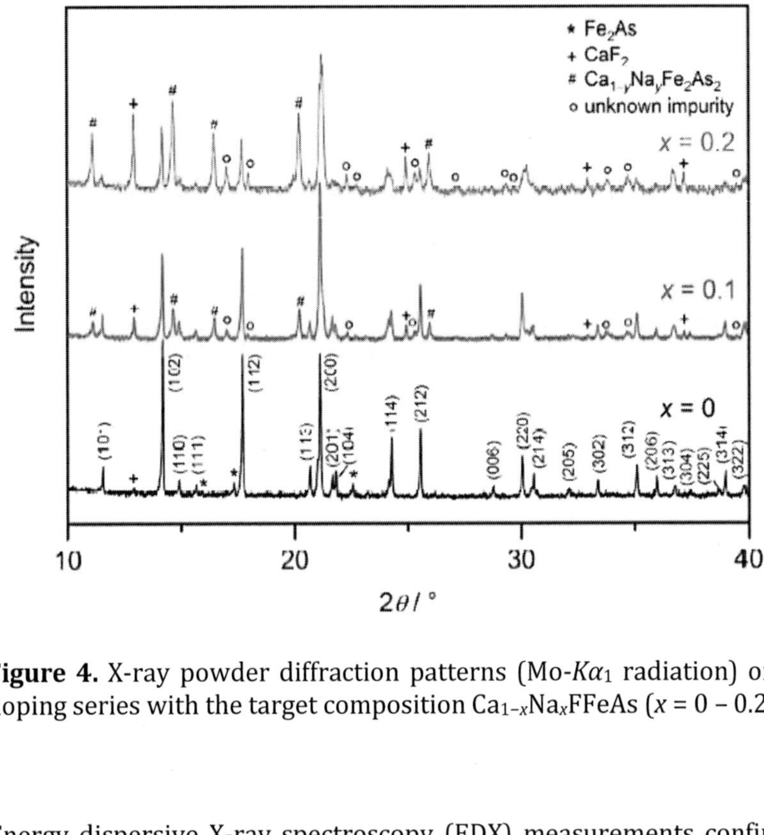


**Figure 4.** X-ray powder diffraction patterns (Mo-$K\alpha_1$ radiation) of the doping series with the target composition $Ca_{1-x}Na_xFFeAs$ ($x = 0 - 0.2$).

Energy dispersive X-ray spectroscopy (EDX) measurements confirm a certain amount of dopants in the target compounds. The crystals of $Ca_{1-x}Na_xFFeAs$ exhibit compositions in a rather narrow range from $Ca_{0.99}Na_{0.01}FFeAs$ to $Ca_{0.86}Na_{0.14}FFeAs$, depending on the sodium concentration used in the synthesis. Single crystals of the series $Ca_{1-x}Na_xFFeAs$ can be structurally refined with the compositions from the EDX analyses (Table 1). Figure 5 shows the development of the lattice parameters, $a$ and $c$, as well as the molar volume, $V_m$, of $Ca_{1-x}Na_xFFeAs$ for $x = 0$, 0.03, 0.08 and 0.14 as a function of the sodium content. The ionic radii with a coordination number of eight are 112 and 118 pm for $Ca^{2+}$ and $Na^{+}$, respectively [19]. The expected increase of the lattice parameters and volume for a rising sodium content is only slightly realized for the *a*-axes, while the increase in the *c*-axes and, subsequently, the volumes are more visible. For the compound $Ca_{0.86}Na_{0.14}FFeAs$ with the highest degree of substitution, the *a*-axis increases by 0.02% and the *c*-axis by 0.17%, as compared to the unsubstituted compound. An almost unchanged *a*-axis has been observed previously in the case of hole-doped $La_{0.87}Sr_{0.13}OFeAs$ [15], whereas the unit-cell size even decreases with increasing sodium content in the case of $Ca_{1-x}Na_xFe_2As_2$ [20]. The observed increase of the *c*-axis for $Ca_{1-x}Na_xFFeAs$ lies in the similar range as in $La_{0.87}Sr_{0.13}OFeAs$ (0.34%). The rather small rise of the *c*-axis length, as well as the molar volume may indicate a slight overestimation of the sodium content, particularly for the crystal with highest degree of substitution ($x = 0.14$).

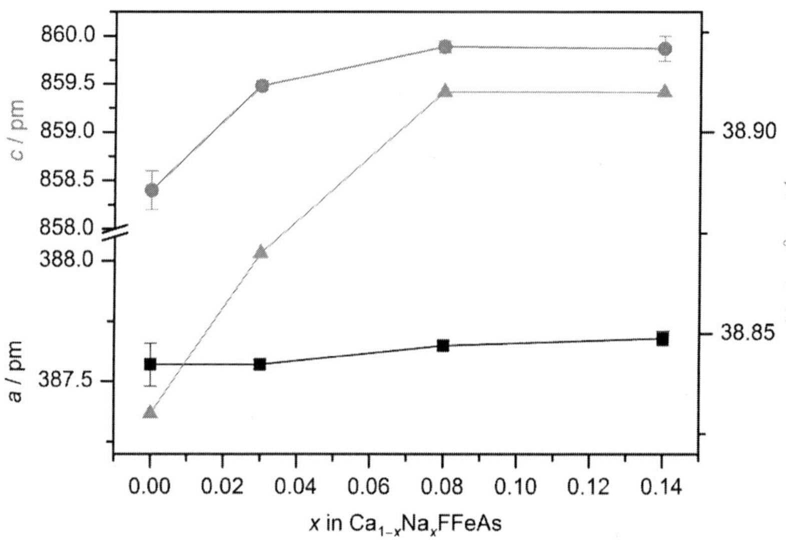


**Figure 5.** Lattice parameters and molar volumes of $Ca_{1-x}Na_xFFeAs$ phases*versus* the sodium doping grade *x* (from energy dispersive X-ray spectroscopy (EDX)).

A closer look at the equivalent isotropic displacement parameters (see the Supplementary Information, Table S2) shows that the ones for the Ca/Na position remain nearly constant for $x > 0$. The parameters of the other atoms increase with rising sodium content. This effect could be explained with an incremental static displacement of the F, Fe and As atoms for an increasing Na level via the statistical occupation of the 2*c* position (Ca/Na). The atomic coordinates of the $Ca_{1-x}Na_xFFeAs$ crystals do not indicate a significant change with rising substitution grade. This is the reason for the virtually unmodified distances, e.g., *d*(Ca/Na–F) = 233 – 234 pm, *d*(Ca/Na–As) = 316 – 317 pm, *d*(Fe–As) = 240 pm and *d*(Fe–Fe) = 274 pm, and angles, e.g., ∢(As–Fe–As′) = 107.7° – 108.8° and 110.3° – 103.4°, within the series (see Table S4 of the Supplementary Information for details). Therefore, it seems that the variation of the electron count of iron via doping is exclusively responsible for the appearance of superconductivity in this system.

## 2.3. Magnetization and Resistivity Measurements on $Ca_{1-x}Na_xFFeAs$

Magnetization measurements on $Ca_{1-x}Na_xFFeAs$-containing powder samples indicate the occurrence of superconductivity. In the range of nominal composition of $x$ = 0.1 – 0.2, a transition temperature to superconductivity of 34 K was observed, regardless of the doping level (Figure 6).

The superconducting volume fraction in the samples with $x$ = 0.1 and 0.2 can be given by about 10% and 24%, respectively. However, highly substituted $Ca_{1-x}Na_xFe_2As_2$ detected as a side phase in all samples was previously reported to exhibit superconductivity with a nearly identical $T_c$ up to 34 K [20]. For this reason, the superconducting transition observed in the powder samples cannot be ascribed definitively to the target phase. However, it seems that it is at least contributing to the content of the superconducting material, because of the observation that in samples with a higher sodium amount of $x$ = 0.3, where an excess of the side phases according to Figure 4 is present and the target phase becomes the minority component, the superconducting volume fraction decreases considerably. Already known hole-doped oxide compounds of the 1111 class show lower transition temperatures (for example, $La_{0.87}Sr_{0.13}OFeAs$: $T_c$ = 25.6 K [15] and $Pr_{0.75}Sr_{0.25}OFeAs$: $T_c$ = 16.3 K [21]).

To investigate the characteristics of phase pure sodium-doped 1111, phase resistivity, as well as directional physical measurements of the magnetic susceptibility have been performed on a larger single crystal of $Ca_{0.89}Na_{0.11}FFeAs$ grown in NaCl flux (Figure 7). Single crystal XRD results in a sodium doping grade of 11(3)% and the following parameters: tetragonal, *P4/nmm*, $Z = 2$, $a = 387.8(1)$ pm, $c = 859.9(2)$ pm, $R_1 = 0.034$, $wR_2 = 0.079$, $GooF = 1.027$. The critical temperature apparent at about 34.5 K in both measurements shows that Na-doped CaFFeAs with a sodium content of 11% incidentally exhibits almost the same $T_c$ as $Ca_{1-x}Na_xFe_2As_2$ ($x \approx 0.66$) [20].

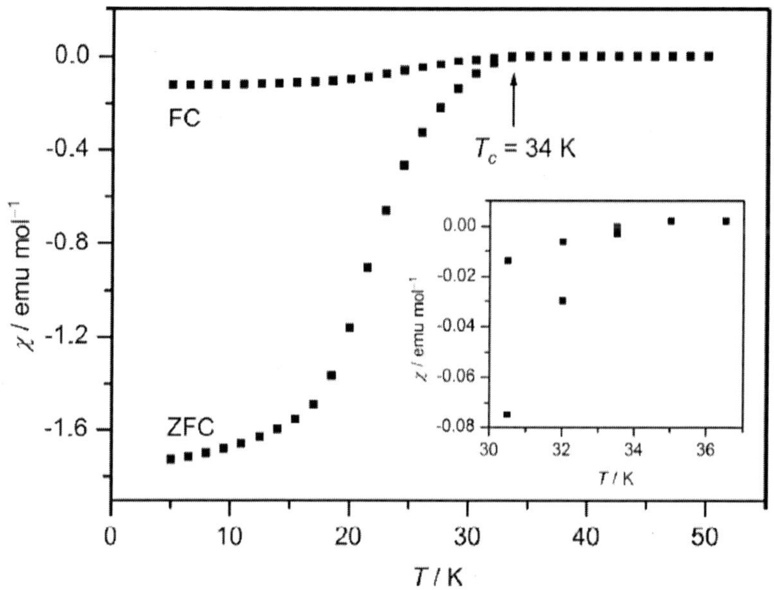


**Figure 6.** Magnetic susceptibility of polycrystalline $Ca_{0.9}Na_{0.1}FFeAs$-containing powder samples in a magnetic field of 100 Oe with a close-up of the transition region.

The anisotropic properties of $Ca_{0.89}Na_{0.11}FFeAs$ are available from the temperature dependence of the upper critical fields, which is shown in Figure 8. The anisotropy factor $\gamma = H_{c2||}(0)/H_{c2\perp}(0)$ can be calculated from the estimated orbital critical fields, $H_{c2\perp}(0)$ and $H_{c2||}(0)$, at 0 K, derived from the Werthamer-Helfand-Hohenberg formula [22], which uses the slope of the $H_{c2}$ curves at $T_c$. In the case of $Ca_{0.89}Na_{0.11}FFeAs$ single crystals, this leads to $H_{c2\perp}(0) = 75.6$ T and $H_{c2||}(0) = 370.7$ T, which results in $\gamma \approx 5$. The received value for the anisotropy factor is in accordance with the literature data for 1111 compounds [23,24]. In contrast 122 phases exhibit smaller values of about two [25,26]. More details on the superconducting properties of $Ca_{1-x}Na_xFFeAs$ ($x$ = 0.11) will be discussed in a forthcoming publication [27].

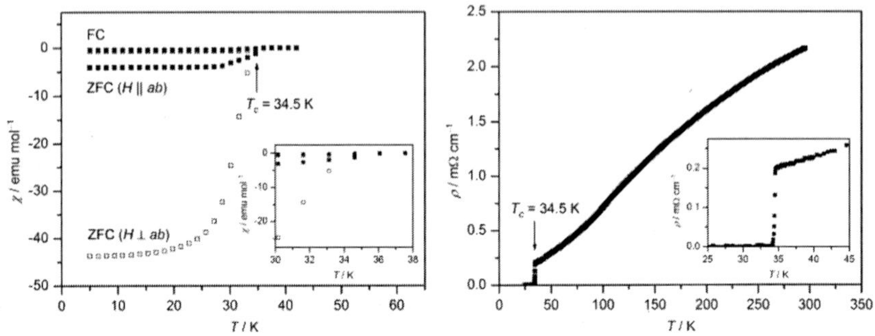


**Figure 7.** Temperature dependence of the zero-field cooled (ZFC) and field cooled (FC) DC magnetic susceptibility ($\mu_0 H$ = 100 Oe) (**a**) and DC resistivity (**b**) of an oriented single crystal of $Ca_{0.89}Na_{0.11}FFeAs$. (The insets are showing close-ups of the transition regions.)

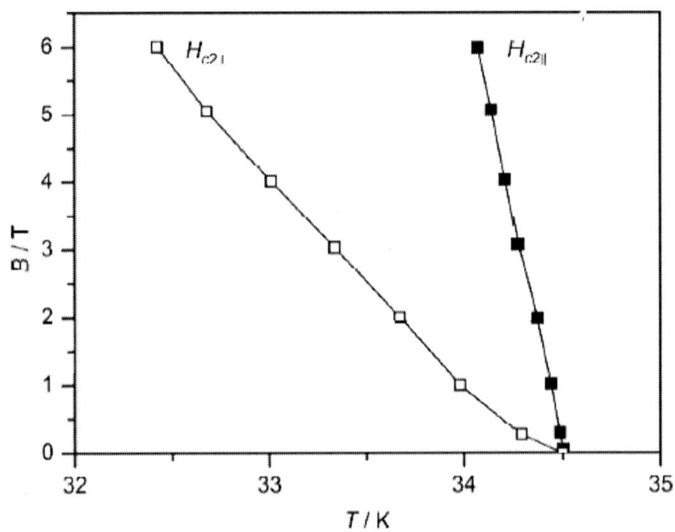


**Figure 8.** *H–T* phase diagram of $Ca_{0.89}Na_{0.11}FFeAs$ single crystals. ($H_{c2\perp}$ and$H_{c2||}$: upper critical fields perpendicular and parallel to the *ab* plane, respectively.)

# 3. EXPERIMENTAL SECTION

The approaches to synthesize $Ca_{1-x}Na_xFFeAs$ were performed via classical solid-state reactions at 1000 °C for 24 h, according to the following equation ($x = 0 - 0.2$).

$$(1-x)\,CaF_2 + 2x\,NaF + (1-x)\,CaAs + 2\,Fe + (1+x)\,As \rightarrow 2\,Ca_{1-x}Na_xFFeAs \quad (1)$$

The precursors were ground to fine powders and pressed to pellets prior to reaction. Previously, CaAs was synthesized by the reaction of calcium pieces with arsenic powder at 900 °C for 10 h in evacuated silica ampoules. All preparation procedures were carried out in an argon-filled glove box ($O_2$, $H_2O$ < 1 ppm). Small black crystals with a disc-like shape of the target compound were obtained. Single crystals with a size of up to 2 × 2 $mm^2$ were available via flux growth in sodium chloride. For this purpose, the educts ($CaF_2$, NaF, CaAs, Fe and As), according to Equation (1), are heated with NaCl in an $Al_2O_3$ crucible at 950 °C for 6 h, held at an intermediate temperature of 850 °C for the same duration and, subsequently, cooled to ambient temperature in a period of 24 h (cooling rate: 34.6 $°C\cdot h^{-1}$). These crystals could be separated from the flux manually.

X-ray diffractometry has been performed with the use of Mo-$K\alpha$ radiation ($\lambda$ = 71.07 pm) on a Nonius KappaCCD (Bruker AXS, Karlsruhe, Germany) and a STOE STADI P (Stoe & Cie, Darmstadt, Germany) with a DECTRIS MYTHEN 1K detector for the single-crystal and the powder X-ray measurements, respectively. The single crystals were sealed in glass capillaries, and the powder samples were measured in transmission geometry as thin films on an amorphous adhesive foil. Energy dispersive X-ray spectroscopy (EDX) was carried out via a Thermo Scientific NORAN System 7 X-ray Microanalysis System (Thermo Fisher Scientific, Waltham, MA, USA) with a UltraDry detector. DC-magnetization measurements of oriented single crystals were performed using a Quantum Design MPMS SQUID magnetometer (Quantum Design, San Diego, CA, USA) in the temperature range 5–50 K. $H_{c2}$ values were determined from DC magnetization in different magnetic fields for field-cooled (FC) and zero-field-cooled (ZFC) measurements in the reversible regime. The electrical resistivity was measured by a standard four-probe DC-method in the temperature range 5 K < $T$ < 300 K.

# 4. CONCLUSIONS

The results presented here show that it is possible to substitute the fluorine-containing 1111 class of iron pnictides at the alkaline-earth metal position with sodium and to obtain $Ca_{1-x}Na_xFFeAs$ phases. Crystals of $Ca_{0.89}Na_{0.11}FFeAs$ exhibit superconductivity at 34.5 K, which is intermediate between the observed temperatures for the electron-doped compounds $Ca_{1-x}Co_xFFeAs$ [7] with $T_c$ = 22 K for $x$ = 0.1 and $Ca_{1-x}Nd_xFFeAs$ [16] with $T_c$= 56 K for $x$ = 0.6. An estimation of the orbital critical fields, $H_{c2\perp}(0)$ and $H_{c2||}(0)$, for $Ca_{0.89}Na_{0.11}FFeAs$ single crystals results in values of 75.6 and 370.7 T, respectively, which yields an anisotropy ratio of $\gamma \approx 5$. This outcome could lead to the discovery of further compounds that are substituted with less valence electron containing elements, which have been very rare, so far. Promising candidates for this purpose could be SrFFeAs and EuFFeAs, with possible substitutes, like sodium and potassium, from an ionic radii point of view. It seems that the variation of the electron count of iron via doping is exclusively responsible for the appearance of superconductivity within this system, because there is no significant change in the interatomic distances and angles with increasing sodium content.

# ACKNOWLEDGMENTS

This work was supported by the German Research Foundation (Deutsche Forschungsgemeinschaft, DFG, Bonn, Germany) as part of the priority program SPP1458. We would also like to thank Martin Dressel and his group from the Physics Department at the University of Stuttgart for the collaboration within the aforementioned program, in particular for enabling us to carry out magnetic measurements on his SQUID machine.

# AUTHOR CONTRIBUTIONS

K.W.W.: Idea, preparation, crystal structure determination and refinement, crystallographic part, physical characterization (X-ray powder diffraction and EDX), writing of the manuscript. L.S.: Single crystal growth, physical characterization (magnetic measurements and interpretation). M.B.: $Ca_{0.89}Na_{0.11}FFeAs$ single crystal structure determination. E.R.: Single crystal resistivity measurement. R.N.: Valuable discussion, contribution to the crystallographic part, proofreading of the manuscript. T.S.: Basic idea, valuable discussion, final crystal structure refinement, proofreading of the manuscript.

## REFERENCES


1. Kamihara, Y.; Hiramatsu, H.; Hirano, M.; Kawamura, R.; Yanagi, H.; Kamiya, T.; Hosono, H. Iron-based layered superconductor: LaOFeP. *J. Am. Chem. Soc* **2006**, *128*, 10012–10013.

2. Kamihara, Y.; Watanabe, T.; Hirano, M.; Hosono, H. Iron-based layered superconductor La[$O_{1-x}$ $F_x$]FeAs ($x$= 0.05–0.12) with $T_c$ = 26 K. *J. Am. Chem. Soc* **2008**, *130*, 3296–3297.

3. Johrendt, D.; Pöttgen, R. Pnictide oxides: A new class of high-$T_c$ superconductors. *Angew. Chem. Int. Ed***2008**, *47*, 4782–4784.

4. Bednorz, J.G.; Müller, K.A. Possible high $T_c$ superconductivity in the Ba-La-Cu-O system. *Z. Phys* **1986**, *B64*, 189–193.

5. Quebe, P.; Terbüchte, L.J.; Jeitschko, W. Quaternary rare earth transition metal arsenide oxides *RT* AsO ($T$ = Fe, Ru, Co) with ZrCuSiAs type structure. *J. Alloys Compd* **2000**, *302*, 70–74.

6. Nitsche, F.; Jesche, A.; Hieckmann, E.; Doert, T.; Ruck, M. Structural trends from a consistent set of single-crystal data of *R* FeAsO ($R$ = La, Ce, Pr, Nd, Sm, Gd, and Tb). *Phys. Rev. B* **2010**, *82*, 134514:1–134514:8.

7. Matsuishi, S.; Inoue, Y.; Nomura, T.; Yanagi, H.; Hirano, M.; Hosono, H. Superconductivity induced by co-doping in quaternary fluoroarsenide CaFeAsF. *J. Am. Chem. Soc* **2008**, *130*, 14428–14429.

8. Tegel, M.; Johansson, S.; Weiß, V.; Schellenberg, I.; Hermes, W.; Pöttgen, R.; Johrendt, D. Synthesis, crystal structure and spin-density-wave anomaly of the iron arsenide-fluoride SrFeAsF. *Europhys. Lett* **2008**, *84*, 67007:1–67007:5.

9. Tegel, M.; Rotter, M.; Weiß, V.; Schappacher, F.M.; Pöttgen, R.; Johrendt, D. Structural and magnetic phase transitions in the ternary iron arsenides $SrFe_2As_2$ and $EuFe_2As_2$. *J. Phys. Condens. Matter* **2008**, *20*, 452201:1–452201:5.

10. Johrendt, D.; Hosono, H.; Hoffmann, R.-D.; Pöttgen, R. Structural chemistry of superconducting pnictides and pnictide oxides with layered structures. *Z. Kristallogr* **2011**, *226*, 435–446.

11. Rotter, M.; Pangerl, M.; Tegel, M.; Johrendt, D. Superconductivity and crystal structures of $(Ba_{1-x}K_x)$ $Fe_2As_2$ ($x = 0 - 1$). *Angew. Chem. Int. Ed* **2008**, *47*, 7949–7952.

12. Takahashi, H.; Igawa, K.; Arii, K.; Kamihara, Y.; Hirano, M.; Hosono, H. Superconductivity at 43 K in an iron-based layered compound $LaO_{1-x}F_x$FeAs. *Nature* **2008**, *453*, 376–378.

13. Ren, Z.-A.; Lu, W.; Yang, J.; Yi, W.; Shen, X.-L.; Li, Z.-C.; Che, G.-C.; Dong, X.-L.; Sun, L.-L.; Zhou, F.; Zhao, Z.-X. Superconductivity at 55 K in iron-based F-doped layered quaternary compound $Sm[O_{1-x}F_x]$FeAs.*Chin. Phys. Lett* **2008**, *25*, 2215–2216.

14. Fujioka, M.; Denholme, S.J.; Ozaki, T.; Okazaki, H.; Deguchi, K.; Demura, S.; Hara, H.; Watanabe, T.; Takeya, H.; Yamaguchi, T.; *et al.* Phase diagram and superconductivity at 58.1 K in $\alpha$-FeAs-free $SmFeAsO_{1-x}F_x$. *Supercond. Sci. Technol* **2013**, *26*, 085023:1–085023:4.

15. Wen, H.-H.; Mu, G.; Fang, L.; Yang, H.; Zhu, X. Superconductivity at 25 K in hole-doped $(La_{1-x}Sr_x)$OFeAs.*Europhys. Lett* **2008**, *82*, 17009:1–17009:5.

16. Cheng, P.; Shen, B.; Mu, G.; Zhu, X.; Han, F.; Zeng, B.; Wen, H.-H. High-$T_c$ superconductivity induced by doping rare-earth elements into CaFeAsF. *Europhys. Lett* **2009**, *85*, 67003:1–67003:4.

17. Sheldrick, G.M. *SHELX-97: Program Package for the Determination of Crystal Structures by Single Crystal X-ray and Neutron Diffraction*; University of Göttingen: Göttingen, Germany, 1997

18. Sheldrick, G.M. A short history of SHELX. *Acta Crystallogr* **2008**, *A64*, 112–122.

19. Shannon, R.D. Revised effective ionic radii and systematic studies of interatomic distances in halides and chalcogenides. *Acta Crystallogr* **1976**, *A32*, 751–767.

20. Zhao, K.; Liu, Q.Q.; Wang, X.C.; Deng, Z.; Lv, Y.X.; Zhu, J.L.; Li, F.Y.; Jin, C.Q. Doping dependence of the superconductivity of $(Ca_{1-x}Na_x)Fe_2As_2$. *Phys. Rev. B* **2011**, *84*, 184534:1–184534:7.

21. Mu, G.; Zeng, B.; Zhu, X.; Han, F.; Cheng, P.; Shen, B.; Wen, H.-H. Synthesis, structural, and transport properties of the hole-doped superconductor $Pr_{1-x}Sr_xFeAsO$. *Phys. Rev. B* **2009**, *79*, 104501:1–104501:6.

22. Werthamer, N.R.; Helfand, E.; Hohenberg, P.C. Temperature and purity dependence of the superconducting critical field, $H_{c2}$. III. Electron spin and spin-orbit effects. *Phys. Rev* **1966**, *147*, 295–302.

23. Jaroszynski, J.; Hunte, F.; Balicas, L.; Jo, Y.; Raicevic, I.; Gurevich, A.; Larbalestier, D.C.; Balakirev, F.F.; Fang, L.; Cheng, P.; *et al.* Upper critical fields and thermally-activated transport of $NdFeAsO_{0.7}F_{0.3}$ single crystal. *Phys. Rev. B* **2008**, *78*, 174523:1–174523:9.

24. Lee, H.-S.; Bartkowiak, M.; Park, J.-H.; Lee, J.-Y.; Kim, J.-Y.; Sung, N.-H.; Cho, B.K.; Jung, C.-U.; Kim, J.S.; Lee, H.-J. Effects of two gaps and paramagnetic pair breaking on the upper critical field of $SmFeAsO_{0.85}$and $SmFeAsO_{0.8}F_{0.2}$ single crystals. *Phys. Rev. B* **2009**, *80*, 144512:1–144512:8.

25. Yuan, H.Q.; Singleton, J.; Balakirev, F.F.; Baily, S.A.; Chen, G.F.; Luo, J.L.; Wang, N.L. Nearly isotropic superconductivity in $(Ba,K)Fe_2As_2$. *Nature* **2009**, *457*, 565–568.

26. Haberkorn, N.; Maiorov, B.; Jaime, M.; Usov, I.; Miura, M.; Chen, G.F.; Yu, W.; Civale, L. Effect of doping on structural and superconducting properties in $Ca_{1-x}Na_x$ Fe2As2 single crystals (x = 0.5, 0.6, 0.75). *Phys. Rev. B* **2011**, *84*, 064533:1–064533:10.

27. Shlyk, L.; Wolff, K.K.; Bischoff, M.; Rose, E.; Schleid, Th.; Niewa, R. Crystal structure and superconducting properties of hole-doped $Ca_{0.89}Na_{0.11}FFeAs$ single crystals. *Supercond. Sci. Technol* **2014**. in press.

5

# INVESTIGATING DOUBLE-REGIME CROSSOVER IN $Y_1BA_2CU_{3-X}R_XO_{7-\Delta}$ SUPERCONDUCTORS

**Mohamed H. Badr*, Lubna M. Sharaf El-Deen, Mohamed El-Hofy, Aly A. Haswa**

Physics Department, Faculty of Science, Minufiya University, Shebin El-Koom, Egypt

## ABSTRACT

The effect of Pr and Gd doping on the transport properties of Cu-deficient YBCO superconductors has been studied. Two series of $Y_1Ba_2Cu_{3-x}R_xO_{7-\delta}$, where R = Pr or Gd, were prepared by the conventional solid-state reaction technique. Resistance measurements showed a suppression of $T_c$ with increasing of Prand Gd-contents in addition to a normal-state metal-to-insulator transition. Moreover, a superconductor-to-insulator transition has been observed at ambient pressure for temperatures less than 50 K for Pr with x = 0.3 and for Gd with x > 0.3. The overall complex behaviours of the resistivity data have been preliminary explained in terms of localization of charge carriers, structural disorders, and magnetic ordering of magnetic moments.

## KEYWORDS

High Tc Superconductivity; Metal-Insulator Transition; Superconductor-Insulator Transition

# 1. INTRODUCTION

Elemental substitutions play an important role in understanding the nature of superconductivity in high $T_c$ superconductors. Substitutions of Cu by X ions in RBCXO system, where R is a rare earth and X is Mn, Cr, Fe, Co, Ni, Al, and Ga, have been studied [1] -[11] . The studies have revealed that substitutions with nonmagnetic ions caused a huge decrease in the critical transition temperature as compared to magnetic ions. Many non-magnetic mechanisms have been proposed to explain the phenomena in addition to pair-breaking due to magnetic impurities in accordance with Abrikosov and Gorkov theory. The dependence of resistivity on temperature proved to be difficult to explain not only for different compounds, but within the same family as well [12] . Metal to insulator transition (MIT) has been observed upon substitutions at Y-, Ba-, and Cu-sites in diverse superconducting systems [13] -[21] . Moreover, evidence of MIT due to other reasons such as changes in oxygen content, pressure, film thickness, and doping at grain boundaries has been reported [22] -[26] . In this work we investigated the effect of doping magnetic ions of R = Pr and Gd on Cu-deficient $Y_1Ba_2Cu_{3-x}R_xO_{7-\delta}$ system with x = 0.1, 0.2, and 0.3 and x = 0.1, 0.2, 0.3 and 0.4, respectively. Our aim is to study and compare the MIT and the observed incomplete superconductor to insulator (SIT) crossovers in the two systems and investigate its origin.

# 2. EXPERIMENTAL DETAILS

The polycrystallized samples were prepared by the conventional solid state reaction method. Raw powders of $Y_2O_3$ (99.99%, Aldrich Chemical Co. Ltd.), $Gd_2O_3$ (99.9%, Aldrich Chemical Co. Ltd.), $Pr_6O_{11}$ (99.9%, Sigma Chemical Co.), $BaCO_3$ (99%, NTL Nentech Ltd.), and CuO (99%, Aldrich Chemical Co. Ltd.), were mixed according to the nominal chemical formula $Y_1Ba_2Cu_{3-x}Pr_xO_{7-\delta}$with x = 0, 0.1, 0.2, and 0.3 and $Y_1Ba_2Cu_{3-x}Gd_xO_{7-\delta}$ with x = 0, 0.1, 0.2, 0.3, and 0.4. The powders were dried for 10 hr at 120°C and mixed according to the required stoichiometric compositions. Then, they were manually ground using an agate mortar for $1/2$ hr followed by mechanical milling for 5 hr to ensure homogeneity. Calcination of homogenous mixtures took place at 900°C for 10 hr in porcelain crucibles. Then, the calcined powders were subjected to a second cycle of grinding under same conditions mentioned above. Using a hydraulic press with a pressure of 6 tons/cm$^2$, we obtained tablets of 13 mm in diameter and 2 - 3 mm in thickness. The obtained tablets were sintered at 950°C for 10 hr with an oxygen flow during the cooling process. Finally, the sintered tablets were reground manually for 1/2 hr followed by mechanical milling for 3 hr. The sintered powders were pressed again and retreated thermally under same pervious conditions. The X-ray diffraction (XRD) patterns were recorded at room temperature using a Bruker D8 Advance diffractrometer with Cu target, and it's Kα radiation. The 2θ scanning speed of detector was 3°/min with a step width of 0.02° and running condition of 40 KV and 40 mA. The X-ray diffraction patterns were analyzed using the XPowder software [27] . The DC resistance was measured using the conventional four-point probe method in the temperature range 40 - 300 K using a closed cycle helium cryostat (APD Cryogenics, USA), programmable current source (Keithley 224), sensitive digital voltmeter (Keithley 182), and autotuning temperature controller (LakeShore 321).

# 3. RESULTS AND DISCUSSION

The X-ray diffraction patterns between 20° and 80° for $Y_1Ba_2Cu_{3-x}Pr_xO_{7-\delta}$ and $Y_1Ba_2Cu_{3-x}Gd_xO_{7-\delta}$ were shown in Figures 1 and 2, respectively. The intensity of peaks for different values of Prand Gd-contents, x, were normalized and shifted for clarity purposes. As seen, the pure sample, x = 0.0, showed the well known Y123 pattern with all peaks indexed in the orthorhombic structure. The calculated a, b, and c lattice parameters for x = 0.0 were given as 3.8374, 3.8867, and 11.6650 Å, respectively. It is worth mentioned to observe the extra two peaks, denoted by (+), appeared for x ≥ 0.1 in both series. These peaks matched the insulating $Y_2BaCuO_5$ phase with characteristic planes (311) and (112). Table 1 showed the dependence of lattice constants, orthorhombicity (tendency towards orthorhombic phase related to the quantity$^{(a-b)/(a+b)}$), difference in peak position of (020) and (200) reflections (Δpp), and grain-size diameter (D) with Prand Gd-contents.

Since the ionic radius of Pr (1.13 Å) is greater than that of Gd (1.05 Å) and Cu (0.93 Å), one should expect a systematic increase in lattice parameters with the increase of x if both occupied the Cu sites. Moreover, the expected increase in the parameters should be stronger for Pr doping (denoted as Pr(Cu)) than that of Gd(Cu). Referring to Table 1 and checking the expected increase, it is noticed that such expectations did not exist. The absence of such expectations is an established knowledge that both dopants would occupy the A-sites rather than Cu-sites. This is consistent with the formation of the impurity insulating $Y_2BaCuO_5$ phase and its increase with increasing of doping content. Linear fitting have beenused to investigate the overall variation of lattice parameters with Prand Gd-contents in the range x = 0 - 0.3. The rate of change of parameters with x was given for Pr(Cu) as 0.08, 0.004, and 0.014 for a-, b-, and c-parameters, respectively. Whereas for Gd(Cu) it was given as 0.113, 0.016, and 0.043 for a-, b-, and c-parameters, respectively. These rates confirm that the lattice constant b is nearly invariant for both systems and that the effect of Gd(Cu) on lattice parameters is stronger than that of Pr(Cu) even though the ionic radius of the latter is greater than the former. For Gd(Cu) with x = 0.4, it was noticed that the a-parameter decreased sharply which could be attributed to the observed strong formation of the insulating phase, Figure 2.

The two peaks denoted by (*) in Figures 1 and 2 referred to the (020) and (200) indices that describe the orthorhombic nature of the structure. The peak positions of these diffraction peaks have been measured and the variation of resulting position difference (Δpp) with Pr and Gd contents was given in Table 1. It is clear that there is a lattice distortion for x > 0.1 for Pr(Cu) and x > 0.0 for Gd(Cu) that was mainly due to changes in a-parameter, since b-parameter was less affected by the substitution. This behavior was reflected on the resistivity measurements as would be discussed later. Furthermore, the decrease in Δpp with increasing of x-content reflected the tendency of the structure towards the tetragonal phase, and a consequent decrease in the critical transition temperature with increasing of x could be expected. Also, Table 1 lists the dependence of grain size on Prand Gd-contents. As observed, the grain size diameter decreased from 280 to 190 Å for Pr(Cu) and from 270 to 180 Å for Gd(Cu) as x increased from 0.1 to 0.3. This decrease will be reflected on the nature of the transition width as will be discussed later.

Figure 3 shows the variation of normalized resistance with temperature for $Y_1Ba_2Cu_{3-x}Pr_xO_{7-\delta}$samples with x = 0.0, 0.1, 0.2, and 0.3. As shown, the x < 0.3 samples still showed a metallic-normal state behaviour $(d(R/R_o)/dT > 0)$ as expected for polycrystalline Y123 system. This could be attributed to weak holes localization (as discussed below) due to the assumption that Pr unoccupied the Cu sites, as mentioned above. The increase in the normal state resistance as x increased from 0 to 0.2 could be attributed to the corresponding absence of Cu ions. Whereas, the sample with x = 0.3 showed a normal-state metallic to insulating transition (MIT) (with $d(R/R_o)/dT < 0$) down to about 82 K. Then, a superconducting behaviour existed (with a decrease of normalized resistance by about 58% from its value at the onset of transition) in the small temperature range 82 - 50 K. This incomplete superconducting state, no $T_c$(offset), was lost for T < 50 K and a crossover from superconducting to insulating transition (SIT) took place in the form of a twofold increase in normalized resistance as the temperature decreased below 50 K by just 16 K. The dependence of the critical transition temperatures and transition width on Pr-content, x, was shown in Figure 4. Moreover, the pure sample, x = 0, showed a transition width of about 1.1 K which confirms the well coupling of grains in the sample. Furthermore, increasing x resulted in a decrease in critical transition temperatures with an increase in the transition width, $\Delta T_c$.

On the other hand, the variation of normalized resistance with temperature of $Y_1Ba_2Cu_{3-x}Gd_xO_{7-\delta}$ samples with x = 0.0, 0.1, 0.2, 0.3, and 0.4 was shown in Figure 5. Unlike Pr(Cu), the normal-state MIT $(d(R/R_o)/dT < 0)$ existed for all values of x > 0 and the insulating nature increases systematically with the increase of x. This could be attributed to strong holes localization of Gd(Cu). The sample with x = 0.4 showed an onset of superconductivity at around 83 K and the state existed down to about 50 K with a strong reduction in normalized resistance by about 96% in the temperature range 83 - 50 K. The incomplete superconducting state was lost for T < 50 K and SIT appeared with the gain of about 95% of the value of normalized resistance at the onset of transition over the small temperature range 50 - 40 K. To confirm the SIT in our system, we have prepared and measured the sample with x = 0.5, as shown in Figure 5. As expected, this sample followed the normal-state nonmetallic pattern with MIT tendency in addition to SIT at lower temperatures. It is important to observe that the normalized resistance at the onset of superconducting state decreased with just 87% as compared to 96% for x = 0.4 over the same temperature range. This suggests the tendency of the system to show less (or loss of) superconducting behaviour with further increase in x due to the gradual decrease in charge carriers concentration. The variation of the critical transition temperatures and transition width with Gd-content, x, was shown in Figure 6. As seen, increasing the Gd-content resulted in a decrease in critical transition temperatures with an increase in the transition width, $\Delta T_c$. The transition width was more affected by Gd(Cu) than Pr(Cu), Figure 4, which could be attributed to very small grain coupling due to participation of the insulating $Y_2BaCuO_5$ phase in these grain boundaries, Figures 2 and 3.

**Table 1**. The variation of lattice parameters (a, b, and c), orthorhombicity, difference in peak position of (020) and (200) reflections (Δpp), and grain-size diameter (D) with Prand Gd-contents, x.

| Pr- & Gd- contents | a (Å) | b (Å) | c (Å) | $(b-a)/(b+a)$ $\times 10^{-3}$ | pp (°) | D (Å) |
|---|---|---|---|---|---|---|
| Pr: 0.1 | 3.8240 | 3.8886 | 11.6710 | 8.4 | 0.92 | 280 |
| 0.2 | 3.8507 | 3.8843 | 11.6663 | 4.3 | 0.82 | 240 |
| 0.3 | 3.8553 | 3.8896 | 11.6711 | 4.4 | 0.66 | 190 |
| Gd: 0.1 | 3.8364 | 3.8893 | 11.6719 | 6.8 | 0.80 | 270 |
| 0.2 | 3.8543 | 3.8923 | 11.6802 | 4.9 | 0.70 | 170 |
| 0.3 | 3.8691 | 3.8911 | 11.6767 | 2.8 | 0.70 | 180 |
| 0.4 | 3.8373 | 3.8951 | 11.6886 | 7.5 | 0.70 | 180 |

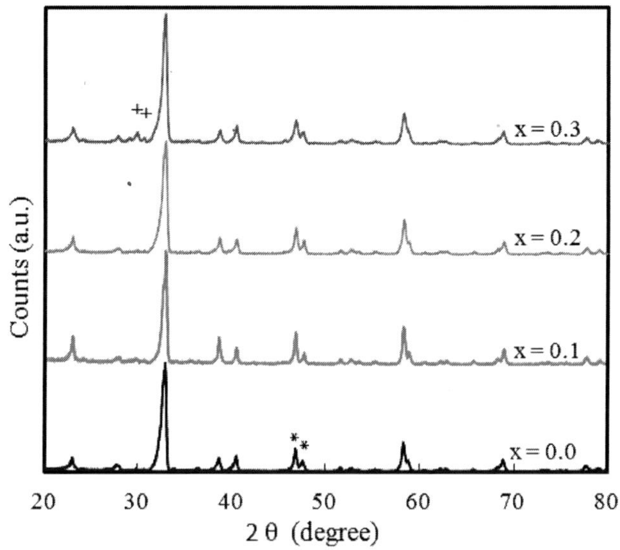


**Figure 1**. Normalized XRD pattern of $Y_1Ba_2Cu_{3-x}Pr_xO_{7-\delta}$ with x = 0.0, 0.1, 0.2, and 0.3. The peaks denoted by (+) referred to $Y_2BaCuO_5$ phase while (*) peaks referred to (020) and (200) indices.

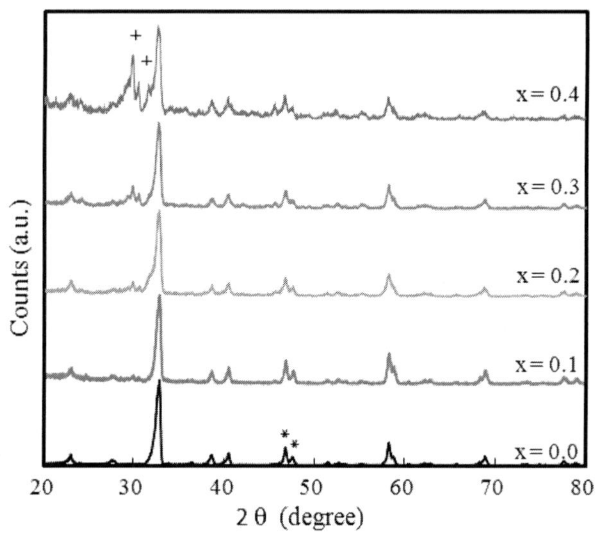


**Figure 2**. Normalized XRD pattern of $Y_1Ba_2Cu_{3-x}Gd_xO_{7-\delta}$ samples with x = 0.0, 0.1, 0.2, 0.3, and 0.4. The peaks denoted by (+) referred to $Y_2BaCuO_5$ phase while (*) peaks referred to (020) and (200) indices.

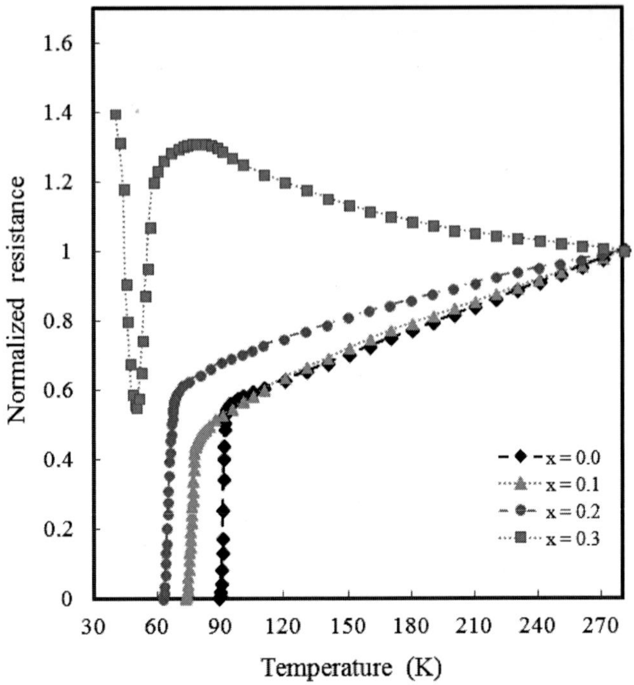


**Figure 3**. Variation of normalized resistance with temperature of $Y_1Ba_2Cu_{3-x}Pr_xO_{7-\delta}$ samples with x = 0.0, 0.1, 0.2, and 0.3.

Some points should be clarified, namely, the tendency of the normal-state to transit from metallic to insulating transition (MIT) with increasing of Prand Gd-contents, the incomplete superconducting states, and the loss of the incomplete superconducting transition (SIT) for x = 0.3 in case of Pr(Cu) and for x > 0.3 for Gd(Cu). As known, high-temperature superconductors (HTS) cuprates with optimal doping are highly anisotropic with twodimensional (2D) charge carriers concentration in the normal and superconducting states. A shift from optimal doping makes it easier for carriers to hope in 3D (rather than the 2D ab-plane) through the "hard" c-axis where the effective mass anisotropy $\Gamma = m_c^*/m_{ab}^* = 26$ for Y123 [28]. This involves a suppression of charge carriers with a consequent normal-state 2D to 3D crossover, namely, metal to insulator transition, MIT. This means that HTS cuprates stretch out to the metal-insulator edge [29] . Let us now study the origin of the shift in optimal structure in our systems. First, since the possibility of Pr and Gd to occupy the Cu-sites is minimal, that would cause a structure distortion leading to a shift in oxygen content from its optimal value, and therefore a decrease in mobile hole concentration, i.e., metallic to insulator transition. The deficiency in oxygen content is reflected in the tendency of the structure towards the tetragonal phase, Table 1. This view agrees with the work of Chang et al. [30] who concluded that mobile holes concentration decreases quite linearly with the increase of rare earth ionic radius.

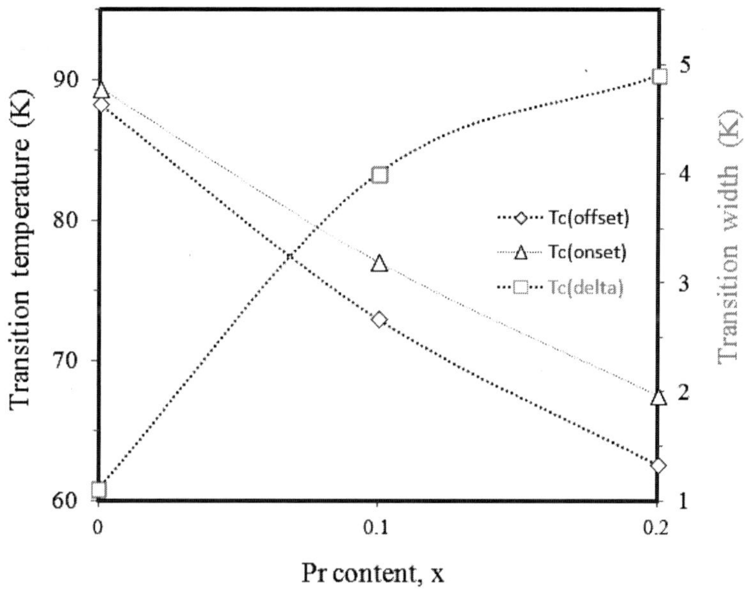


**Figure 4**. Dependence of the critical transition temperatures and transition width on Pr-content, x.

Secondly, let us comment on the observed incomplete superconductivity for high Pr and Gd contents. Ahn et al. [31] have studied the effect of the polarization field formed by ferroelectric oxide lead zirconate titanate layer on $GdBa_2Cu_3O_{7-x}$ HTS thin film. They observed a dip in the normal-state nonmetallic behaviour started at around 50 K and have been attributed to nonuniformity in the thickness of the HTS thin film. The observed incomplete superconducting state in our samples could be explained as follows. These samples are characterized by the strong development of the insulating $Y_2BaCuO_5$ phase, Figures 1 and 2, and wide transition width, Figures 4 and 6. We suggest here that both characters are responsible for the absence of zero-resistance where the wide transition width is a signature of increasing lattice imperfections and the number of grain sizes [32] . In our samples, the lattice distortion and the intensity of the insulating phase increased with increasing of x. Moreover, the decrease in grain size with increasing of x, Table 1, implies an increase in the number of grain sizes. This would suppress the superconducting tunneling between grains.

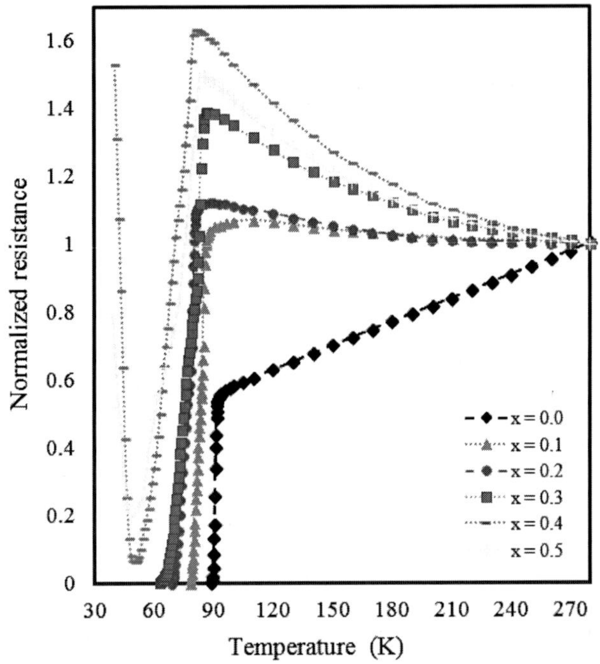


**Figure 5**. Variation of normalized resistance with temperature of $Y_1Ba_2Cu_{3-x}Gd_xO_{7-\delta}$ samples with x = 0.0, 0.1, 0.2, 0.3, 0.4, and 0.5.

Thirdly, as indicated by Figures 4 and 6, we observe that the offset critical temperature of Pr(Cu) is less than Gd(Cu). Furthermore, while Pr(Cu) with x = 0.3 showed incomplete superconducting state, the same content showed a complete transition in case of Gd(Cu) substitution, Figures 3 and5. These observations could not be explained purely in terms of the magnetic nature of Pr and Gd ions. Whereas, the reported critical temperaturesoffset are greater than that required for ordering magnetic moments. The observed negative impact of Pr(Cu) as compared to Gd(Cu) could be attributed to the $Pr^{3+}$-$Pr^{4+}$ mixed valent state and $Gd^{3+}$ [16] . These electronic configurations require the depletion of charge carriers (holes localization) in the $CuO_2$ planes due to Pr(Cu) to be stronger than that of Gd(Cu).

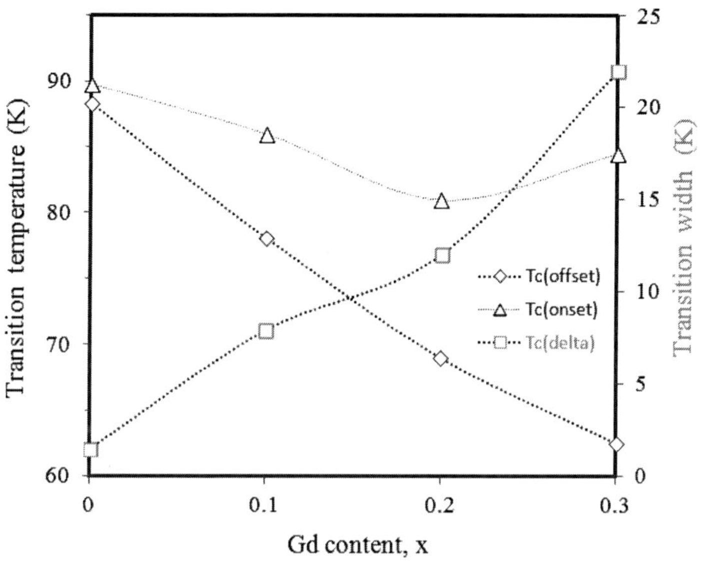


**Figure 6.** Dependence of critical transition temperatures and transition width on Gd-content, x, of prepared $Y_1Ba_2Cu_{3-x}Gd_xO_{7-\delta}$ samples.

Finally, let us examine the superconducting to insulator transition (SIT) observed at low temperatures. Kobayashi et al. [33] have observed SIT in λ-$BETS_2Fe_xGa_{1-x}Cl_4$ (λ-type BETS alloys; BETS = bis(ethylenedithio) tetrathiafulvalene) with x = 0.43 at ambient pressure and with x = 0.55 at 1 k bar. The observed SIT have been considered as an intrinsic property of the crystals. Explanation of the observed superconductor-insulator transition below 50 K in our samples with high x contents is a difficult task. It could be attributed to the established high crystal structural disorder along with the rule played by grain boundaries, such as segregation of Pr and Gd ions at grain boundaries [34] . Another reason might be due to ordering of magnetic moments of Pr and Gd ions [35] [36] . Further work is needed to identify the nature of the observed transitions in these systems.

# 4. CONCLUSION

We have investigated the effect of partial doping of Pr and Gd on the electrical resistance of Cu-deficient Y123 superconductors. Resistance measurements showed a suppression of the critical transition temperature with increasing of Gd-content and this suppression increased in case of Pr doping. A normal-state metal-to-insulator crossover has been observed with $x > 0$ and $x = 0.3$ for Gd(Cu) and Pr(Cu), respectively. Moreover, Pr(Cu) with $x = 0.3$ showed incomplete superconducting transition, while the same content showed a complete transition in case of Gd(Cu). Furthermore, a superconductor-to-insulator crossover has been observed for Pr with $x = 0.3$ and for Gd with $x > 0.3$ at temperatures less than 50 K. To the best of our knowledge, no SIT has been reported in the studied two systems before. The observed crossovers were preliminary explained in terms of localization of charge carriers, structural disorders, and magnetic ordering of magnetic moments.

# REFERENCES

1. Jamadar, T.A. and Ghosh, A.K. (2009) Metal to Insulator Transition in Mn Substituted Underdoped $NdBa_2Cu_3O_{7-y}$: Evidence of Strong Charge Localization. Physica C, 469, 1971-1976.

2. Babu, T.G.N. and Greaves, C. (1993) Critical Hole Density for Superconductivity in Sr-Containing Phases Related to $YBa_2Cu_3O_7$: Structure, Superconductivity and Cation Substitutions in $YSr_2Cu_{2.8}Cr_{0.2}O_7$. Physica C, 207, 44-50.

3. Skakle, J.M.S. (1998) Crystal Chemical Substitutions and Doping of $YBa_2Cu_3O_x$ and Related Superconductors. Materials Science and Engineering: R: Reports, 23, 1-40.

4. Tarascon, J.M., Greene, L.H., Barboux, P., McKinnon, W.R. and Hull, G.W. (1987) 3d-Metal Doping of the HighTemperature Superconducting Perovskites La-Sr-Cu-O and Y-Ba-Cu-O. Physical Review B, 36, 8393-8400.

5. Mendels, P., Bobroff, J., Collin, G., Alloul, H., Gabay, M., Marucco, J.F., Blanchard, N. and Grenier, B. (1999) Normal-State Magnetic

Properties of Ni and Zn Substituted in $YBa_2Cu_3O_{6+x}$: Hole-Doping Dependence. Europhysics Letters, 46, 678-684.

6. Tarascon, J.M., Barboux, P., Miceli, P.F., Greene, L.H., Hull, G.W., Eibschutz, M. and Sunshine, S.A. (1988) Structural and Physical Properties of the Metal (M) Substituted $YBa2\ Cu_{3-x}M_xO_{7-y}$ Perovskite. Physical Review B, 37, 7458- 7469.

7. Xiao, G., Streitz, F.H., Gavrin, A. and Chien, C.L. (1988) Superconductivity and Magnetism in Transition Element Substituted $YBa_2Cu_3O_7$ Compounds. Journal of Applied Physics, 63, 4196.

8. Xiao, G., Streitz, F.H., Gavrin, A., Du, Y.W. and Chien, C.L. (1987) Effect of Transition-Metal Elements on the Superconductivity of Y-Ba-Cu-O. Physical Review B, 35, 8782-8784.

9. Kistenmacher, T.J. (1988) Substitution for Copper in $YBa_2\ Cu_3O_y$: The First 3%. Physical Review B, 38, 8862-8867.

10. Goldschmidt, D., Direktovitch, Y., Knizhnik, A. and Eckstein, Y. (1996) Similar Depression of Tc by Zn and Ni Impurities in1:2:3 $(Ca_{0.4}La_{0.6})(Ba_{1.35}La_{0.65})(Cu_{1-z}Mz)_3O_y$. Physical Review B, 54, 13348-13351.

11. García, S., Cobas, R., Musa, J.E. and Baggio-Saitovitch, E.M. (2000) Normal-State Behavior of the Resistivity in the Superconducting $YBa_2(Cu_{1-x}Li_x)_3O_{7-x}$ System: An Evidence of Scattering by the Spin Fluctuations. Physica C, 341, 1895-1896.

12. Ito, T., Takagi, H., Ishibashi, S., Ido, T. and Uchida, S. (1991) Normal-State Conductivity between Cupric Oxide Planes in Copper Oxide Superconductors. Nature, 350, 596-598.

13. Beschoten, B., Quitmann, C., Kelley, R., Onellion, M. and Güntherodt, G. (1996) Metal-Insulator Transition and Electronic Structure in Pr-Doped $Bi_2Sr_2(Ca_z\ Pr_{1-z})Cu_2O_{8+y}$. Physica B, 223, 519-521.

14. Tarascon, J.-M., Barboux, P. and Hull, G.W. (1989) Bismuth Cuprate High-Tc Superconductors Using Cationic Substitution. Physical Review B, 39, 4316-4326.

15. Clayhold, J., Hagen, S.J. and Ong, N.P. (1989) Approaching the Mott-Hubbard Insulator in the 85-K Superconductor $Bi_2(Sr,Ca)_3Cu_2O_{8+d}$ by Doping With Tm. Physical Review B, 39, 7320-7323.

16. Akhavan, M. (1995) Normal Electrical Resistivity of the Superconductor-Insulator $d_{1-x}Pr_xBa_2Cu_3O_{7-y}(0\leqslant x\leqslant 1)$ System. Physica C, 250, 25-29.

17. Tomkowicz, Z., Lunkenheimerb, P., Knebelb, G., Bałandac, M., Pacynac, A.W. and Zaleskid, A.J. (2000) InsulatorMetal Transition by the Substitution of Ho, Y or Ca for Pr in $PrBa_2Cu_3O_{7-\delta}$. Physica C, 331, 45-56.

18. Ando, Y., Boebinger, G.S., Passner, A., Wang, N.L., Geibel, C., Steglich, F., Kimura, T., Okuya, M., Shimoyama, J., Kishio, K., Tamasaku, K., Ichikawa, N. and Uchida, S. (1997) Low-Temperature Normal-State Resistivity of High-Tc Cuprates. Physica C, 282, 240-243.

19. Prabhakaran, D. and Subramanian, C. (1997) Metal-Insulator Transition in the Pr Substituted Bi-2212 Bulk Textured Crystals. Physica C, 291, 73-78.

20. Chattopadhyay, M.K., Ghatak, S.K. and Dey, T.K. (1997) Metal-Insulator Transition in $YBa_{2-x}Ca_xCu_3O_{7-\delta}$ System. Physica C, 281, 211-217.

21. Cao, X.W., Tang, Y.J. and Ho, J.C. (1996) Superconducting $T_c$ and Normal-State Resistivity in $Nd_{1-x}Pr_xBa_2Cu_3O_{7-\delta}$ System. Physica C: Superconductivity, 259, 361-364.

22. Semba, K., Matsuda, A. and Mukaida, M. (2000) Carrier-Concentration-Drivensuperconductor-to-Insulator Transition in $YBa_2Cu_3O_{6+x}$. Physica B: Condensed Matter, 281-282, 904-905.

23. Li, X., Sun, X., Wu, W., Chen, Q., Shi, L., Zhang, Y., Kotaka, Y. and Kishio, K. (1997) Variations of Normal State Resistivity and $Cu^{2+}$ Localized Spin Moment in the Single Crystal $Bi_2Sr_2$ $CaCU_2O_{8+x}$. Physica C: Superconductivity, 279, 241-245.

24. Fujita, T., Hori, J., Iwata, S., Yoshino, Y., Kurisaki, H., Goko, T., Yamane, K. and Nakamura, F. (2001) Superconductor-Insulator Transition in Under-Doped Cuprates. Physica C: Superconductivity, 364-365, 274-277.

25. Goldman, A.M. (2003) Superconductor-Insulator Transitions in the Two-Dimensional Limit. Physica E: Low-Dimensional Systems and Nanostructures, 18, 1-6.

26. M'chirgui, A., Zouaoui, M., Ben Azzouz, F., Yangui, B. and Ben Salem, M. (2003) Electron Localization Induced by Grain Boundary Disorder in the Normal State of High-$T_c$Superconductors. Physica C: Superconductivity, 125, 71-76.

27. Martin, J.D. (2005) Using XPowder: A Software Package for Powder X-Ray Diffraction Analysis.

28. Ferrell, D., Williams, C., Wolf, S., Bansel, N. and Kogan, V. (1988) Experimental Evidence for a Transverse Magnetization of the Abrikosov Lattice in Anisotropic Superconductors. Physical Review Letters, 61, 2805-2805.

29. Mandal, P., Poddar, A., Ghosh, B. and Choudhury, P. (1991) Variation of $T_c$ and Transport Properties with Carrier Concentration in Yand Pb-Doped Bi-Based Superconductors. Physical Review B, 43, 131 02-13111.

30. Chang, B.C., Yang, C.Y., Hsu, Y.Y. and Ku, H.C. (2007) Origin of Metal-Insulator Transition in the Weak-Ferromagnetic Superconductor System $RuSr_2RCu_2O_8$ (R = Rare Earths). Physica C: Superconductivity, 460-462, 503-505.

31. Ahn, C., Gariglio, S., Paruch, P., Tybell, T., Antognazza, L. and Triscone, J. (1999) Electrostatic Modulation of Superconductivity in Ultrathin $GdBa_2Cu_3O_{7-x}$ Films. Science, 248, 1152-1155.

32. Shaltiel, D., Bezalel, M., Golosovsky, M., Kwok, W.K. and Fendrich, J.A. (1999) Induced Microwave Absorption by Magnetic Modulation in Untwinned and Twinned YBaCuO Crystals and Its Comparison with Resistivity Measurements. Physica C: Superconductivity, 315, 23-35.

33. Kobayashi, H., Akutsu, H., Ojima, E., Sato, A., Tanaka, H., Kobayashi, A. and Cassoux, P. (1999) Supercondutor-toInsulator Transition of $\lambda$-$BETS_2FexGa_{1-x}Cl_4$. Synthetic Metals, 103, 1837-1838.

34. Durrell, J.H., Eom, C.-B., Gurevich, A., Hellstrom, E.E., Tarantini, C., Yamamoto, A. and Larbalestier, D.C. (2011) The Behavior of Grain Boundaries in the Fe-Based Superconductors. Reports on Progress in Physics, 74, 124511- 124549.

35. Ishikawa, M. and Fischer, Ø. (1977) Destruction of Superconductivity by Magnetic Ordering in $Ho_{1.2}Mo_6S_8$. Solid State Communications, 23, 37-39.

6

# HOLE-PAIR FORMATION IN CUPRATE SUPERCONDUCTORS DESPITE ANTIFERROMAGNETIC FLUCTUATIONS

**Rana Janardon Singh[1], Shakeel Khan[2]**

[1]Physics Department, Aligarh Muslim University, Aligarh, India

[2]Department of Applied Physics, Aligarh Muslim University, Aligarh, India

## ABSTRACT

We have earlier proposed models of preformed hole pairs based on the results of our electron paramagnetic resonance experiments. A hole doped in a cuprate superconductor causes ferromagnetic alignment of the spins of the holes of 4 $Cu^{2+}$ ions of the plaquette $(CuO)_4$ in which it enters. Spin alignments undergo oscillations from vertically upward to vertically downward of the $CuO_2$ plane. Vertical projections of spins go on changing when they pass through different plaquettes going to zero when they pass through the $CuO_2$plane. Ferromagnetic alignments of spins produce magnetic fields on the plane proportional to their vertical projections. When two holes travelling in $CuO_2$ plane come across each other at a certain distance between them, they are attracted towards each other by Heisenberg exchange interaction and their path is decided by the magnetic field produced due to spin alignments. Their path is similar to $3dx^2 - y^2$ atomic orbital. Y-123 has been chosen as an example. Due to plethora of evidence of antiferromagnetic fluctuations in cuprates, hole-pair formation has been tried in Y-123 assuming antiferromagnetic fluctuations in it. It has been found that hole-pair formation in spite of AFM fluctuations can be explained on the same lines as done earlier. Hole-pair formation was tried in Tl-2201 to test whether the same rules apply in cuprates with very high coherence lengths. Coherence length in Tl-2201 = 52 Å, whereas in Y-123 = 15 - 20 Å in CuO2 plane. It has been reported that in Tl-2201 the CuO2 plane is very flat and smooth. From this it was concluded that high coherence length is the result of the smoothness of the plane. Further it was concluded that the smoothness of the CuO2 plane depends upon the nature of the near neighbors of the CuO2 plane. Near neighbors of Y-123 and Tl-2201 have been compared.

## KEYWORDS

Hole-Pair Formation in Cuprate Superconductors, Buckling Angle in $CuO_2$ Plane, Coherence Length in a-b-Plane $(\xi ab)$

## 1. INTRODUCTION

@NolistTemp# In the absence of consensus regarding the mechanism of superconductivity in cuprates, people are thinking of preformed hole-pairs [1] [2] which below a certain temperature undergo Bose-Einstein condensation and cause superconductivity. We have proposed models of preformed hole-pairs [3] -[5] for resistanceless current flow in the a-b plane and along c-axis of cuprate superconductors. For better understanding of this paper, it should be read in conjunction with the paper in Ref. [4] because the detailed treatment given in [4] cannot be reproduced here. However, some salient points of the paper [4] are given here which may be of help to readers. As prepared, cuprate superconductors are electron paramagnetic resonance (EPR) silent because of antiferromagnetic (AFM) coupling of $Cu^{2+}$ ions in the all-important $CuO_2$ plane. When cuprate superconductors are deoxygenated, they show EPR spectra of Cu-monomer, dimmer, tetramer and octamer, but more frequently $(CuO)_4$ tetramer. $(CuO)_4$ tetramer is the unit cell of two-dimensional (2-D) $CuO_2$ plane and by understanding its properties, one can understand the properties of $CuO_2$plane which is the seat of superconductivity. Spectra of $(CuO)_4$ and other fragments [6] -[13] given above are observed also in constituents of superconductors as $BaCuO_2$, $SrCuO_2$, $CaCuO_2$, $Bi_2CuO_4$, CuO etc. On deoxygenation, $(CuO)_4$ unit gets magnetically isolated from the bulk by breaking 8 Cu-O bonds situated at its 4 corners. In [14] , it has been found that Cu-O bonds are $\approx 87.5\ \%$ ionic and $\approx 12.5\ \%$ covalent. Breaking of 8 bonds amounts to loss of one electron from the $(CuO)_4$ plaquette, because oxygen is more electronegative than copper. Loss of one electron means the introduction of one hole in the $(CuO)_4$ plaquette. This situation can be represented by the following equation. An isolated $(CuO)_4$ unit $=$ $(CuO)_4$ unit of continuous sheet of $CuO_2$ plane + a hole inside it. Observation of 4 fine structure signals which corresponds to total electronic spin $=2$ and 13 components in each fine structure line which corresponds to total nuclear spin $I=6$ suggest that advent of a hole in a $(CuO)_4$ entity causes ferromagnetic (FM) coupling of spins of 4 $Cu^{2+}$ ions resulting into total electronic spin $=2$ and 13 hyperfine components in each fine structure signal suggest that the advent of the hole in $(CuO)_4$ entity causes 4 holes of the 4 $Cu^{2+}$ ions tocirculate around the $(CuO)_4$ framework and each hole comes into contact with all 4 $Cu^{2+}$ ions. As discussed above, on introduction of a hole in a $(CuO)_4$ plaquette, spins of all 4 $Cu^{2+}$ ions align ferromagnetically and should continue to fluctuate with the same frequency corresponding to 41 meV antiferromagnetic signal seen in Y-123 which we have chosen as an example. 41 meV equated to hν gives $\nu$ to be equal to $10^{13}$ hertz or time-period $T = 10^{-13}$ sec. But the question will be in which direction? It has been noted in [1] that modest magnetic field when it is perpendicular to $CuO_2$ plane suppresses AFM resonance in $YBa_2Cu_3O_{6.6}$more significantly than when applied horizontally. It suggests that the FM coupled spins of holes of 4 $Cu^{2+}$ ions fluctuate vertically above and below the $CuO_2$ plane. Starting from a position where FM aligned spins stand vertically upward of the $CuO_2$ plane, they have maximum resultant spin angular momentum. When they cross $CuO_2$ plane, spins find themselves spread in $CuO_2$ plane. The diagonally opposite ones directed in opposite directions with resultant spin angular momentum equal to zero after a time $=T/4$. After next $T/4$, they get FM aligned vertically downward of the $CuO_2$ plane; after another $T/4$, the yagain find themselves spread in $CuO_2$ plane with zero spin angular momentum. Yet after another $T/4$, they go back to the original positions or vertically aligned above the $CuO_2$ plane. The alignment of spins of $Cu^{2+}$ holes and their circulation around the $(CuO)_4$ plaquette produce magnetic field on the $CuO_2$ plane in proportion to resultant spin angular momentum and locations of holes during their circulation. The magnetic field produced has been calculated [4] to vary from $\approx +5300 \rightarrow 0 \rightarrow -5300$ gauss. When 2 holes wandering in the $CuO_2$ sheet of a superconductor come across a column or row of $(CuO)_4$ plaquettes within a certain distance, they are attracted towards each other by Heisenberrg exchange interaction and follow a path guided by the exchange interaction between them and magnetic field generated by the directions of spins of $Cu^{2+}$ holes and their locations during circulation around a $(CuO)_4$plaquette in which they are situated at that instant. The path executed by the pair of holes is similar to $3d_{x2-y2}$ atomic orbital which has been supported by many theories and experiments. The wandering hole hops from one oxygen ion to the other forcing the holes of $Cu^{2+}$ ions to circulate along the periphery of the $(CuO)_4$ plaquette. The spin angular momentum of wandering hole does not combine with spin or angular momentum of $Cu^{2+}$ holes. To avoid confusion, let it be clear that the holes wandering in the $CuO_2$ plane due to doping will be called wandering holes and those associated with $Cu^{2+}$ ions will be called $Cu^{2+}$ holes.

## 2. FORMATION OF HOLE-PAIRS

A brief description of hole-pair formation is given below. Mathematical treatment of hole-pair formation in cuprates has been given in [4] . The shape of the hole-pair has been shown in figure 3 of [4] . This figure has been reproduced in this paper as figure 1. Full mathematics is avoided here, but some portions are explained here which will be of help in grasping the main idea and final equation describing the order parameter. Application of the equation in describing the shape of the hole-pair has been shown through a shortened table which is a part of table 1 in [4] . The shortened table is numbered 1 in this paper.

In figure 1, the squares A, B, C, D, E, F, G. H. I are the unit cells of $CuO_2$ 2-D plane, each taken to be a square of side 38.4 Å, which is one side of a plaqette in a representative Y-123 superconductor (though in actual case there is a very small difference between a and b sides). When 2 wandering holes enter figure 1, Hole 1 from A side and Hole 2 from E side, they move towards each other under the effect of Heisenberg exchange interaction. Their velocities are modified by the magnetic field present in each $(CuO)_4$ plaquette they traverse.

Here we will explain some symbols and terms and the final equation which determines the path of holes in the preformed hole-pairs. Table 1 in this paper which shows position of holes at different angles along their paths in figure 1 will be discussed for further clarification.

The velocity of a charged particle (here hole) does not change by magnetic field; only its direction is changed and $V^2$ is always equal to $V_x^2+V_y^2$. Motion of charged particle of charge "e" in $XY$-plane under the action of magnetic field $Hz$ (in $z$-direction) is given by the following equations:

$$\frac{dVx}{dt}=\left(\frac{e}{mc}\right)H_zV_y \tag{1}$$

$$\frac{dVy}{dt}=-\left(\frac{e}{mc}\right)H_zV_x \tag{2}$$

$$\frac{dVz}{dt}=0 \tag{3}$$

where $m$ is the mass of the charged particle. Equation (3) goes to zero and need not be considered further.

In the above equations, the three quantities $H_z$, $V_x$ and $V_y$ have been used in the following forms:

$H_z=H_z\cos wt$; $V_x=V_x\sin wt$; $V_y=(-)V_y\cos wt$, because magnitude of these quantities depends on the positions of holes in the figure which depends on the angle ωt. $V_y$ has been expressed as $V_y\cos wt$ with a negative sign, because in this problem, the origin has been chosen at the center of the cell A and for all positions of the holes except at the origin, $y$-coordinate is always negative.

For angular velocities also, two notations have been used: $\omega'$ and $\omega$. $\omega'$ corresponds to the Euclidean angle according to which the total angle in going round a circle is $=360^\circ$ or $=2\pi$.But for the spins of $Cu^{2+}$ holes, one oscillation is completed for Hole 1 in going from A to E and for Hole 2 in going from E to A. Both these angles are equal to 180° according to Euclidean geometry. This is why on the circumference of the circle in figure 1, both the angles, $\omega'$ and ω have been shown. It means that $\omega'=2\omega$. With this much introduction, it will be easy to appreciate the full meaning of the final formula in Ref. [4] defining the positions of holes at different angles as shown in figure 1. The final formula in [4] is shown below and in this paper, it is numbered 4.

$$R \cdot \omega t = \pm \frac{1}{2} \cdot R \cdot \ln \left| \tan\left( \frac{\omega t}{2} + \frac{\pi}{4} \right) \right| \tag{4}$$

In equation (4), $R$ is some length used for plotting this equation. $R$ has been given a nominal value of 7.68 Å, which is twice the side (38.4 Å) of the unit cell of 2-dimensional $CuO_2$ plane of Y-123 superconductor. The R.H.S. of Equation (4) indicates the position of a wandering hole at an angle $\omega t$. The magnitude of the R.H.S. of Equation (4) is shown by a straight line from the coordinate of the angle $\omega t$ on the circumference of the circle towards the center of the circle. The tips at the end of lines (shown by dots) represent the positions of wandering holes. The magnitudes of the R.H.S. of Equation (4) for angles varying from $0^\circ$ to $360^\circ$ have been given in Table1 In the range of angles $0^\circ$ to $180^\circ$ all the values are positive, but in the range $180^\circ$ to $360^\circ$ all the values are negative. But in the range of $180^\circ$ to $360^\circ$, the direction of the magnetic field is also negative, that is , downward of the $CuO_2$ plane, whereas in the range $0^\circ$ to $180^\circ$, it is upward of the $CuO_2$ plane. In Equation (4) also, there is ± sign. It means in both ranges, the holes are attracted towards the center of the circle or the center of the cell C. With the help of Table 1 and figure 1, meaning of the Equation (4) can be understood. So far discussion has been mainly for Hole 1. But same is true for Hole 2 which starts from E and completes $360^\circ$ of journey ($360^\circ$ to $720^\circ$). In figure 1, Holes 1 and 2 always occupy positions diagrammatically opposite to each other.

Going through Table 1, one finds that at certain angles $(\omega t)$ the position of Hole 1 lies beyond the cell C. From $\omega t = 80^\circ$ to $100^\circ$, the position of Hole 1 lies beyond the cell C. At $\omega t = 90^\circ$, the position of Hole 1 is indeterminate according to Equation (4) or effectively very far from the cell C. At $\omega t = 74.6^\circ$, the position of Hole 1 is just at the center of the cell C. The position of Hole 2 at $\omega t = 74.6^\circ + 360^\circ = 434.6^\circ$ is also at the center of the cell C. Two holes cannot be present at the same place at the same time. Due to Coulomb repulsion, they cannot come very near to each other in the cell C.

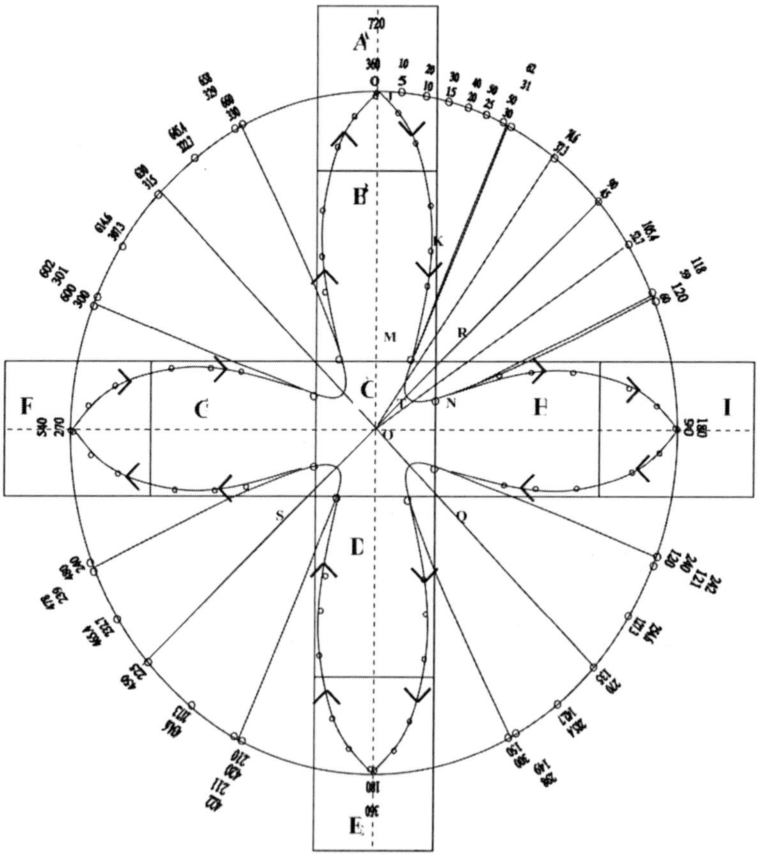

**Figure 1**. Formation of hole-pairs, both the holes traverse the same path continuously.

Hole 1 remains inside the cell C within angles $\omega t = 62^\circ$ to $118^\circ$. Similarly Hole 2 remains inside the cell C within angles $\omega t = (360 + 62 = 422^\circ)$ to $(360 + 118 = 478^\circ)$. Inside the cell C, holes repel each other, their velocities are reduced and Heisenberg exchange interaction becomes ineffective due to such a small separation between the two charged particles. From the cell C, Hole 1 is deflected towards cells H-I and Hole 2 towards the cells G-F. When Hole 1 reaches cell I and Hole 2 reaches cell F, they experience maximum magnetic field because of vertical alignment of all the 4 $Cu^{2+}$ spins in respective cells. They are turned back from these cells due to magnetic mirror effect. Just to refresh memory, magnetic mirror effect is that force that causes the ions in the ionosphere to oscillate between the north pole and the south pole of the earth due to highest strength of the magnetic field at the two poles. The path of both the holes has been shown in figure 1. The holes are indistinguishable and both holes follow the path A-C-I-C-E-C-F-C-A. Ultimately they circulate along a path similar to the shape of the atomic orbital $3d_{x2-y2}$. Complete derivation of Equation (4) has been given in [4].

## 3. HOLE-PAIR FORMATION IN SPITE OF ANTIFERROMAGNETIC FLUCTUATIONS

It has been observed that in high $T_c$ cuprate superconductors [15] -[22] and in heavy Fermion systems ($UPd_2Al_3$, $CeCoIn_5$) [23] [24] and also in iron oxide superconductor ($Ba_{0.6}K_{0.4}Fe_2As_2$) [25] , inelastic neutron scattering exhibit AFM fluctuations dominated by a resonance signal in single layered $CuO_2$ superconductors [21] [22] , there appears only one resonant signal; in two-layered superconductors [15] -[20] , two signals are observed of odd and even parity where the modes differ in symmetry with respect to exchange between adjacent $CuO_2$ layers. Odd parity signal which is resonance signal is stronger and occurs at smaller energy and the even parity signal is weaker and occurs at higher energy. A universal relation between AFM resonance signal and superconducting gap $(2\Delta)$ has been demonstrated according to which energy of resonance signal $E_r$ is proportional to 2Δ, but always less than $2\Delta$ [26] . McDonald et al. [27] have pointed out that the experimentally determined Cooper pair wave function in cupratesmaps directly on the spin fluctuation disturbance responsible for the AFM peaks measured in inelastic neutron scattering. Large number of works cited above showing AFM excitations, commensurate or incommensurate suggest that these fluctuations are intrinsic property of cuprate superconductors and so should be integral part of any theory. But these fluctuations have failed to explain superconductivity.

**Table 1** . *wt* = angle which the spin vector of the hole makes with a direction perpendicular to the $CuO_2$ plane. In the rotation of $360^\circ$ in the Cartesian system, the spin vector completes two oscillations and so ωt varies from 0-720 The expression 3.84. $\ln\left|\tan\left(\frac{\omega t}{2}+\frac{\pi}{4}\right)\right|$ gives the position of the hole at the angle ωt in the (3).

| $\omega t$ (1) | $3.84\cdot\ln\cdot\left|\tan\left(\frac{\omega t}{2}+\frac{\pi}{4}\right)\right|$ (2) | (1) | (2) |
|---|---|---|---|
| 0 | 0.000 | 180 | 0.000 |
| 20 | 1.367 | 200 | −1.367 |
| 40 | 2.927 | 220 | −2.927 |
| 60 | 5.053 | 240 | −5.053 |
| 62 | 5.333 | 242 | −5.333 |
| 74.6 | 7.680 | 254.6 | −7.680 |
| 80 | 9.354 | 260 | −9.354 |
| 89 | 18.207 | 269 | −18.207 |
| 90 | Indeterminate | 270 | Indeterminate |
| 91 | 18.207 | 271 | −18.207 |
| 100 | 9.354 | 280 | −9.354 |
| 105.4 | 7.680 | 285.4 | −7.680 |
| 118 | 5.333 | 298 | −5.333 |
| 120 | 5.053 | 300 | −5.053 |
| 140 | 2.927 | 320 | −2.927 |
| 160 | 1.367 | 340 | −1.367 |
| 180 | 0.000 | 360 | 0.000 |

**Table 2** .Showing positions of holes in different $(CuO)_4$ plaquette or cells, no. of cells in FM or half FM coupling and no. of cells in AFM coupling.

| S. No. | | No. of cells in FM and half-AFM couplings | Positions of holes in cells, first Hole 1 and then 2 |
|---|---|---|---|
| 1 | A, E | FM = 2 | 7 |
| 2 | B, D | Half – FM = 4 = 2FM | 5 |
| 3 | C, C | No coupling | 8 |
| 4 | H, G | Half – FM = 2 = 1 FM | 7 |
| 5 | I, F | FM = 2 | 7 |
| 6 | H, G | Half – FM = 4 = 2 FM | 5 |
| 7 | C, C | No coupling | 8 |
| 8 | D, B | Half – FM = 2 = 1 FM | 7 |
| 9 | E, A | FM = 2 | 7 |
| Total of couplings | | FM = 12 | AFM = 61 |

## 3.1. Degree of Antiferromagnetism and Ferromagnetism with Doping Level

Kopp et al., [28] noted that any trace of AFM fluctuation is absent in over doped regime of cuprates. Superconductivity arises in Mott insulators after doping level of nearly 5%; attains optimal value at 16%; overdoping starts at 19% and the superconducting dome terminates in the vicinity of 25%. In the overdoped regime, spin susceptibility shows a ferromagnetic upturn. He also suggested that at the end of superconductivity dome, there should be genuine ferromagnetism at zero temperature. Sonier et al. [29] observed gradual disappearance of antiferromagnetism on doping of cuprates and onset of static magnetic order in the highly doped regimes. But this magnetism is not of long range order but magnetic moments appear in dilute form. The main point in [28] [29] is that charge doping or hole-doping induces FM order in cuprates and competing ferromagnetic fluctuations are simultaneously present with superconductivity. It has also been concluded that on increasing doping level upto 25%, ferromagnetic fluctuations increase at the expense of antiferromagnetic fluctuations.

## 3.2. EPR Signals Due to Ferromagnetism though Experimentally Only Antiferromagnetic Fluctuations Are Observed

We have proposed models of preformed hole-pairs in cuprates [3] -[5] on the basis of FM spin fluctuation in $(CuO)_4$ plaquette of $CuO_2$ plane, but there is plethora of evidence that AFM fluctuations are intrinsic properties of cuprates. We will now show that hole-pairs can be formed on the same lines as done earlier without consideration of AFM fluctuations. For this, let us consider only A, B, C cells of figure 1, shown separately in figure 2.

We will consider the behaviour of Hole 1 only and the same applies to Hole 2. When the hole1 enters cell A of figure 2, the spins of all the 4 $Cu^{2+}$ ions at its corners align vertically above the $CuO_2$ plane as has been concluded from our EPR experiments and has been used in the formation of hole-pairs infigure 1. It is to be remembered from figure 1 that when a hole goes from cell A to B and then to C, the vertical components of spins gradually decrease and so also the magnetic field produced from them. Presently we are concerned with the states of spins 3, 4, 5, 6 when Hole 1 enters B from A. The spins 3 and 4 will remain in ferromagnetic alignment, but their spins will be tilted from the vertical direction in direct proportion to the fraction of time-period T of spin fluctuation spent by the hole in coming from A to B cell. But what about the spins numbered 5 and 6.Will there be FM exchange between the pair of spins 3 and 4 and the pair of 5 and 6 spins or AFM exchange. But because there is a hole present in the cell B [28] [29] which cause FM alignment of spins in any cell, so all the 4 spins 3, 4, 5, 6 will be in FM arrangement, but all 4 of them more inclined towards the horizontal plane. When the hole reaches the cell C, there will be again realignment of spins with decreased vertical component. When the hole reaches the center of the cell C, the total vertical component will be zero. As has been discussed in the hole-pair formation in figure 1, the hole will never reach the center of the cell C, because at the same time the other partner of the hole pair would also reach the center of the cell C which is not possible because of the Coulomb repulsion between the holes. Thus the whole process of hole-pair formation is in the same way as in the formation of hole-pair described in figure 1.

But here a question arises, why in neutron diffraction experiments on cuprate superconductors, only AFM fluctuations are observed and not FM fluctuations. To answer this question, let us calculate the percentages of FM and AFM alignments of spins during one time-period $T$ of journey of Holes 1 and 2 in figure 1. The calculation of FM and AFM order can be done under 2 assumptions: 1) in whichever 2 cells out of the 9 cells the 2 holes are present, they will be in FM alignment and the rest 7 in AFM alignment. In this way, during one time-period $T$ of journey of two holes in figure 1, $2/9 \times 100\% \approx 22\%$ will be in FM order and $\approx 78\%$ in AFM order.

Under the assumption (2) when Holes 1 and 2 are in cells A and E respectively in the beginning, there will be FM order in cells A and E and in the rest 7, AFM order. When Hole 1 enters cell B and Hole 2 enters cell D, there will be subdued FM order in cells A and B due to Hole 1 and in cells E and D due to Hole 2, because there is common boundary between A and B for Hole 1 and common boundary between E and D for Hole 2. It can be said that there is half FM order both in A and B due to Hole 1 and half FM order both in D and E due to Hole 2. Half FM order is justified, because when both the holes enter cell C, FM order is nearly lost. It can be said that full ferromagnetism is due to the hole1 while in A and no ferromagnetism while in C. Thus when Hole 1 has reached cell B from A, it can be appropriately said that in both cells there is ferromagnetism of half strength only. From the cell C, due to curvature of the paths of holes and Coulomb repulsion, Hole 1 is reflected towards cells H-I and Hole 2 towards G-F. When Holes 1 and 2 reach cell H and G respectively, half FM order is attained, but FM order is built downward of the $CuO_2$ plane. When Holes 1 and 2 reach I and F cells respectively, there will be full FM order downward of the $CuO_2$ plane. When the holes turn back from the cells I and F at the extreme ends, they enter cells H and G respectively, half FM order is obtained in cells I and H due to Hole 1 and in cells F and G due to Hole 2. Again both the holes reach cell C, where there will be negligible or zero magnetic order. From there, Hole 1 will be reflected towards cells D-E and Hole 2 towards cells B-A with the same type of magnetic order as in going from cell C towards cells H-I for the Hole 1 and towards G-F for Hole 2, but FM order now building upward of the $CuO_2$ plane. When Hole 1 has reached cell E and Hole 2 cell A, full FM order will be attained and by this time their journey in one time-period T has been completed. Thereafter they will be following the same path repeatedly.

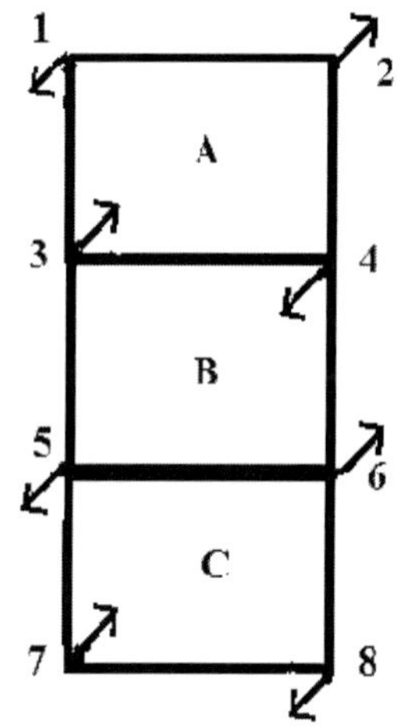


**Figure 2**. Showing spin configuration in AFM alignment. Spins have been numbered.

Calculations by the first assumption gives $\approx 22$ percent of cells in FM spin alignment and $\approx 78$ percent in AFM alignment. Calculation by the second method gives 16% FM order and $\approx 84\%$ AFM order. Calculation of FM and AFM orders have been shown in table 2. Ratio of FM order to AFM order $= 12/73 \times 100\ \% \approx 16\ \%$. Calculation from the second assumption is more realistic. But the above calculations have been done in the most favourable case where at all the possible places of hole-pair formation, holes are present. In actual cases (from underdoped to optimum doped), there will be many patches where hole-pairs would not have been formed because no holes are available. Because of the small percentage of FM alignment in comparison to AFM alignment, the former may be submerged under AFM alignment in actual experiments. AFM coupling of spins of holes of $Cu^{2+}$ spins is unable to explain superconductivity. But FM coupling of $Cu^{2+}$ spins can explain superconductivity [4] but not supported by neutron scattering experiments. The reason may be that under overwhelming AFM order, minority transitory FM order is submerged. The present mechanism of hole-pair formation is supported by theoretical considerations [4] and cannot be rejected by experiments because of its indetectibility of FM order due to its lean and fleeting presence. Time-period of a hole in Y-123 is $10^{-13}$ sec. And in this time a hole has to cross 10 cells (not only 9 cells but 10 cells because C cell will come twice in its path). Time spent in C cell will be greater than in any other cell, because in this cell both of the holes face Coulomb repulsion and their velocities are reduced and practically there is no magnetic order in this cell. So a hole has to pass through 8 cells only with some kind of FM order and it will take less than $10^{-14}$ sec per cell. There is another reason for FM order not to be observed in experiments is that half of time, projections of holes will be above the plane and half of time below the plane. They may cancel each other because changeover is very fast. It may be possible that the transitory FM order in such a shot spell is not detected in neutron scattering experiments. Kopp et al. [28] and Sonier et al. [29] have maintained that in superconductivity dome in the phase diagram, FM order coexists with AFM order.

Formation of hole-pairs in cuprates with large coherence lengths in a-b plane $\xi_{ab}$: We have till now discussed formation of preformed hole-pairs taking Y-123 as a representative example which has coherence length $\xi_{ab}$ of the order of 15 - 20 Å. But there are cuprate superconductors whose $\xi_{ab}$ are much larger than that of Y-123. There are also cuprates whose $\xi_{ab}$ are nearly equal to that of Y-123 or a little smaller. They can be understood on the lines of arguments given for Y-123. The problem is how to explain hole-formation when $\xi_{ab}$ is quite large. A broad view of the properties of cuprate superconductors is given in the table 3with the parameters with which we may be concerned in this paper. From table 3, one thing becomes clear that for a single $CuO_2$ layered superconductor, $\xi_{ab}$ has the largest value, followed by two layered and the smallest ones are for the three layered cuprates. Coherence lengths along c-axis $\xi_c$ are quite small in all cases. For small $\xi_c$ values, Kumar et al. [30] have given a reasonable explanation for highly enhanced resistivity in c-axis transport in normal state. They interpreted the suppression of single particle transport along the c-axis in the normal state due to the blocking of inter-block transport by the intra-block coupling to many-body environments (i.e., entanglement with other electronic degrees of freedom). This mechanism is called Quantum Zeno Effect (QZE). From this it can be said that coherence lengths depend on the velocity of transport of charge carriers. Regarding the values of $\xi_{ab}$ it can be said that when there are more than one $CuO_2$ layer, there is some kind of interaction between layers that reduces the velocity of charge carriers which lowers the value of $\xi_{ab}$ in two $CuO_2$ layered and the lowest value in three layered cuprates. Example of interaction between layers of $CuO_2$ in cuprate superconductors is the occurrence of odd and even AFM excitations in two $CuO_2$ layered cuprates. It explains why $\xi ab$ is highest in single layered cuprates. As in multi-layered cuprates, in single layered cuprates also velocity of holes depends on the buckling of Cu-O-Cu angle in the $CuO_2$ plane. Buckling may arise due to interaction of $3d_{z2-r2}$ and $2p_z$ orbitals of the oxygen ions of nearest layers vertically above or below. Interaction between $3d_{x2-y2}$ orbital of the $Cu^{2+}$ ion and $2px$ and $2py$ orbitals of the planar oxygen when a hole reaches $O^{2-}$ ion may be another cause of buckling.

Chmaisen et al. [31] , observed that for a fixed chemical composition, increased $CuO_2$ plane buckling lowers the transition temperature. Amit Keren [32] while investigating superconductor ($\{Ca_xLa_{1-x}\}$ $Ba_{1.75-x}$ $La_{0.25x}$) concluded that $T_c$ is determined by a single scale set by super exchange interaction J, which in turn is determined by the in-plane Cu-O-Cu buckling angle. Thus buckling angle affects $T_c$, super exchange interaction and also coherence length. As seen from table 3, the in-plane coherence length $\xi_{ab}$ in Tl-2201 is equal to 52 Å (highest in cuprates). It means that in the beginning of hole-pair formation, two holes are situated at a distance of 52 Å which is equivalent to $\approx 13$ $(CuO)_4$ plaquettes. For hole-pair formation the wave function of Hole 1 in the first cell (in a figure similar to figure 1 but with 13 cells in y-direction and 13 cells in the x-direction intersecting each other at the central 7th cell) must overlap with that of Hole 2 in the 13th cell. The odd resonance peak in this superconductor has been found to be equal to 47 meV, which when equated to hν gives $\nu = 1.135 \times 1013$ hertz or time-period $T = 8.81 \times 10^{-14}$ second. As has been shown in [4] that for hole-pair formation in Y-123, each hole has to cover a distance of $\approx 38.5$ Å (equivalent to 10 cell length) in one time-period $=10^{-13}$ sec. For hole-formation in Tl-2201, each hole has to cover a distance of $\approx 100$ Å (equivalent to 26 cell length) in one time-period $=8.81\times10^{-14}$ second.

It has been found that $CuO_2$ plane in TL-2201 is quite flat and smooth. The effect of smoothness of $CuO_2$ plane can also be guessed from comparison of the velocities of hole pairs in Y-123 and Tl-2201. In Y-123, each hole of a hole-pair covers angular distance $\omega t = 360^\circ$ in $10^{-13}$ sec corresponding to linear distance of $10\times3.84=38.4$ Å (each side of the cell taken to be equal to 38.4 Å). In the case of Tl-2201, the linear distance $=26\times3.84=99.84$ Å. Velocity of hole-pair = distance travelled/time period. For Y-123, velocity $=(38.4\times10^{-8})/10^{-13}=3.84\times10^{6}$ cm/sec.

For Tl-2201, velocity $=(99.84\times10^{-8})/8.81\times10^{-14}=11.33\times10^{6}$ cm/sec. Thus velocity of hole pairs in Tl- 2201 is $\approx 2.95$ times more than that in Y-123. It is known that velocity of hole-pairs in cuprates is proportional to $n_s/m^*$ (where $n_s$ superconducting density and $m^*$ = effective mass of hole pair). All the data about Y- 123 and Tl-2201 used here have been taken for optimum doping, hence $n_s$ will be nearly equal in both cases. So velocity of hole-pairs should be proportional to $1/m^*$. Thus it can be concluded that the coherence length of cuprates depends upon the effective mass of hole-pair in superconductors. It may also be mentioned here that effective mass of hole-pairs is less than that of single hole.

Different effective masses of hole-pairs in different cuprates may be attributed to the nature of near neighbours of $CuO_2$ plane in the vertical direction. In Y-123, near neighbours are as follows: CuO-BaO-$CuO_2$-$YCuO_2$-BaO-CuO and for Tl-2201, they are as follows: TlO-TlO-BaO-$CuO_2$-BaO-TlO-TlO. Near neighbours in the two cases are quite different.

Near neighbours control the smoothness and flatness of the plane or buckling of Cu-O-Cu angle which ultimately controls the velocity of holes in the $CuO_2$ plane. One difference that is obvious in the immediate neighbourhood of $CuO_2$ plane in Tl-2201 and Y-123 is that in the former, $Cu^{2+}$ ions of $CuO_2$ plane have bipyramidal coordinations with the $O^{2-}$ ions of BaO plane, whereas in the latter, on one side of $CuO_2$ plane , there is coordination between $Cu^{2+}$ ions of $CuO_2$ plane with $O^{2-}$-ions of BaO plane but on the other side, $Cu^{2+}$ ions of $CuO_2$ plane are not coordinated to any oxygen ion , because in Y layer there is no oxygen ion. In addition to coordination of ions, the atoms of surroundings may affect the properties of $CuO_2$ layers by differences in electronic structure, ionization energy and electronegativity etc.

**Table 3**. Showing for different superconductors crystal structure, coherence lengths in ab-plane and along c-axis in Angstroms, energy of odd and even inelastic neutron neutron scattering peaks in milielectron volts, superconducting energy gap (2Δ) and ratio of (2Δ) and energy of odd peak (2Δ/Eodd).

| Substances | Crystal Structure ($A^{\circ}$) | $T_c$ (K) | $\xi_{ab}$ ($A^{\circ}$) | $\xi_c$ ($A^{\circ}$) | Energy of odd peak (meV) | Energy of even peak (meV) | 2Δ (meV) | 2Δ/Eodd |
|---|---|---|---|---|---|---|---|---|
| La-214 | a = b = 3.77, c = 13.25 | 38 | 21 | - | 18 | - | 35 | |
| Y-123 | a = 3.823, b = 3.885, c = 11.7 | 92 | 13 - 16 | 2 - 3 | 41 | 53 | 79 | |
| Bi-2212 | a = 5.40, b = 5.41, c = 37.01 | 96 | 27 - 38 | 1.6 - 1.8 | 43 | - | - | - |
| Bi-2223 | a = 5.42, b = 5.41, c = 37.01 | 122 | 29 | 0.93 | - | - | - | - |
| Tl-2201 | a = b = 3.85, c = 23.2 | 92 | 52 | 3 | 47 | - | 85 | 1.81 |
| Tl-2212 | a = b = 3.85, c = 23.2 | 118 | 20 - 31 | 0.3 - 6.8 | - | - | - | - |
| Tl-2223 | a = b = 3.85, c = 35.6 | 128 | 11 - 13.6 | - | - | 11 | - | - |
| Hg-1201 | a = b = 3.85, c = 9.5 | 95 | 17 - 34 | - | 56 | - | 88 | 1.57 |
| Hg-1212 | - | 127 | 21.1 | - | - | - | - | - |
| Hg-1223 | a = b = 3.857, c = 15.7 | 133 | 13 | - | 72 | - | 117 | 1.63 |

The general condition for Heisenberg exchange interaction to take place is that the ratio $r_{ij}/r_{3d}$ must be greater than 3 but not much greater, where $r_{ij} =$ distance between atoms i and j atoms and $r_{3d} =$ radius of the 3d orbital. But there are instances where Heisenberg exchange interaction is effective at much larger separations. In a series of free radicals in aqueous solution, the exchange rate [33] is much greater than the reaction rate and the critical exchange distance is between one and three hard sphere encounter distance in agreement with several theoretical predictions. In [33], exchange interaction has been found to be effective at distances $\approx 30$ Å. Velocity of a hole in Tl-2201 is much higher than that in Y-123. Higher velocity means that the surface is flat and smooth which means that there is no or very small variation of electron density along its path or the buckling angle is zero or very small. When intervening space between two holes is smooth or without any variation of electron density, the wave functions of electrons can spread over large distances. Holes situated at much larger distances than $r_{ij}/r_{3d}$ can also be bound by Heisenberg exchange interaction. It can be concluded that hole-pair formation in Tl-2201 (with very large $\xi_{ab}$) and Y-123 (with small or moderate $\xi_{ab}$) can be explained on the same lines by transitory FM order induced by holes wandering in the $CuO_2$ plane.

# 4. SUMMARY

Following are the important points in this paper:

1) We briefly described our EPR work on deoxygenated cuprate superconductors. It was inferred from our work that an isolated $(CuO)_4$ plaquette (after breaking of all its 8 Cu-O bonds at its 4 corners from the surrounding) is equivalent to a $(CuO)_4$ plaquette of continuous $CuO_2$ plane with a hole inside it. In isolated $(CuO)_4$ plaquettes, magnetic field is generated due to the alignment of spins of 4 $Cu^{2+}$ ions in the plaquettes. So it was concluded that a hole on entering a $(CuO)_4$ plaquette of continuous $CuO_2$ sheet will also produce magnetic field caused by the alignment of spins of 4 $Cu^{2+}$ holes.

2) When a hole proceeds along a column or row of plaquettes in a continuous $CuO_2$ plane, magnetic field produced goes on oscillating from a direction vertically upward of the $CuO_2$ plane to vertically downward,

attaining zero while crossing the plane. Also the magnitude of the magnetic field goes on changing when holes pass from one cell to the other. When two holes proceeding towards each other along a column or row of cells in $CuO_2$ plane come at a certain separation between them, are attracted towards each other by Heisenberg exchange interaction. But the direction of their motion is decided by the magnitude and direction of magnetic field produced when holes move from one cell to the other. Two holes moving along $CuO_2$ plane under the effect of exchange interaction and magnetic field form a hole-pair and they carve out a path which is similar to $3d_{x2-y2}$ atomic orbital. For hole-pair formation, Y-123 has been taken as an example.

3) In the above model, no magnetic field was supposed to be present in any cell, until a hole enters a cell. The above model was extended to the case when AFM alignment of spins is present in all the cells except those cells where FM alignment is present due to entry of holes. Again example was Y-123.

4) The model proposed in [4] for Y-123 with planar coherence length $=15-20$ Å in a-b plane has been extended to Tl-2201 with very high coherence length $=52$ Å. The model seems to be successful even in Tl-2201. The model seems to be successful because of the following two reasons. The $CuO_2$ plane in Tl-2201 is more flat and smooth. Due to the smoothness of the path between two holes, the velocity of hole-pairs in Tl-2201 is nearly 2.95 times more than that in Y-123.

5) Flatness or smoothness of the $CuO_2$ plane is attributed to the nature of near neighbours of $CuO_2$plane in the superconductor. Differences between the near neighbours of $CuO_2$ plane in Y-123 and Tl-2201 have been recounted.

## REFERENCES


1. Dai, P.C., Mook, H.A., Aeppli, G., Hayden, S.M. and Dogan, F. (2000) Resonance as a Measure of Pairing Correlations in High Tc Superconductor YBa2Cu3 O6.6. Nature, 406, 965-968.

2. Stankowski, J., Krupski, M. and Roman, M. (2004) Remarks on the Phase Diagram of High Temperature Superconductors. Materials Science-Poland, 22, 175.

3. Singh, R.J. (2009) Preformed Hole-Pairs in Cuprate Superconductors. International Journal of Modern Physics B, 231, 53-76.

4. Singh, R.J. (2011) Model of Preformed Hole-Pairs in Cuprate Superconductors. Journal of Modern Physics, 2, 885897.

5. Singh, R.J. (2012) Model of Preformed Hole-Pairs in c-Axis Transport in Cuprate Superconductors. World Journal of Condensed Matter Physics, 2, 228-236.

6. Punnoose, A., Maurya, B.P., Jilson, M., Umar, M., Haque, M.I. and Singh, R.J. (1993) EPR Observation of Cu2+-Cu2+ Pairs in Cupric Oxide Powder. Solid State Communications, 88, 195-198.

7. Singh, R.J., Alex, P., Mathew, J., Maurya, B.P. and Khan, S. (1994) S = 1 and S = 2 EPR Signals in Modified CuO and BaCuO2. Physical Review B, 49, 1346.

8. Singh, R.J., Ikram, M., Singh, A., Punnoose, A., Maurya, B.P. and Shakeel, K. (1995) Copper Tetramers in High Temperature Superconductors. Physics Letters A, 208, 369-374.

9. Punnoose, A. and Singh, R.J. (1995) EPR Studies of High Tc Superconductors and Related Systems. International Journal of Modern Physics B, 9, 1123.

10. Singh, R.J., Sharma, P.K., Singh, A. and Shakeel, K. (2001) EPR Spectra of Deoxygenated High Temperature Superconductors. Physica C, 356, 285-296.

11. Khan, S., Mohd, I., Arti, S. and Singh, R.J. (1997) EPR Study of Deoxygenated La2CuO4. Physica C, 281, 143-148.

12. Shakeel, K., Arti, S. and Singh, R.J. (1998) EPR Study of La2?xMxCuO4 (M = Sr, Ba). Solid State Communications, 106, 621-626.

13. Shakeel, K., Arti, S. and Singh, R.J. (1999) EPR Study of La1.854Sr0.146CuO4. Physica C: Superconductivity, 325, 165172.

14. Meng, Q.B., Wu, Z.J. and Zhang, S.Y. (1998) Evaluation of the Energy Barrier Distribution in Many Particle Systems Using the Path Integral Approach. Journal of Physics: Condensed Matter, 10, L89.

15. Rossat-Mignod, J., Regnault, L.P., Veltier, C., Bourges, P., Burtlet, P., Bossy, J., Henry, J.Y. and Caperlot, G. (1991) Neutron Scattering Study of YBa2Cu3O6+x System. Physica C, 185-189, 86-92.

16. Yu, G., Li, Y., Zhao, X., Bari?i?, N., Cho, Y., Bourges, P., Hradil, K., Mole, R.A. and Greven, M. (2010) Magnetic Resonance in the Model High-Temperature Superconductor HgBa2CuO4+δ. Physical Review B, 81, Article No: 064518.

17. Pailhès, S., Ulrich, C., Faque, B., Hinkov, V., Sidis, Y., Ivanov, A., Lin, C.T., Keimerand, B. and Bourges, P. (2006) Doping Dependence of Bilayer Resonant Spin Excitations in (Y, Ca)Ba2Cu3O6+x. Physical Review Letters, 96, Article No: 257001.

18. Fong, H.F., Bourges, P., Sidis, Y., Regnault, L.P., Ivanov, A., Gu, G.D., Koshizuka, N. and Keimer, B. (1999) Neutron Scattering from Magnetic Excitations in Bi2Sr2CaCu2O8+δ. Nature, 398, 588-591.

19. Capogna, L., Faque, B., Sidis, Y., Ulrich, C., Bourges, P., Pailhes, S., Ivanov, A., Tallon, J.L., Liang, B., Lin, C.T., Rykov, A.I. and Keimer, B. (2007) Odd and Even Magnetic Resonance Modes in Highly Overdoped Bi2Sr2CaCu2O8+δ. Physical Review B, 75, Article No: 060502.

20. He, H., Bourges, P., Sidis, Y., Ulrich, C., Regnault, L.P., Pailhes, S., Bazigirova, N.S. and Keimer, B. (2002) Magnetic Resonant Mode in the Single Layer High Temperature Superconductor Tl2Ba2CuO6+δ. Science, 295, 1045-1047.

21. Bourges, P., Regnault, L.P., Sidis, Y. and Veltier, C. (1996) Inelastic-Neutron-Scattering Study of Antiferromagnetic Fluctuations in YBa2Cu3O6.97. Physical Review B, 53, 876.

22. Batista, C.D., Ortiz, G. and Balasky, A.V. (2001) Unified Description of the Resonance Peak and Incommensuration in High-Tc Superconductors. Physical Review B, 64, Article No: 172508.

23. Sato, N.K., Aso, N., Miyake, K., Shiima, R., Thalmer, P., Varelogiannis, G., Geibel, C., Steglich, F., Fulde, P. and Komatsubara, T. (2001) Strong Coupling between Local Moments and Superconducting "Heavy" Electrons in UPd2Al3. Nature, 410, 340-343.

24. Stock, C., Broholm, C., Hudis, J., Kang, H.J. and Petrovic, C. (2008) Spin Resonance in the d-Wave Superconductor CeColn5. Physical Review Letters, 100, Article No: 087001.

25. Christiansen, A.D., McMorrow, D.F., R?nnow, H.M., Lake, B., Heiden, S.M., Aeppli, G., Perring, T.G., Mangkorntong, M., Nohara, M. and Takagi, H. (2008) Dispersive Excitations in the High Temperature Superconductor La2?xSrxCuO4. Nature, 456, 930.

26. Yu, G., Li, Y., Motoyama, E.M. and Greven, M. (2009) A Universal Relationship between Magnetic Resonance and Superconductivity Gap in Unconventional Superconductors. Nature Physics, 5, 873-875.

27. McDonald, R.D., Harrison, N. and Singleton, J. (2009) Exact Mapping of the dx2−y2 Cooper-Pair Wavefunction onto the Spin Fluctuations of the Cuprate: The Fermi Surface as a Driver for "High Tc" Superconductivity. Journal of Physics: Condensed Matter, 21, Article No: 012201.

28. Angela, K., Amit, G. and Sudip, C. (2007) Competing Ferromagnetism in High Temperature Copper Oxide Superconductors. Proceedings of the National Academy of Sciences of the United States of America, 104, 6123-6127.

29. Sonier, J.E., Kaiser, C.V., Pacradoum, V., Sabok-Sayer, S.A., Cochrane, C., MacLaughlin, D.E., Komaya, S. and Hussey, N.E. (2010) Direct Search for a Ferromagnetic Phase in a Heavily Overdoped Nonsuperconducting Copper Oxide. Proceedings of the National Academy of Sciences of the United States of America, 107, 17131-17134.

30. Kumar, N. and Jayannavar, A.M. (1992) Temperature Dependence of c-Axis Resistivity of High-Tc Layered Oxides. Physical Review B, 45, 5001.

31. Chmaisen, O., Jorgensen, J.D., Short, S., Knizhnik, A., Eikstein, Y. and Shaked, H. (1999) Scaling of Transition Temperature and CuO2 Plane Buckling in High Temperature Superconductors. Nature, 397, 45-48.

32. Keren, A. (2009) Evidence for Magnetic Mechanism for Cuprate Superconductivity. New Journal of Physics, 11, Article No: 065006.

33. Bartels, D.M., Trifunac, A.D. and Lawler, R.G. (1988) Observation of Heisenberg Spin Exchange between Reactive Free Radicals. Chemical Physics Letters, 152, 109-115.

7

# MODEL OF PREFORMED HOLE-PAIRS IN C-AXIS TRANSPORT IN CUPRATE SUPERCONDUCTORS

**Rana Janardan Singh**

Department of Physics, Aligarh Muslim University, Aligarh, India

## ABSTRACT

Model of hole-pairs in electrical transport along ab plane in cuprate superconductors has already been proposed. It has been found to be in the shape of $3d_{x2-y2}$ orbital of an electron in an atom. This time, model of hole-pairs in transport along c-axis in cuprate superconductors is proposed. In ab-plane, hole-pairs are formed along $CuO_2$ plane; one hole-pair covering 9 - 10 two dimensional $CuO_2$ unit cells in $3d_{x2-y2}$ configuration. In the investigation of c-axis hole-pairs, cuprate superconductors have been sub-divided into three categories depending on the number of $CuO_2$ planes/formula unit. There is a little different treatment for finding out the order parameter in each category. Coherence lengths along ab-planes are of the order of a few tens of Angstroms, whereas along c-axis, they are less than even their a-, b-lattice constants. In cuprates with 2 or 3 $CuO_2$ planes, the order parameter is of $3d_{z2-x2}$ type in zx-plane with lobes along both the axes much constrained. For cuprates with a single $CuO_2$ layer, the order parameter is of $3d_{x2-y2}$ type, but its dimensions are less than a-, b-lattice constants.

## KEYWORDS

Cuprate Superconductors; Preformed Hole-Pairs; Electron Paramagnetic Resonance; 41 meV Peak in Neutron Diffraction

## 1. INTRODUCTION

Holes are Fermionic particles and are charge carriers in cuprate superconductors in the normal state. When two holes combine to form a Bosonic particle, they carry current in cuprate superconductors in superconducting state. Cooper pairs formed by charge carriers involving different quasi-particles have till now failed to explain high temperature cuprate superconductivity. People are veering round the idea that Bosonic pairs of charge carriers are formed much before the transition temperature and these preformed pairs undergo Bose-Einstein condensation at a lower transition temperature resulting into superconductivity.

The model of preformed hole-pairs in ab-plane of cuprate superconductors has already been given by us in the references [1,2]. The model of preformed hole-pairs in electrical conduction along c-axis of cuprate superconductors is proposed on the same lines as used for the model proposed in [2]. In [2], the positions of the two holes forming a pair at different angles (ωt, where ω = angular velocity of holes and t = time) have been expressed mathematically and results shown in a table and also shown in the **Figure 1** of [2]. Ideally, this paper should be read in conjunction with [2].

However, some basic points in the derivation of $3d_{x2-y2}$ order parameter in ab-plane in [2] and also the **Figure 1** will be briefly presented which may be of help in understanding the present paper even in the absence of [2]. **Figure 1** of [2] is reproduced here as **Figure 2** to make the numbering of the rest of the figures continuous. These basic points are:

## 1.1. Electron Paramagnetic Resonance Studies

We had earlier studied electron paramagnetic resonance (EPR) of nearly all cuprate superconductors and also their constituents as CuO, $BaCuO_2$, $CaCuO_2$, $SrCuO_2$, $BiCuO_4$ etc. [3-10] after deoxygenation. It may be mentioned here that cuprate superconductors as such are EPR silent because of antiferromagnetic coupling between $Cu^{2+}$ ions. On deoxygenation (due to loss of oxygen from the bulk) many fragments break away and get magnetically isolated from the bulk. These fragments are Cumonomer, Cu-dimer, Cu-tetramer and Cu-octamer. These fragments were identified on the basis of interpretation of EPR spectra. The most frequently occurring fragment was $(CuO)_4$. The spectra of $(CuO)_4$ showed that the total spin angular momentum of 4 holes associated with 4 $Cu^{2+}$ ions in the $(CuO)_4$ unit = 2, which may result from ferromagnetic coupling of the spins of 4 $Cu^{2+}$ ions (spin of a hole of $Cu^{2+}$ ion = 1/2). As $(CuO)_4$ is the unit cell of 2-dimensional $CuO_2$sheet which is the seat of superconductivity provided a clue for understanding cuprate superconductivity. $(CuO)_4$ fragment is formed when its bonds with 8 surrounding oxygen ions in $CuO_2$ plane are broken. $(CuO)_4$ fragment remains magnetically isolated from the bulk. As Cu-O bond is ≈87.5% ionic and ≈12.5% covalent [11], breaking of 8 bonds amounts to removal of 1 electron from $(CuO)_4$ unit or advent of 1 hole inside it. As oxygen is more electronegative in comparison to copper, negative charge equal to that of 1 electron will be taken away by 8 oxygen ions which have broken away from the isolated $(CuO)_4$. To avoid any confusion between $CuO_2$ and $(CuO)_4$, the difference between the two must be explained. $CuO_2$ is the continuous sheet of copper and oxygen ions in the ab-plane of a cuprate superconductor, where each $Cu^{2+}$ ion is connected to four $O^{2-}$ ions. $(CuO)_4$ entity is a combination of four $Cu^{2+}$ and four $O^{2-}$ ions with four $Cu^{2+}$ ions at the corners of a rectangular figure of dimensions equal to aand blattice constants of the superconducting material and $O^{2-}$ ions placed at the centers of four sides of the rectangle. $(CuO)_4$ remains isolated from every ion/atom in the $CuO_2$ sheet.

It was concluded from above arguments that even in continuous $(CuO)_2$ sheet, if a hole wandering in the $(CuO)_2$ plane on entering a $(CuO)_2$ unit cell will produce the same effect or resulting into ferromagnetic coupling of the spins of the holes of the 4 $Cu^{2+}$ ions. It can be said that the following equation will hold good:

An isolated $(CuO)_4$ plaquette = a $(CuO)_2$ unit cell in the continuous $CuO_2$ sheet + a hole inside it.

It has been reported in [12] that when a hole enters a $CuO_2$ unit cell, the spins align vertically above the $CuO_2$. The field produced will be equal to $4\ \mu/r^3$, where $\mu$ = magnetic moment of a hole and r = distance from the holes. Its maximum value at the center of the plaquette ≈ 1900 gauss. The [12] also suggests a relation between the direction of entry of the hole in a $CuO_2$ plane and the direction of the magnetic field produced. The direction of the magnetic field produced remains at 90º to the direction of entry of the hole and inclined to it in clockwise direction.

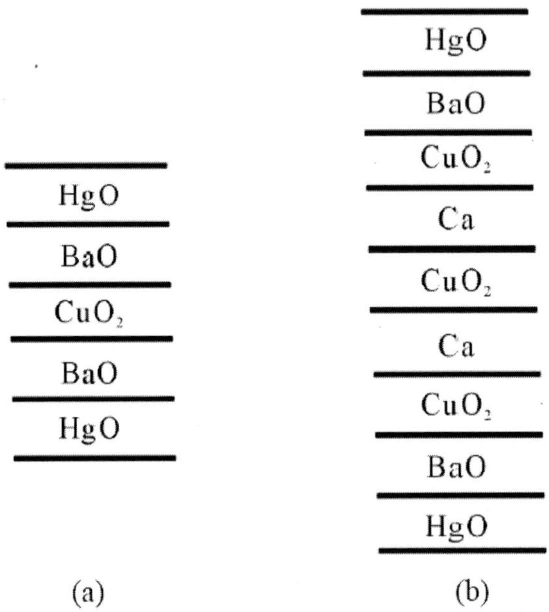


**Figure 1**. Layering schemes of (a) Hg-1201; (b) Hg-1223.

## 1.2. 41 meV Peak Observed in Neutron Diffraction and Nuclear Magnetic Resonance

A peak at 41 meV (with a little difference in energy from one to other superconductor) is observed in neutron diffraction and nuclear magnetic resonance spectra of high temperature superconductors. This peak is associated with spin angular momentum. 41 meV when equated to hν, ν comes to be equal to $10^{13}$ hertz or time-period T = $10^{-13}$ sec. We associate this frequency with oscillations of spins of 4 $Cu^{2+}$ holes, which means that they oscillate with this frequency from vertically upward direction to the vertically downward direction. The manner of oscillation is that when the spins are aligned vertically upward, they produce maximum magnetic field, then they distribute themselves in $CuO_2$ plane in completely dephased manner producing zero magnetic field, and again they align in the vertically downward direction producing maximum magnetic field in the opposite direction. Again they come back to the $CuO_2$ plane producing zero magnetic field and so on.

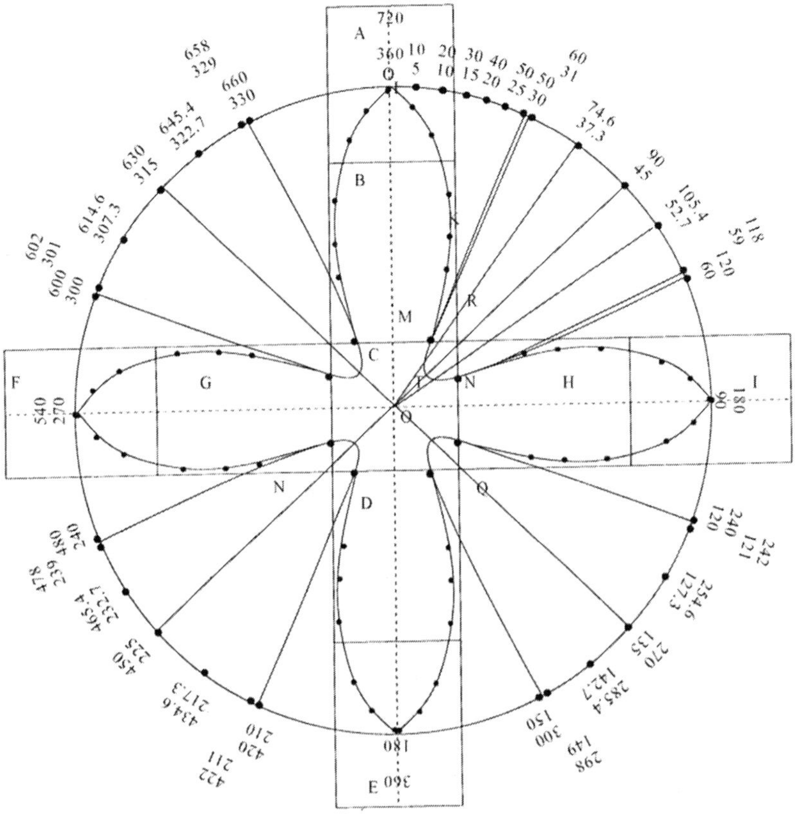


**Figure 2**. Model of hole-pairs in ab-plane. Both the holes traverse the same path but at any instant they are on the opposite sides of the central point. This figure is reproduction of **Figure 1** from [2]. It has been reproduced to get a clearer picture of preformed hole-pairs.

## 1.3. Exchange Interaction

There is always exchange interaction between two Fermions when they exist within certain separation. In the $CuO_2$ plane when two holes come near each other from opposite sides at less than a certain distance, they are bound as a hole-pair. This pair of holes under the effect of magnetic field generated by four $Cu^{2+}$ holes in the $CuO_2$ unit cells move together in a path which is like a $3d_{x2-y2}$ electron orbital in a atom. When the separation between the holes is very small the Coulomb repulsion between the holes will supersede exchange attraction. In $3d_{x2-y2}$ path when the separation between the holes becomes large, the holes are turned back due to magnetic mirror effect. To elucidate the concept of magnetic mirror, it may be mentioned that the pole regions of the earth serve as magnetic mirrors that reflect the incoming charged particles of ionosphere and they move back and forth between two poles. The two poles of the earth act as a magnetic bottle. Mathematically also, for a charged particle moving in a magnetic field, we have $mv^2/r = q\ vxH$ or $r = mv/qH$, where m = mass of the particle, v = velocity of the particle, q = charge on the particle, r = radius of curvature and H = strength of magnetic field. The relation $r = mv/qH$ can be written as $r = v/\alpha\ H$, where $\alpha = q/m$ = specific charge, which is a constant quantity. In the $3d_{x2-y2}$ path of the holes v is constant (components $v_x$ and $v_y$ may change but not v by application of magnetic field). Hence from the relation, $r = v/\alpha\ H$, greater is H, smaller is r. At the farthest distance between the holes in the $3d_{x2-y2}$ configuration the magnetic field is the highest and so the radius of curvature smallest. Therefore, the holes turn back from these positions. Then they face decreasing magnetic field and continue their journeys forward.

## 1.4. Preformed Hole-Pairs Formed at Temperatures Higher than Transition Temperature

It may be mentioned here that preformed hole pairs are formed at much higher temperatures than $T_c$. As the temperature is lowered their number goes on increasing. At a particular low temperature ($T_c$), they undergo BoseEinstein condensation resulting into superconducting state.

For better understanding of the above ideas, the **Figure 1** of [2] is reproduced below as **Figure 2**in the present paper to maintain the continuity of the numbering of figures. There are many similarities and also some dissimilarities with the model we are going to give for c-axis transport.

Now we will show how the preformed hole-pairs are formed in electrical conduction along c-axis of cuprate superconductors. The shape and size of the paired charge carriers (holes) are essentially derived on the same lines as in the ab-plane as given in [2], but in details they are quite different. If one walks along the ab-plane, one finds uniform landscape consisting of $CuO_2$ unit cells only. But if one travels along the c-axis or in a direction perpendicular to ab-plane, one has to cross through different types of terrains at nearly equal intervals. For example, if one travels along c-direction inside the Tl-2212 crystal, one has to cross through the patches consisting of $CuO_2$, Ca, $CuO_2$, BaO, TlO, TlO, BaO, $CuO_2$ unit cells as shown in the **Figure 3**(b) and repetitions of such patches. It has been reported in [13-16] that c-axis resistivity is $\approx 10^2$ to $10^5$ times larger than resistivity in ab-plane. Caxis resistivity sometimes even exceeds maximum metal resistivity of Mott-Ioffe-Regel limit. Maximum metallic resistivity is attained when mean free path of electrons = interatomic distance. Crossing this limit means that mean free path is less than interatomic distance. Though c-axis resistivity $\rho_c$ is much higher than $\rho_{ab}$ in the normal state but the transition temperature $T_c$remains the same in ab-plane or along c-axis. The origin of very high c-axis resistivity in normal state is that c-axis transport is metallic at higher temperatures, semiconducting at lower temperatures and again metallic before $T_c$ [16] but in abplane, it is metallic throughout upto $T_c$. In the superconducting state, the transport in ab-plane is through preformed hole-pairs and along c-axis also, transport between planes is through Josephson's tunnelling of preformed hole-pairs. These two kinds of hole-pairs along c-axis and in ab-plane are different in size and shape but are Bosonic particles and do not encounter any resistance in transportation.

Kumar et al. [13-15] gave a very reasonable explanation for highly enhanced resistivity in c-axis transport in normal state. They interpreted the suppression of single particle transport along the c-axis in the normal state as blocking of the inter-block transport by the intra-block coupling to the many body environments (i.e., entanglement with other electronic degrees of freedom). This mechanism is called Quantum Zeno Effect (QZE). However as stated earlier, the Bosonic preformed hole-pairs at $T_c$ do not suffer any scattering from the above or any other mechanism.

Now we will take up the construction, shape and size of the preformed hole-pairs in the electric transport along c-axis of cuprate superconductors that undergo BoseEinstein condensation at the transition temperature $T_c$, which is the main aim of this paper. Broadly, there are 5 types of high $T_c$cuprate superconductors: 1) La-214 ($La_{2-x}$-

$M_xCuO_4$, M = Ba, Sr); 2) Y-123 ($YBa_2Cu_3O_{7-\delta}$); 3) ($Tl_2 Ba_2Ca_{n-1}Cu_nO_{2n+4}$ with n = 1, 2, 3); 4) ($Bi_2Sr_2Ca_{n-1}Cu_nO_x$ with n = 1, 2, 3); and 5) ($HgBa_2Ca_{n-1}Cu_nO_y$ with n = 1 and 3).

The properties of the last three superconductors are more anisotropic than the first two. $CuO_2$ is common among all of them. The layering schemes of all the superconductors have been given in the Figures 1 and 3-5. The main properties of superconductors have been given in the **Table 1**.

It is well known that in superconductors, $CuO_2$ layers are the seat of superconductivity and they govern most of the properties. So depending upon the number of $CuO_2$ layers in the unit cell of a superconductor, they have been divided into three categories: 1) Having 3 layers of $CuO_2$ as in $Tl_2Ba_2Ca_2Cu_3O_x$, $Bi_2Sr_2Ca_2Cu_3O_y$, $HgBa_2Ca_2Cu_3O_z$; 2) 2 layers of $CuO_2$ as in $Tl_2Ba_2CaCu_2O_x$, $Bi_2Sr_2CaCu_2O_y$, $YBa_2Cu_3O_{7-\delta}$; 3) 1 layer of $CuO_2$ as in $Tl_2Ba_2CuO_x$, $HgBa_2CuO_z$, $La_{1-x}M_xCuO_4$ (where M = Ba or Sr).

Before going further with the discussion of hole-pair formation, we will establish clear distinction between two kinds of holes which have been mentioned earlier in the text and will often come in our discussion to avoid any confusion. One kind of hole is that which is generated in the superconductors due to the level of doping and which is the electric current carrier in superconductors and it will be called lattice hole. The other kind of holes is that which is associated with $Cu^{2+}$ ions in $CuO_2$ network and which generates magnetic field on arrival of a lattice hole in the $CuO_2$ unit cell. This kind of holes will be called $Cu^{2+}$ holes.

First we discuss the representative case of $Tl_2Ba_2CaCu_2O_x$, in the category (2) consisting of 2 layers of $CuO_2$. In all our discussions, we will be mainly interested in giving the shape, size and symmetry of the preformed hole-pairs (i.e., order parameter) in transport along c-axis and estimating the value of $\xi_c$ (coherence length along c-axis). In Tl-2212, there are two $CuO_2$ layers or blocks and in between them there is one layer containing Ca as shown in the **Figure 3**(b) [Layer is the common term but it is generally used for 2-dimensional objects. Along the c-axis, there are 3-dimensional blocks of TlO, BaO, $CuO_2$ etc. heaped one over the other. So in our discussion, both the terms can be used as synonyms]. As $CuO_2$ layers are most active layers, therefore, the hole pair is expected to be formed in between these two layers or in the block containing Ca.

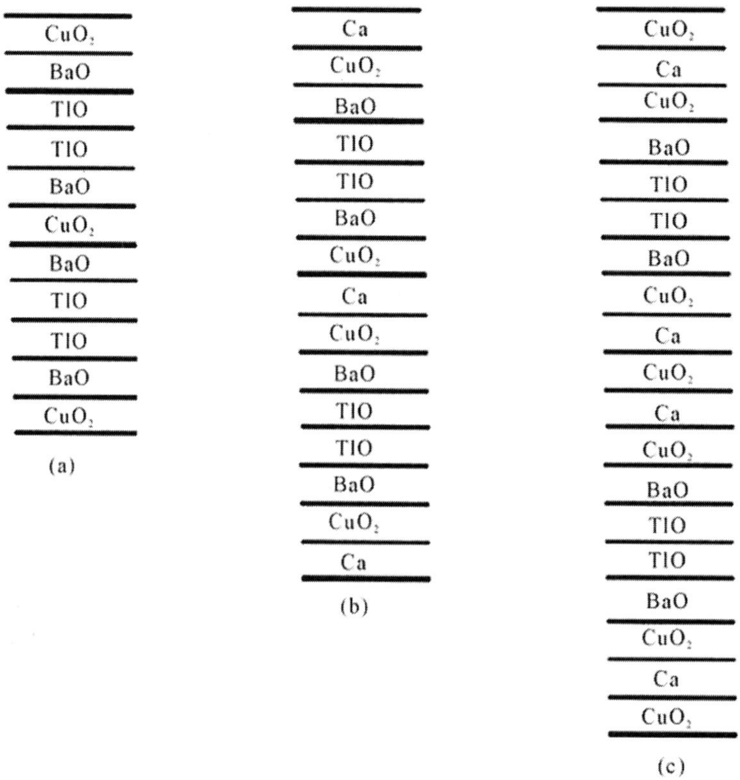


**Figure 3**. Layering schemes of (a) Tl- 2201, (b) Tl-2212, (c) Tl-2223. Note: Layering schemes of Bi-2201, Bi-2212 and Bi-2223 are similar to those of their Tl-counterparts with the difference that instead of Tl and Ba in Tl-superconductors, Bi and Sr are substituted in Bi-superconductors. Another important difference occurring in Bi-superconductors is that in the packing of layers along c-axis, two layers of BiO separated by 3Å are shifted relative to each other along the diagonal direction of the perovskite sub-cell.

We now explain the formation of hole-pairs shown in the **Figure 6**(a). The coordinate system for the **Figure 6** has been given in the **Figure 6**(c).

There are two lattice holes, one approaching from above and the other from below. Only the motion of the hole 1 starting from position A in the figure will be described. Same kind of description is true for the hole 2 starting from E in the figure. The hole 2 is always situated diagrammatically opposite to the hole 1 because of starting from the opposite side. The motion of the two holes is guided by the three effects mentioned in the beginning, 1) magnetic field generated by the advent of a hole inside a $CuO_2$ block; 2) oscillations of spins of holes of four $Cu^{2+}$ ions with a frequency of $10^{13}$ hertz which produce varying magnetic field with the same frequency. **Figure 6**(b) shows the variation of magnetic field from a maximum to zero, then from zero to the maximum but in opposite direction and then again to zero as shown in Parts A, B, C, D respectively; 3) Both the holes are always under the effect of exchange interaction which varies with separation between the holes.

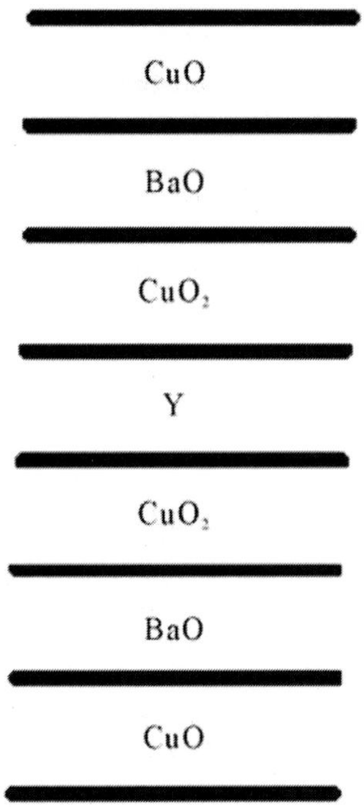


**Figure 4**. Layering scheme of Y-123.

It needs to be mentioned here that the motion of a charged hole in a magnetic field is governed by right hand rule. Hole 1 starting from A travels with its thermal velocity towards the Ca-block but not in straight line but along the curved path AB (**Figure 6**(a)) due to the magnetic field direction shown in the **Figure 6**(b)(A). Then it is deflected along the path BC due to Coulomb repulsion from $Ca^{2+}$ ion and also from the hole 2 which has simultaneously reached near the $Ca^{2+}$ ion at F on the opposite side of hole1. At this instant the magnetic field seems to have nearly vanished (**Figure 6**(b)(B)) and the Coulomb repulsion seems to have superseded attractive exchange interaction. Charge distribution on the $Ca^{2+}$ ion seems to act as reflecting planes for holes 1 and 2 when they reached the positions at B and F respectively. From B, the hole 1 reaches C under the effect of increasing magnetic field in the downward direction. From C the hole 1 reverses its direction of motion and follows the path C to D. At C the magnetic field is maximum but vertically downward and at D, the magnetic field becomes nearly zero. The important question is how the path of the hole 1 is reversed? The reason is the magnetic mirror effect. Magnetic mirror effect has been explained earlier. At D, again the Coulomb repulsion supersedes exchange interaction and charge distribution on $Ca^{2+}$ acts like a reflecting plane and directs hole 1 towards E. From D, the hole 1 reaches E under the effect of magnetic field increasing vertically upwards. In the time the hole 1 has reached E through the path ABCDE, the hole 2 has reached A through the path EFGHA. The hole 2 experiences same kind of forces as the hole 1. Both the holes are always situated diametrically opposite to each other in the path ABCDEFGHA. The two holes continue moving along the path ABCDEFGHA and form a hole-pair.

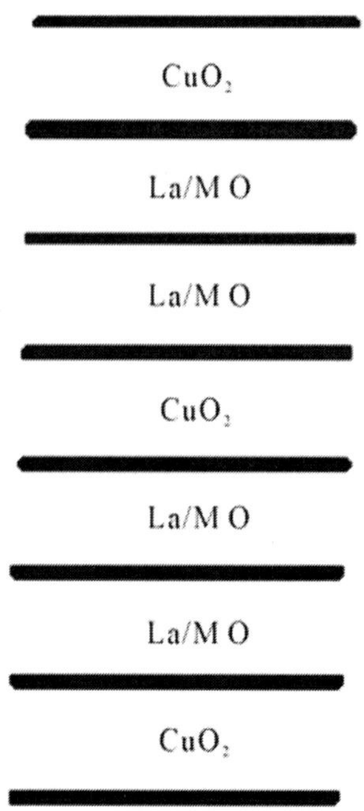


**Figure 5**. Layering scheme of La-214.

**Table 1**. Some important data about superconductors. $^{36}Kr_{84} = 1s^2 2s^2 2p^6 3s^2 3p^6 3d^{10} 4s^2 4p^6$. $^{54}Xe_{131} = 1s^2 2s^2 2p^6 3s^2 3p^6 3d^{10} 4s^2 4p^6 4d^{10} 5s^2 5p^6$. $O_x$ denotes partial filling.

| Superconductor | $T_c$ (K) | $\zeta ab$ (Å) | $\zeta c$ (Å) | Lattice constants | Type1/Type2 Reservoirs/formula unit | Electronic structure of some cations |
|---|---|---|---|---|---|---|
| La-214 | 38 | 21 | - | a = b = 3.77<br>c = 13.25 | La/SrO-La/SrO $CuO_2$ | $^{57}La_{139}$ = (Xe)$5d^1 6s^2$ |
| Y-123 | 92 | 13 - 16 | 2 - 3 | a = 3.823<br>b = 3.885<br>c = 11.7 | BaO-CuO-BaO $CuO_2$-Y-$CuO_2$ | $^{39}Y_{89}$ = (Kr)$4d^1 5s^2$ |
| Bi-2212 | 96 | 27 - 38 | 1.6 - 1.8 | a = 5.40<br>b = 5.41<br>c = 30.90 | SrO-BiO-BiO-SrO $CuO_2$-Ca-$CuO_2$ | $^{83}Bi_{209}$ = (Xe)$4f^{14} 5d^{10} 6s^2 6p^3$ |
| Bi-2223 | 122 | 29 | 0.93 | a = 5.42<br>b = 5.41<br>c = 37.01 | SrO-BiO-BiO-SrO [$CuO_2$-Ca-$CuO_2$-Ca-$CuO_2$] | |
| Tl-2201 | 92 | 52 | 3 | a = b = 3.86<br>c = 11.6 × 2 | BaO-TlO-TlO-BaO $CuO_2$ | $^{81}Tl_{204}$ = (Xe)$4f^{14} 5d^{10} 6s^2 6p^1$ |
| Tl-2212 | 118 | 20 - 31 | 0.3 - 6.8 | a = b = 3.85<br>c = 14.6 x 2 | BaO-TlO-TlO-BaO $CuO_2$-Ca-$CuO_2$ | |
| Tl-2223 | 128 | 11 - 13.6 | - | a = b = 3.85<br>c = 17.8 x 2 | BaO-TlO-TlO-BaO [$CuO_2$-Ca-$CuO_2$-Ca-$CuO_2$] | |
| Hg-1201 | 95 | 17 - 34 | - | a = b = 3.85<br>c = 9.5 | BaO-$HgO_x$-BaO $CuO_2$ | $^{80}Hg_{201}$ = (Xe)$4f^{14} 5d^{10} 6s^2$ |
| Hg-1223 | 135 | 13 | - | a = b = 3.857<br>c = 15.7 | BaO-$HgO_x$-BaO [$CuO_2$-Ca-$CuO_2$-Ca-$CuO_2$] | |

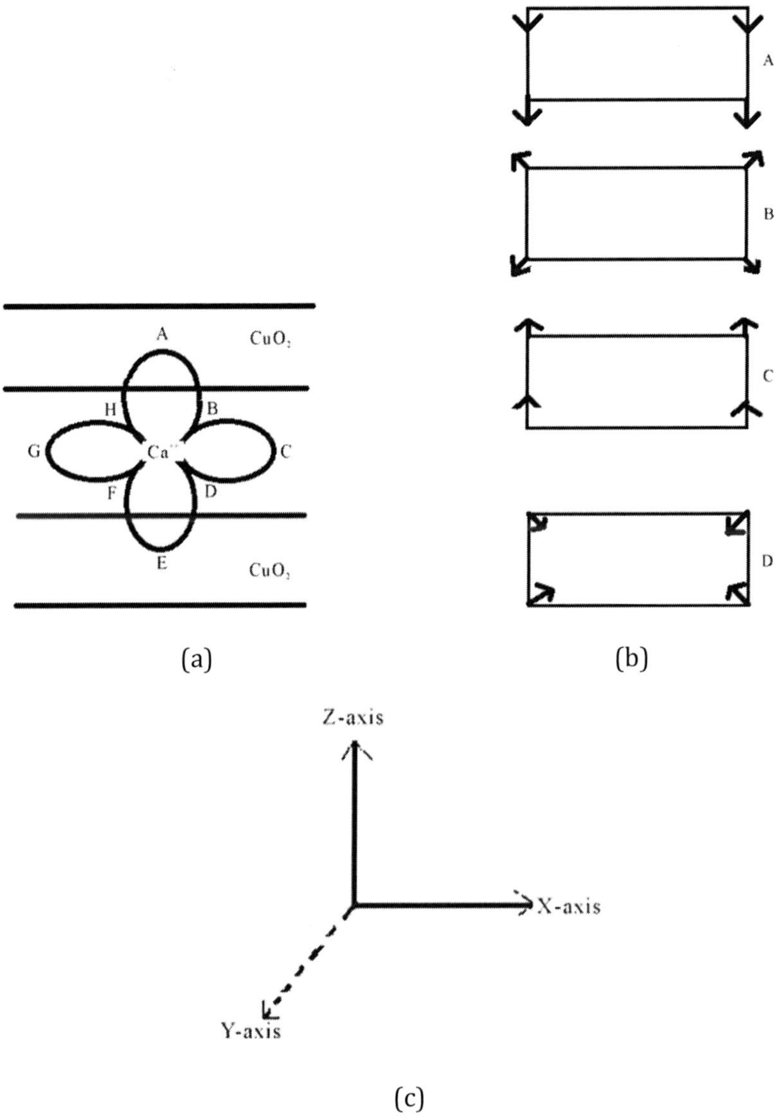


**Figure 6**. (a) Model of hole-pairs in c-axis transport; (b) Directions of magnetic field produced by spin alignments of 4 $Cu^{2+}$ ions of $CuO_2$ unit cell. From A to B to C to D to A again, the spin arrangements have been shown at intervals of T/4. Time-period of spin oscillation (T) = $10^{-13}$ sec; (c) Coordinate system for all figures under **Figure 6**.

In the journey of hole 1 from A, there are 4 stations at B, C, D and at E. At each station, there is drastic change in the path of the hole 1. The question is what is special about these stations? As have been mentioned above, these are the positions from where magnetic field changes from maximum at A to minimum or zero at B, then maximum at C but in the opposite direction, then zero at D and again maximum at E in the same direction as at A. In the journey of lattice hole from A to E, the spins of $Cu^{2+}$ holes have completed one full oscillation. It means that time taken by the hole 1 is $10^{-13}$sec. as the frequency of oscillations of spins of $Cu^{2+}$ holes which cause magnetic field = $10^{13}$ hertz. Therefore, the time for the hole 1 to complete each segment of its journey A to B; B to C ; C to D and D to E should be the same and equals to $1/4 \times 10^{-13}$ sec. For completing each segment of its journey, it travels with its thermal velocity and for equal time of $1/4 \times 10^{-13}$ sec. Hence all the segments should also be of the same length. The same is true for the hole 2. Thus it can be said that in coming from A to E, spins have completed one oscillation or tilted by 360°, though the change of geometrical angle =180° only as seen from the **Figure 6**(a) and also **Figure 2**.

The shape of the hole-pair shown in the **Figure 6**(a) is $3d_{z2-x2}$ because the coordinates involved are z-axis and x-axis. It is not of usual $3d_{x2-y2}$ type obtained for the hole pair in ab-plane of superconductors. The dimensions of $3d_{x2-y2}$ hole pairs in ab plane is usually a few tens of Angstroms. But along c-axis, the dimensions are usually less than aand b-lattice constants. The reason for the small dimension of hole-pairs along c-axis transport may be explained as follows. We have seen that the length of every limb of every $3d_{z2-x2}$ hole pair is decided by the distance travelled by a hole in time = $1/4 \times 10^{-13}$ sec. or equals to velocity of the hole x$1/4 \times 10^{-13}$ sec. Because of incoherent in-plane scattering [13-15], the velocity of hole in the x-direction is much reduced. Also the velocity of the hole in the z-direction or along the c-axis is also much reduced due to uneven terrain and also due to the Quantum Zeno Effect (QZE) [13-15] which suppresses c-axis transport due to intra-block incoherent scattering. It may be possible that limbs of hole-pairs in zand x-directions may be of different length depending upon the resistivity along the zand x-directions. It is well known that greater is the resistivity of a material, less is the velocity of charge carriers. Because the velocities of holes along xor z-direction are not known exactly, it can not be said which limb is bigger and which shorter. But it is known that both limbs will be constrained in comparison to $3d_{x2-y2}$ hole-pairs in ab-plane.

Now we discuss $Tl_2Ba_2Ca_2Cu_3O_x$ (others in the class being $Bi_2Sr_2Ca_2Cu_3O_y$ and $HgBa_2Ca_2Cu_3O_z$) which has three bocks of $CuO_2$ and two intervening blocks of Ca as shown in the **Figure 3**(c). In this case there are two pairs of $CuO_2$ layers each containing a Ca layer in between. One pair consists of upper and middle $CuO_2$ layer with a Ca layer in between, the other pair consists of the middle and the lower $CuO_2$ layer with another Ca layer in between. Two holes coming from opposite sides, one appearing in the upper $CuO_2$ layer and the other appearing in the middle layer simultaneously will form a hole-pair in the intervening Ca block in the same manner as explained in the case of Tl-2212 discussed above. In the other pair of $CuO_2$ layers consisting of the middle and the lower $CuO_2$ blocks with a Ca block in between, a hole-pair will be formed in the same way as in the first pair. Definitely, there is more probability of hole-pairs in Tl-2223 than in Tl-2212, which should ensure higher $T_c$ for Tl-2223 than for Tl-2212, which is corroborated by experiment. $T_c$ of Tl-2223 = 128 K and $T_c$ for Tl-2212 = 118 K.

Next we discuss $Tl_2Ba_2CuO_x$ (or Tl-2201). Others in this class are $HgBa_2CuO_z$ and $La_{2-x}M_xCuO_4$, where M = Ba or Sr. In this system , there is only one $CuO_2$ layer bounded on both sides by four blocks of BaO, TlO, TlO, BaO as shown in the **Figure 3**(a). If two holes, one from above and one from below reach the $CuO_2$ block simultaneously, both will induce the spins of four $Cu^{2+}$ holes in the $CuO_2$ unit to align in the A-configuration [12] of the **Figure 6**(b). and then oscillate from one configuration to the other (i.e., A to B to C to D to A). In the same way as in Tl-2212, the two holes under the action of the magnetic field produced from the fluctuating spins of the four $Cu^{2+}$ ions and exchange effects will follow a closed path in the plane of the $CuO_2$ layer, i.e., in the ab-plane or xy-plane. The symmetry of the order parameter will be of $3d_{x2-y2}$ and not $3d_{z2-x2}$ as in Tl-2212. The hole-pairs along c-axis transport in other cases discussed here have been found to be of $3d_{z2-x2}$type but in the case of Tl-2201 and its associates, it is of type $3d_{x2-y2}$. It has been found experimentally that transport in Tl-2201 is metallic both in ab-plane and along c-axis. There arises an obvious question. If in Tl-2201, the order-parameter both in ab-plane and along c-axis is of $3d_{x2-y2}$ type and transport in both case being metallic, why there should be so much difference between $\xi_{ab}$ (= 52 Å) and $\xi_c$ (= 3 Å). It should be because of different paths taken in current flow in ab-plane and along c-axis. Conduction in the ab-plane is smooth consisting of homogeneous $CuO_2$unit cells, but conduction along c-axis is through heterogeneous slabs of BaO, TlO, TlO, BaO and their repetitions. Moreover, Quantum Zeno Effect (QZE) discussed earlier will be effective in reducing the velocity of hole in x-direction due to incoherent scattering.

From the **Table 1**, it is noticed that $\xi_{ab}$ are of the order of tens of Angstroms, but $\xi_c$ is less that aand b-lattice constants except in Tl-2212, where it has been reported to vary from 0.3 to 6.8 Å. Such large variation of $\xi_c$ seems to be due to some kind of experimental error. In [13-18], it has been pointed out that it is quite possible for any measurement of resistivity $\rho_c$ to pick up some in-plane component of resistive tensor—perhaps externally due to misalignment of contacts or internally due to the randomly distributed defects.

There are two more points worth considering: 1) Why $T_c$ of a superconductor increases when number of $CuO_2$ layers per unit cell increases (see **Table 1**) from one to three. It is because more the number of $CuO_2$ layers, greater will be the number of hole-pairs formed and transition temperature will be reached at higher temperature. Legget [19] has shown that for identically doped members of the same homologous series, as a result of the inter-plane Coulomb interaction, the difference $\Delta T_c^{(n)}$ of the transition temperature for the nth member from the single layer value is given by $\Delta T_c^{(n)}$ = Constant (1 – 1/n). This value of $\Delta T_c^{(n)}$ is to be added to the single layer $T_c$.

Ratan Lal et al., [20] on the basis of phenomenological arguments and structural features have inferred that Cu $3d_{3z^2-r^2}$ orbitals are involved in the transport along c-axis. This kind of orbital may be correct for single particle transport in normal state. But the question with which we are concerned here is to find out what is the symmetry of the hole-pair in the electrical transport along c-axis at $T_c$ and below. It has been found out here that symmetry of the hole-pair carrying current along c-axis at $T_c$ or below is of $3d_{z^2-x^2}$ type for n (number of $CuO_2$ layers/unit cell) =2 and 3 and of $3d_{x^2-y^2}$ type for n = 1.

## 2. SUMMARY

The model of preformed hole-pairs in ab-plane of cuprate superconductors has been proposed earlier, which is of symmetry $3d_{x^2-y^2}$. The model of hole-pairs in the electrical transport along c-axis at $T_c$ and below has been proposed in this paper. The models for c-axis transport have been proposed in three sections corresponding to the number of layers n = 1, 2, 3 per formula unit. For n = 2 and 3, the symmetry of the preformed pairs is of $3d_{z^2-x^2}$ type (i.e., in zx-plane). But for n = 1, the symmetry of the order parameter is of $3d_{x^2-y^2}$ type. The reasons leading to very small coherence length in c-axis transport have been given. It is suggested to carry out angle-dependent experimental studies on cuprate superconductor single crystals as thermal conductivity, specific heat, STM imaging etc to verify the proposed hole-pair models.

# REFERENCES


1. R. J. Singh, "Preformed Hole Pairs in Cuprate Superconductors," International Journal of Modern Physics B, Vol. 23, No. 1, 2009, pp. 53-76.

2. R. J. Singh, "Model of Preformed Hole-Pairs in Cuprate Superconductors," Journal of Modern Physics, Vol. 2, No. 8, 2011, pp. 53-76.

3. A. Punnoose, B. P. Maurya, J. Mathew, M. Umar, M. I. Haque and R. J. Singh, "EPR Observation of $Cu^{2+}$-$Cu^{2+}$ Pairs in Cupric Oxide Powders," Solid State Communications, Vol. 88, No. 3, 1993, pp. 195-198.

4. R. J. Singh, A. Punnoose, J. Mathew, B. P. Maurya, et al., "S = 1 and S = 2 EPR Signals in Modified CuO and $BaCuO_2$," Physical Review B, Vol. 49, No. 2, 1994, pp. 1346-1349.

5. R. J. Singh, M. Ikram, A. Punnoose, B. P. Maurya and S. Khan, "Copper Tetramers in High-Temperature Superconductors," Physics Letters A, Vol. 208, No. 4, 1995, pp. 369-374.

6. A. Punnoose and R. J. Singh, "EPR Studies of High-$T_c$ Superconductors and Related Systems," International Journal of Modern Physics, Vol. 9, No. 10, 1995, pp. 1123-1157.

7. S. Khan, M. Ikram, A. Singh and R. J. Singh, "EPR Study of Deoxygenated $La_2CuO_4$," Physica C, Vol. 281, No. 2-3, 1997, pp. 143-148.

8. S. Khan, A. Singh and R. J. Singh, "EPR Study of $La_{2-x}Sr_xCuO_4$ [M = Ba,Sr]," Solid State Communications, Vol. 106, No. 9, 1998, pp. 621-626.

9. S. Khan, A. Singh and R. J. Singh, "EPR Study of $La_{1.854}Sr_{0.146}CuO_4$," Physica C, Vol. 325, 1999, pp. 165-172.

10. R. J. Singh, P. K. Sharma, A. Singh and S. Khan, "EPR Spectra of Deoxygenated High Temperature Superconductors," Physica C, Vol. 356, No. 4, 2001, pp. 285-296.

11. Q. B. Meng, Z. J. Wu and S. Y. Zhang, "Evaluation of the Energy Barrier Distribution in Many-Particle Systems Using the Path Integral Approach," Journal of Physics: Condensed Matter, Vol. 10, No. 5, 1998, pp. L85-L88.

12. P. C. Dai, H. A. Mook, G. Aeppli, S. M. Hayden and F. Dogan, "Resonance as a Measure of Pairing Correlations in the High-$T_c$ Superconductor $YBa_2Cu_3O_{6.6}$," Nature, Vol. 406, No. 6799, 2000, pp. 965-968.

13. N. Kumar, T. P. Pareek and A. M. Jayannavar, "Normal State c-Axis Resistivity of the High-$T_c$ Cuprate Superconductors," 1997.

14. N. Kumar, "The Zeno Effect and Interlayer Pairing Mechanism for High Temperature Superconductivity in Layered Materials," 2000.

15. N. Kumar and A. M. Jayannavar, "Temperature Dependence of the c-Axis Resistivity of High-$T_c$ Layered Oxides," Physical Review B, Vol. 45, No. 9, 1992, pp. 5001- 5004.

16. K. Takenaka, M. Mizuhasi, H. Takagi and S. Uchida, "Interplane Charge Transport in $YBa_2Cu_3O_{7-y}$," Physical Review B, Vol. 50, No. 9, 1994, pp. 6534-6537.

17. Y. C. Ma, J. W. Liu, H. W. Lu and H. L. Zheng, "Out of Plane Temperature Dependent Resistivity Studies on TlBased Superconductors," Journal of Physics: Condensed Matter, Vol. 19, No. 18, 2007, pp. 186203-186209.

18. R. Jin, D. P. Grandatto and H. R. Rott, "Normal State Resistivity of Superconducting $Bi_{1.95}Sr_{1.65}La_{0.4}CuO_{6+\delta}$," Physica C, Vol. 250, No. 3, 1995, pp. 395-402.

19. A. J. Legget, "Cuprate Superconductivity: Dependence of $T_c$ on the c-Axis Layering Structure," Physical Review Letters, Vol. 83, No. 2, 1999, pp. 392-395.

20. R. Lal, Ajay, R. L. Hotta and S. K. Joshi, "Model for c-Axis Resistivity in Cuprate Superconductors," Physical Review B, Vol. 75, No. 10, 1998, pp. 6126-6136.

8

# DEVELOPMENT AND PRESENT STATUS OF ORGANIC SUPERCONDUCTORS

**Gunzi Saito and Yukihiro Yoshida**

[1] Faculty of Agriculture, Meijo University, Shiogamaguchi 1-501 Tempaku-ku, Nagoya, Japan

## 1. INTRODUCTION

Since the first observation of superconductivity by K. Ones with a critical temperature of superconductivity ($T_c$) of 4.2 K on mercury (1911), many researchers have persuaded such exciting system on organic materials with vain. Even metallic behavior was hardly seen on the organic materials. Little's theoretical proposal (1964) for high $T_c$ superconductivity ($T_c$> 1000K) was based on a polymer system having both a conduction path and highly polarizable pendants, which mediate the formation of Cooper pairs in the conduction path by electron-exciton coupling [1]. There are at least two inorganic polymer superconductors without doping (graphite and diamond are superconductors by doping: see Section 5), poly(sulfur nitride) $(SN)_x$ (1975, $T_c$≤ 3 K) [2] and black phosphorus (1984, $T_c$~ 6 K at 16 GPa and 10.7 K at 29 GPa) [3], with crystalline forms.However, so far no organic polymers have been confirmed to show superconductivity which is easily destroyed by a variety of disorder. Only crystalline polymers were reported to exhibit metallic behavior: a doped polyaniline by chemical oxidation of monomers[4] and $MC_{60}$(Scheme 1) having linearly polymerized $C_{60}^{\bullet-}$with one-dimensional (M = Rb, Cs) or three-dimensional (M = K) metallic behavior[5].

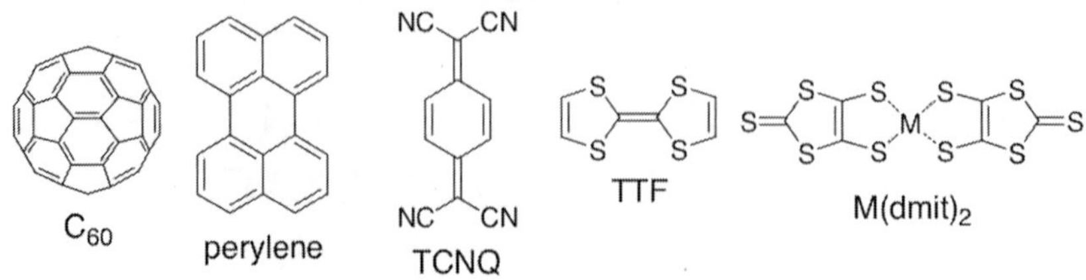


**Scheme 1.**

The Little's model accelerated the exploration of the conducting organic materials of low molecular weight, that had been started by the finding of highly conductive perylene•halides charge-transfer (CT) solids ($10^{0}$–$10^{-3}$ $Scm^{-1}$) in early 1950s [6] and TCNQ CT solids from the 1960s [7].The first metallic CT solid TTF•TCNQ appeared in 1973 [8] based on the two main requirements for the conductivity, namely, (1) a uniform segregated stacking of the same kind of component molecules, and (2) the fractional CT state (uniform partial CT) of the molecules. Since TTF•TCNQ has a low-dimensional segregated stacking, it showed a metal-insulator (MI) transition (Peierls transition) below about 60 K. For TTF•TCNQ, the Peierls transition occurs by the nesting of the one-dimensional Fermi surface causing lattice distortion associated with the strong electron-phonon interaction and forms charge density wave (CDW). There are also several one-dimensional organic metals which show MI transitions by the formation of spin density wave (SDW) when the periodicity of the SDW coincides with the nesting vector of Fermi surface and no lattice distortion occurs in this case. An increase in the electronic dimensionality is inevitable to prevent the nesting of Fermi surfaces and develop superconductors. Several attempts have been made through "pressure", "heavy atom substitution", or "peripheral addition of alkylchalcogen groups" (Fig. 1). The latter two correspond to the enhancement of the self-assembling ability of the molecules.

Appropriate examples taking TTF derivatives are shown in Fig. 1. Based on TMTSF molecules several superconductors under pressure have been prepared with warped one-dimensional Fermi surface since 1980 (a in Fig. 1)[9–14]. In general, the ratio of transfer energies ($t_{//}$ / $t\perp$) is larger than 3 for one-dimensional Fermi surface and a closed two-dimensional Fermi surface is formed when $t_{//} \leq 3t\perp$, where$t_{//}$ and $t\perp$are the transfer energies along the directions of the largest and second largest intermolecular interactions. The BO (BEDO-TTF) molecules afforded stable two-dimensional metals having two-dimensional Fermi surface (b in Fig. 1) owing to the strong self-assembling ability by intermolecular S···S and hydrogen-bonds [15],and only two superconductors are known since 1990 ($T_c$≤ 1.5 K). The substitution of an ethylenedioxy group with an ethylenedithio group (BO→ ET (BEDT-TTF)) destabilized the metallic state of BO compounds and provided unstabletwo-dimensional conductors (c in Fig. 1).Consequently, variety of superconductors and other functional solids have been developed based on two-dimensional metals of ET since 1982 ($T_c$≤ 13.4 K) [16–20] and its analogues ($T_c$≤ 10 K) [21].

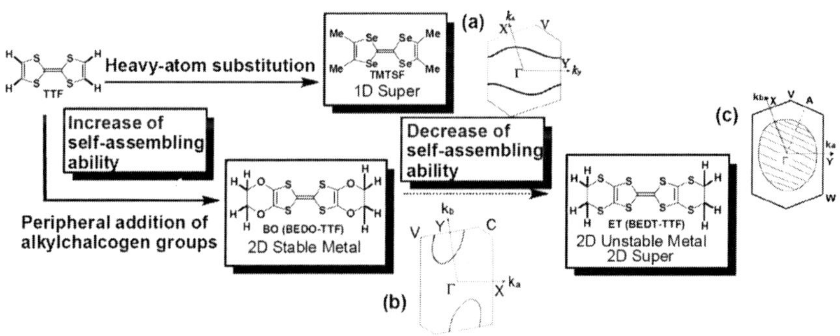


**Figure 1.** Strategy for chemical modification of the TTF molecule to increase (arrows) or decrease (dotted arrow) the electronic dimensionality by the aid of enhancement or suppression of the self-assembling ability of the molecules, respectively[**16**]. Typical Fermi surfaces of TMTSF (a: $(TMTSF)_2NbF_6$), BO (b: $(BO)_{2.4}I_3$), and ET (c: $\beta$-$(ET)_2I_3$) CT solids are depicted.

In this review, we mainly introduce the development and present status of organic superconductors of CT type based on electron donor molecules such as ET, electron acceptor molecules such as $C_{60}$(highest $T_c$ = 38 K at 0.7 GPa), aromatic hydrocarbons (highest $T_c$ = 33 K at ambient pressure (AP)), $M(dmit)_2$ (highest $T_c$ = 8.4 K at 0.45 GPa //*b*)and graphite (highest $T_c$ = 15.1 K at 7.5 GPa), carbon nanotube ($T_c$~ 15 K in zeolite), and B doped diamond ($T_c$ = 11 K at AP). Besides those, single component organic compounds show superconductivity ($T_c$≤2.3 K at 58 GPa). The most reported $T_c$values of CT solids of $C_{60}$, aromatic hydrocarbons, those recently prepared, and those under pressure are the on-set values that are approximately 0.5–1 K higher than the mid-point $T_c$ values. All donor based superconductors are stable in open air, however, only $M(dmit)_2$ superconductors are stable among the acceptor based superconductors.

Most of the superconducting phases of TMTSF, ET, and $C_{60}$materials and also of oxide superconductors reside spin-ordered phases such as SDW and antiferromagnetic (AF) phases. We briefly describe the recent development of superconductors having superconducting phase next to spin-disorder state (quantum spin liquid state).

# 2. PREPARATION OF ORGANIC SUPERCONDUCTORS

CT solids are prepared mainly by the following three redox reactions: (1) electrocrystallization (galvanostatic and potentiostatic), (2)direct reaction of donors (D) and acceptors (A) in the gaseous, liquid, or solid phase, and (3) metathesisusually in solution (D•X + M•A → D•A + MX, M: cation, X: anion). In the latter two cases, single crystals are produced by the diffusion, concentration, slow cooling, or slow cosublimation methods.

Electrocrystallization (main procedures in detail and corresponding references are described in Section 11 of Ref. 17) is performed with a variety of glass cells, as shown in Fig. 2. Strictly speaking, the potentiostatic method is the proper way, in which a three-compartments cell is employed and one of the compartments contains the reference electrode, such as saturated calomel or Ag/AgCl electrode. However, this method is troublesome when a large number of crystal-growth runs are performed for a long period of time due to the following: 1) the contamination through the use of a reference electrode cell, and 2) the limited space for the experiment. The galvanostatic method is much more convenient than the potentiostatic one from these points of view. An H-cell (20 ml or 50 ml capacity) and an Erlenmeyer-type cell (100 ml) with a fine-porosityglass-frit equipped with two platinum wire electrodes (1–5 mm in diameter) have been used (Fig. 2).

There are many factors and tricks to grow single crystals of good quality. The important factors besides both the purity and the concentration of the component materials are the kinds of solvent and electrolyte, the surface of the electrode, the current (0.5–5 μA), and temperature. THF (tetrahydrofuran), $CH_2Cl_2$, TCE (1,1,2-

trichloroethane), chlorobenzene, $CH_3CN$, and benzonitrile are commonly utilized solvents. The addition of 1–10 v/v% ethanol occasionally accelerates the crystal growth. As for the electrolyte, solubility in organic solvent is an important factor and usual electrolytes are tetrabutylammonium (TBA) or tetraphenylphosphonium salt of anion X. Sometimes, the electrolyte is a combination of soluble and insoluble materials. For example, single crystals of $\kappa$-$(ET)_2Cu(NCS)_2$ were prepared using 1) CuSCN + KSCN + 18-crown-6-ether, 2) TBA•SCN + CuSCN, or 3) $Cu(NCS)_2$•K(18-crown-6-ether). Low solubility of the components of the electrolyte in the specific solvent usually retarded single crystal growth. Ionic liquids such as 1-ethyl-3-methylimidazolium (EMI, Scheme 2) salts of X were found to afford single crystals of high quality, recently. Regarding the surfaces of electrodes each research group has special treatments such as burning (but not melting) or polishing with very fine powder. The electrode surface can be treated by applying a current to switch the polarity in a 1 M $H_2SO_4$ solution. When the radical species are unstable in solution, CT solids can be grown by applying a high current at very low temperatures; *e.g.*, salts of fluoranthene (–30°C, 2 mA, Ni electrode), naphthalene, and azulene.

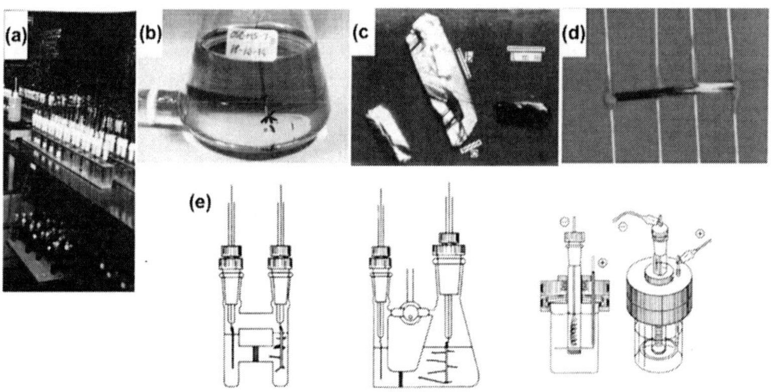


**Figure 2.** a) Galvanostatic electrocrystallization using 20 ml cells on the desk. Under the desk the diffusion method is seen. (b) Single crystals of $\kappa$-$(ET)_2Cu(NCS)_2$ on the electrode in 100 ml cell and (c) showing two-dimensional conducting plane (*bc*). (d) Single crystal of $(TMTSF)_2ClO_4$ with four gold wires connected by gold paste. (e) Typical glass cells for electrocrystallization.

EMI

**Scheme 2**

Besides the electrocrystallization, superconducting single crystals of good quality were prepared by direct chemical oxidation of ET with iodine in gas or with TBA•$I_3$ or TBA•$IBr_2$ in solution. Better-quality single crystals of (TTF)[Ni(dmit)$_2$]$_2$ were obtained by the diffusion method of the metathesisreaction rather than electrocrystallization. No single crystals of the electron acceptor based superconductors were obtained except for the M(dmit)$_2$ system.

The earliest routefor reductive intercalation of $C_{60}$ solids by alkali or alkaline-earth metals is the vapor-solid reaction by vacuum annealing. Almost all superconductors based on graphite and polyaromatic hydrocarbons have been obtained in accordance with this synthetic route. Besides pristine alkali metals, sodium mercury amalgams, sodium borohydride, alkali azides, and alkali decamethylmanganocene have been utilized for source of alkali metal vapors to reduce $C_{60}$ solids. Disproportionation reaction between $C_{60}$ and $M_6C_{60}$ (M: alkali metal) has been sometimes utilized to obtain superconducting $M_3C_{60}$ or non-superconducting $M_4C_{60}$. For a low-temperature solutionroute, liquid ammonia and methylamine have been sometimes utilized for the reaction media. Especially, $Cs_3C_{60}$ with the highest $T_c$ among the $C_{60}$ superconductors can be obtained only when the stoichiometric amounts of cesium metal and $C_{60}$ were reacted in thedissolved methylamine media (see Section 3-2), while the conventional vapor-solid reaction gives energetically stable $Cs_1C_{60}$ and $Cs_4C_{60}$instead of $Cs_3C_{60}$ with nominal composition.

# 3. STRUCTURES AND PROPERTIES

## 3.1. Superconductors Based On Electron Donors

### 3.1.1.One-Dimensionalsuperconductors. (Tmtsf and Tmttf Families)

TMTSF [9–14] has provided eight quasi-one-dimensional superconductors; $(TMTSF)_2X$ with highest$Tc$~ 3 K[11] (Table 1). Most of them were prepared by electrocrystallization using TBA•X as electrolyteexcept the $NbF_6$ salt which can be only prepared by using ionic liquid EMI·$NbF_6$ [12]. They are isostructural to each other and the crystal structure of $(TMTSF)_2NbF_6$ is depicted in Fig. 3, where TMTSF molecules form a zigzag dimer that forms a segregated column along the face-to-face direction (*a*-axis) with no short Se···Se atomic contacts (Fig. 3a,3b). Along the side-by-side direction (*b*-axis), rather short Se···Se atomic contacts were seen (Fig. 3c), however, those less than the sum of the van der Waals radii (3.80 Å) are present only for X = $ClO_4$ and $FSO_3$. For the $PF_6$ salt, $t_a$and $t_b$were estimated to be 0.25–0.30 eV and 0.031 eV, respectively. Consequently, the Fermi surface of $(TMTSF)_2X$ is not closed, but open with fair warping due to the lack of adequate side-by-side transfer interactions (Fig. 1a).

Salts with octahedral anions exhibited MI transition at 11–17 K at AP due to SDW, and a superconductivity appeared with on-set $T_c$ of ca. 1 K at 0.6–1.2 GPa (Fig. 3d). Salts with (pseudo)tetrahedral anions, on the other hand, exhibited order-disorder (OD) transition of anion molecules when the superlattice created was coincide with the nesting vector of the warped Fermi surface ($2a \times 2b$, Fig. 1a).

The isomorphous $(TMTTF)_2X$ (TMTTF: Scheme 3) salts displayed superconductivity under high pressure of 2.6–9 GPa with $T_c$ less than 3 K for X = Br, $BF_4$,$PF_6$, and $SbF_6$[22–25]. Table 1summarizes superconductors of $(TMTSF)_2X$ and $(TMTTF)_2X$ showing $\sigma_{RT}$, temperature at which conductivity shows maximum ($T_{max}$) due to an MI transition, critical pressure to induce superconductivity ($P_c$), $T_c$, phenomena which cause the MI transition (transition temperature), and superlattice after the ordering of the anion molecules.

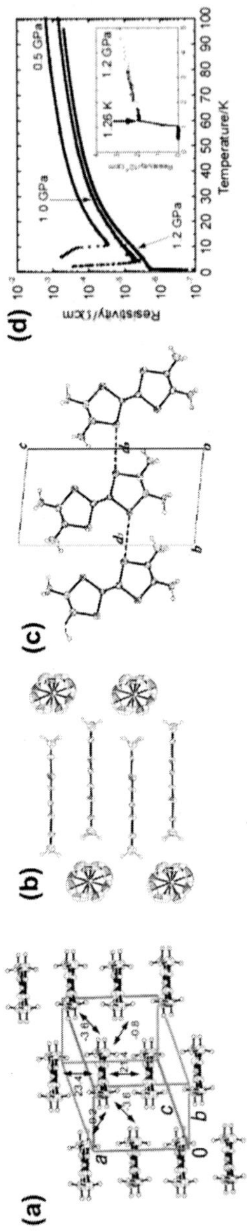


**Figure 3.** Crystal structure of $(TMTSF)_2NbF_6$ [12]. (a) Segregated column of TMTSF molecules. The numbers indicate the overlap integrals in $10^{-3}$ units. (b) Zigzag stacking of the TMTSF column. (c) Se···Se atomic contacts ($d_7$, $d_9$) along the side-by-side direction. (d) Temperature dependence of resistivity under pressure.

$(TMTSF)_2ClO_4$ is the only AP superconductor among them and shows no Hebel-Slichter coherence peak[26], which should be observed just below $T_c$ for a normal BCS-type superconductor having an isotropic gap[27], in the early measurements of relaxation rate of $^1H$ NMR absorption. Later, the thermal conductivity suggested a fully-gapped order parameter [28]. The superconducting coherent lengths are 710 (//*a*), 340 (//*b*), and 20 Å (//*c*), indicating a quasi-one-dimensional character. The application of magnetic field breaks the superconducting state and induces a sequence of SDW (field-induced SDW: FISDW) states above 3 T. Upper critical magnetic field $H_{c2}$of the $PF_6$ salt ($H_{c2}$ = 6 (//*b*'), 4 (//*a*) T at 0.1 K) is far beyond the Pauli limit ($H_{Pauli}$) for the BCS-type superconductor with weak coupling [10].A generalized phase diagram including $(TMTSF)_2X$ and $(TMTTF)_2X$ indicates that the superconducting phase neighbors the magnetic SDW phase (Fig. 4)[29].

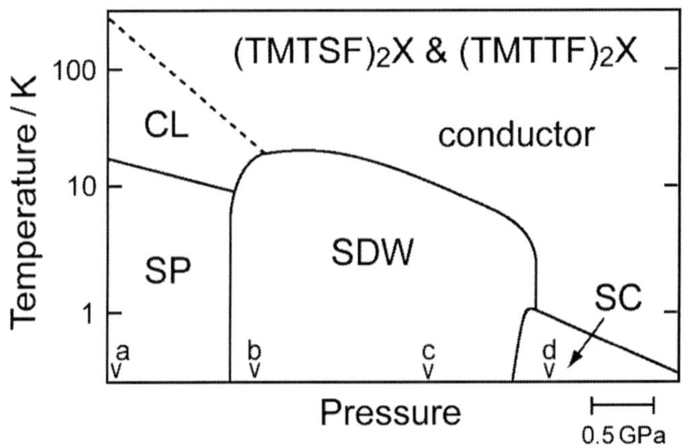


**Figure 4.** Generalized phase diagram for the $(TMTSF)_2X$ and $(TMTTF)_2X$[29]. CL, SP, and SC refer to charge-localized (which corresponds to charge-ordered state), spin-Peierls, and superconducting states, respectively. The salts a–d at AP locates in the generalized diagram. a: $(TMTTF)_2PF_6$, b: $(TMTTF)_2Br$, c: $(TMTSF)_2PF_6$, d: $(TMTSF)_2ClO_4$.

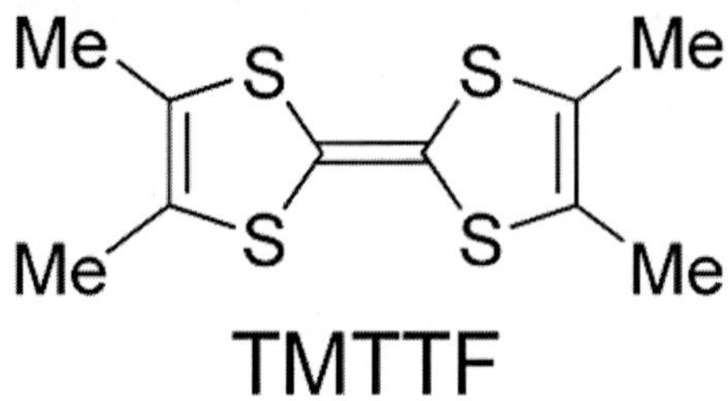

**Scheme 3.**

**Table 1.** Organic superconductors of $(TMTSF)_2X$ and $(TMTTF)_2X$

| X | Symmetry | $\sigma_{RT}$ / S cm$^{-1}$ | $T_{max}$[a)] / K | $P_c$ / GPa | $T_c$[b)] / K | Characteristics[c)] |
|---|---|---|---|---|---|---|
| **TMTSF system** | | | | | | |
| $PF_6$ | octahedral | 540 | 12-15 | 0.65 | 1.1 | SDW (12 K), FISDW |
| $AsF_6$ | octahedral | 430 | 12-15 | 0.95 | 1.1 | SDW (12 K, $J$ = 604 K) |
| $SbF_6$ | octahedral | 500 | 12-17 | 1.05 | 0.38 | SDW (17 K) |
| $NbF_6$ | octahedral | 120 | 12 | 1.2 | 1.26 | SDW (12 K) |
| $TaF_6$ | octahedral | 300 | 15 | 1.1 | 1.35 | SDW (11 K) |
| $ClO_4$ | tetrahedral | 700 | – | 0 | 1.4 | OD (24 K, $a \times 2b \times 2c$), FISDW, ($\gamma$ = 10.5, $\beta$ = 11.4, $\Theta$ = 213) |
| | | – | 5 | – | – | SDW (5 K) by rapid cool |
| $ReO_4$ | tetrahedral | 300 | ~182 | 0.95 | 1.2 | OD (177 K, $a \times 2b \times 2c$) |
| $FSO_3$ | pseudo-tetrahedral | 1000 | ~88 | 0.5 | 3 | OD (88 K, $a \times 2b \times 2c$) |
| **TMTTF system** | | | | | | |
| $PF_6$ | octahedral | 20 | 245 | 5.2-5.4 | 1.4-1.8 | spin-Peierls (15 K) |
| $SbF_6$ | octahedral | 8 | 150 | 5.4-9 | 2.8 | charge-order, AF (8 K) |
| $BF_4$ | tetrahedral | 50 | 190 | 3.35-3.75 | 1.38 | OD (40 K), SDW and SC (coexist) |
| Br | spherical | 260 | 100 | 2.6 | 1.0 | AF (15 K) |

a) *T*max: temperature at maximum conductivity. b) *T*c: on-set. c) SDW: spin density wave, OD: order-disorder transition of anion and newly formed superlattice. FISDW: field-induced SDW. γ and β are important quantities experimentally determined to obtain $D(\varepsilon_F)$ (eq. 1) and Θ (eq. 2) which are related with *T*c by eq. 3 for the BCS-type superconductors.

$\gamma$: Sommerfeld coefficient, mJ mol$^{-1}$ K$^{-2}$

$$\gamma = \pi^2 k_B^2 D(\varepsilon_F)/3 \qquad (1)$$

$\beta$: mJ mol$^{-1}$ K$^{-4}$

$$\beta = 48\pi N k_B / 5\Theta^3 \qquad (2)$$

$\Theta$: Debye temperature, K

$$T_c \propto \Theta \exp(-1/V_{el\text{-}ph} D(\varepsilon_F)) \quad (3)$$

$k_B$: Boltzmann constant, $g$: coupling constant, $V_{el\text{-}ph}$: electron-phonon coupling potential, $D(\varepsilon_F)$: density of states at Fermi level per one spin.

## 3.1.2.Two-Dimensionalsuperconductors. (Bo, Et, And Bets Families)

### *3.1.2.1. BO superconductors*

TTF derivatives with "peripheral addition of alkylchalcogeno groups"were found to be effective to increase dimensionality of CT solids and suppress the Peierls-type MI transition for many BO [15] and ET [16–20]conductors. The robust intermolecular interactions in the BO complexes have provided a metallic state even in the strongly disordered systems. The strong two-dimensionality in the BO complexes hardly exhibited any phase transition including the superconductivity (only two superconductors with $T_c \leq$ 1.5 K were found) [30,31].

### *3.1.2.2. ET superconductors*

The first ET two-dimensional organic metal down to low temperatures is $(ET)_2(ClO_4)(TCE)$[32]. Since then, hundreds of ET solids have been prepared. ET molecules tend to pile up one after the other with sliding to each other so as to minimize the steric hindrance caused by the terminal ethylene groups. A neutral ET molecule is non-planar and becomes almost flat on formation of the partial CT complex except the terminal ethylene groups which are thermally disordered at high temperatures. Segregated packing of such molecules leaves cavities along the molecular long axis, where counter anions and sometimes solvent molecules occupy. It was pointed out that the ethylene conformation is one of the key parameters determining the physical and structural properties including the super conductivity [33]. ET molecule also has a strong tendency to form proximate intermolecular S···S contacts along the side-by-side direction leading to an increment of the side-by-side transfer integrals $t\perp$(Fig. 5). The ET conductors are composed of alternating structures of two-dimensional conducting layer and insulating anion layer. Significant donor···anion interactions arise from the short atomic contacts between the ethylene hydrogen atoms of ET and anion atoms around the anion openings in the anion layer as schematically shown in Fig. 5b [34].

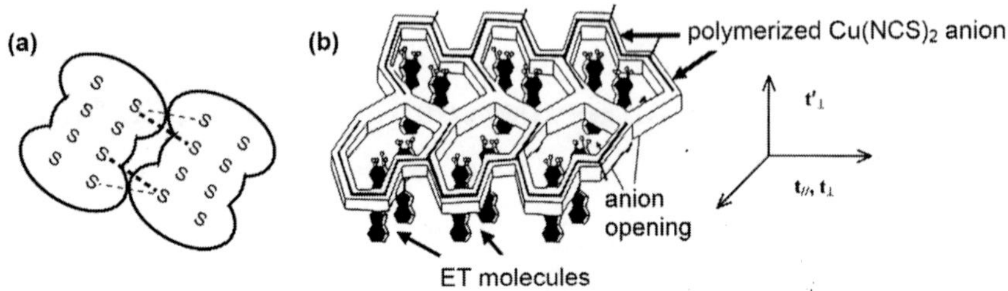


**Figure 5.** a) Schematic figure of an example of S···S atomic contacts observed in ET salts. Thick dotted lines: $S_{in}$···$S_{out}$. Thin dotted lines: $S_{out}$···$S_{out}$. (b). Schematic view of $\kappa$-$(ET)_2Cu(NCS)_2$ indicating anion openings and transfer interactions ($t_{//}$, $t\perp$, and $t'\perp$)[34]. ET dimers are nearly orthogonally aligned ($\kappa$-type packing). For $\kappa$-type salt, $t_{//} \sim t\perp >> t'\perp$.

The steric hindrance exerted by bulky six-membered rings of ET molecules prevents the formation of intermolecular $S_{in}$···$S_{in}$ contacts ($S_{in}$: sulfur atom in the TTF skeleton, Fig. 5a). No particular patterns of intermolecular $S_{in}$···$S_{out}$ ($S_{out}$: sulfur atom in the six-membered ring) are favorable as well. As a consequence various kinds of S···S contacts are produced depending on the donor packing patterns ($\alpha$-,$\beta$-, $\theta$-, $\kappa$-phases, and so forth, see Section 3-1-2-5), and are comparable to the other intermolecular interactions; *i.e.*, face-to-face ($\pi$-$\pi$) and donor···anion interactions. Any interactions could not solely determine the donor packing picture. It is thus much more difficult to predict the donor packing pattern for the ET system compared to those for TMTSF, especially for salts with small and discrete anions such as $I_3$, $ClO_4$, $PF_6$, etc., where polymorphic isomers are frequently afforded. In the salts with discrete linear anions such as $I_3$ and $I_2Br$, the component molecules have great freedom of motion and the donor packing pattern can be changed by thermal or pressure treatment[35,36]. Scanning tunneling microscope (STM) measurements[37–39] revealed that the surface structure of $\beta$-$(ET)_2I_3$ crystals contains many defects, voids, and reconstruction of donor packing attributed to the unstable structure of the anion layers, while the surface structures of salts with polymerized anions such as $\kappa$-$(ET)_2Cu(NCS)_2$ (Fig. 7b in Section 3-1-2-3) and $\alpha$-$(ET)_2MHg(SCN)_4$ (M = $NH_4$ and K) are stable with no defects.

The polymerized anions in $\kappa$-$(ET)_2$ Cu $(NCS)_2$ form the insulating layer having openings as seen in Fig. 5b. Two ET molecules form a dimer unit which fits into each opening. In more accurate description, the hydrogen atom of one ethylene group of ET molecule fits into the core created by anion molecules, like a key-keyhole relation. The position of such an ethylene hydrogen atom projected onto the anion cores produces unique patterns; called $\alpha$-type (5 superconductors), $\beta$-type (6 superconductors), $\theta$-type (1 superconductor), and $\kappa$-type (about 31 superconductors) [40]. It means that the ET molecules arrange according to the anion core or opening pattern created by polymerized anions.

Different kinds of ET ET ($\pi$-$\pi$, S S) and ET anion (hydrogen bonds) intermolecular interactions, large conformational freedom of ethylene groups, flexible molecular framework, fairly narrow bandwidth ($W$), and strong electron correlations, which are represented by on-site Coulomb repulsion energy $U$, gave a rich variety of complexes with different crystal and electronic structures ranging from insulators to superconductors (corresponding references are cited in Ref. 16): Mott insulators (including spin-Peierls systems, antiferromagnets, spin-ladder systems, and quantum spin liquid), one-dimensional metals with CDW transition, two-dimensional metals with CDW transition, two-dimensional metals with FISDW transition, charge-ordered insulators, monotropic complex isomers,and two-dimensional metals down to low temperatures.

About 60 ET superconductors have so far been known. Table 2 summarizes selected ET superconductors and related salts. They are classified into three classes based on the transport behavior at AP: 1) Salts in Class I are metallic down to rather low $T_c$. 2) Salts in Class II are close to a Mott insulator and a poor metal showing $T_c$> 10 K. 3) Salts in Class III are insulators (Mott, CDW, or charge order). Figure 6 compares the temperature dependence of resistivity for several $\kappa$-$(ET)_2X$ with that of $\beta$-$(ET)_2AuI_2$ (6, Class I) which exhibited metallic behavior down to $T_c$at 4.9 K [48]. $\kappa$-$(ET)_2Cu(CN)[N(CN)_2]$ [49] (1, Class II) showed a monotonous decrease of resistivity with upper curvature down to$T_c$. $\kappa$-$(ET)_2Cu[N(CN)_2]Br$ [55] (3, Class II) exhibited similar behavior to that of $\kappa$-$(ET)_2Cu(NCS)_2$[51](2, Class II) except a metallic regime near RT in 2. They have a semiconductive region down to 70–80 K followed by a metallic behavior down to $T_c$. $\kappa$-$(ET)_2Cu$ $[N(CN)_2]Cl$ [57–60] (4, Class III) ia a Mott insulator and showed a semiconductor ($\varepsilon_g$ = 24 meV)-semiconductor($\varepsilon_g$ = 104 meV) transition at ca. 42 K due to an AF fluctuation resulting in a weak ferromagnetbelow 27 K (Néel temperature $T_N$ = 27 K). Under a low pressure, it showed a similar temperature dependence to that of κ-$(ET)_2Cu[N(CN)_2]Br$.κ-$(ET)_2Cu_2(CN)_3$ [61–65] (5, Class III) is semiconductive (a Mott insulator) and under pressure it also behaves similarly to 4 (semiconductor-metal-superconductor).

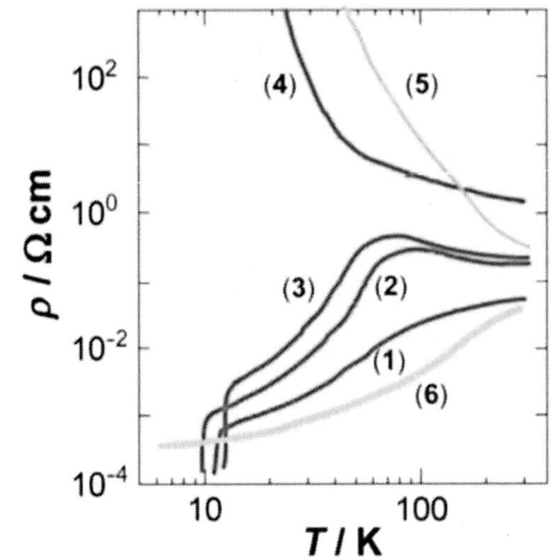


**Figure 6.** Temperature dependences of resistivity of 10 K class superconductors κ-$(ET)_2Cu(CN)[N(CN)_2]$ (1), κ-$(ET)_2Cu(NCS)_2$ (2), κ-$(ET)_2Cu[N(CN)_2]Br$ (3), and κ-$(ET)_2Cu[N(CN)_2]Cl$ (4), which are compared with that of a good metal with low $T_c$β-$(ET)_2AuI_2$ (6) and a Mott insulator κ-$(ET)_2Cu_2(CN)_3$ (5) at AP.

Table 2 summarizes $T_c$ of H- and D-salts (salt using $h_8$-ET and $d_8$-ET, respectively; Scheme 4). The calculated $U/W$values close to unity suggestthat those salts have strong electron correlation.Currentlyβ'-($h_8$-ET)$_2$ICl$_2$ (on-set $T_c$ = 14.2 K at 8.2 GPa[68], mid-point $T_c$ of 13.4 K is estimated) and D-salt of 4 ($T_c$ = 13.1 K at 0.03 GPa)[58] show the highest $T_c$ under pressure, while both are Mott insulators at AP. At AP, D-salt of 1 shows the highest $T_c$ of 12.3 K[50] followed by H-salt of 3 ($T_c$ = 11.8 K) [55,56].These salts are electronically clean metals as evidenced by the observation of quantum oscillations (Shubnikov-de Haas (SdH), de Haas-van Alphen (dHvA)), and geometrical oscillations: angular dependent magnetoresistance oscillation (AMRO), which afford topological information for the Fermi surface [19,20,70].

**Table 2.** Selected ET conductors and superconductors. Except ET•TCNQ, the compoundis represented by*Greek alphabet*-$(ET)_2X$ (Greek alphabet: type of donor stacking, $L_1$, $L_2$: ligand). a) Class I : good metal with low $T_c$, II: 10 K class AP superconductor, III: Mott insulator. 1–6 are the numbers inFig. 6.

| Class | Salt | $CuL_1L_2$ $L_1$, $L_2$ | $\sigma_{RT}$/ S cm$^{-1}$ | $T_c^{a)}$ / K H-salt | D-salt | $t'/t$ | $U/W$ | Ground state at AP$^{b)}$ | Characteristics, Quantum oscillations | Ref |
|---|---|---|---|---|---|---|---|---|---|---|
| I | $\kappa$-$(ET)_2I_3$ | | 30 | 3.6* | – | 0.58 | | SC | $\gamma$ = 18.9, $\beta$ = 10.3, $\Theta$ = 218, SdH, dHvA | [41] |
| | $\beta$-$(ET)_2I_3$ | | 60 | 1.5, 2.0, 8.1 | – | 0.55 | | SC | SdH, dHvA, AMRO | [42–47] |
| | 6 $\beta$-$(ET)_2AuI_2$ | | 20–60 | 4.9 | – | – | | SC | SdH, dHvA | [48] |
| II | 1 $\kappa$-$(ET)_2Cu(CN)[N(CN)_2]$ | CN, $N(CN)_2$ | 5–50 | 11.2* | 12.3* | 0.66–0.71 | 0.87 | SC | AMRO | [49,50] |
| | 2 $\kappa$-$(ET)_2Cu(NCS)_2$ | SCN, NCS | 5–40 | 10.4* | 11.2* | 0.82–0.86 | 0.94 | SC | $\gamma$ = 25, $\beta$ = 11.2, $\Theta$ = 215, SdH, dHvA, AMRO | [51–54] |
| | 3 $\kappa$-$(ET)_2Cu[N(CN)_2]Br$ | $N(CN)_2$, Br | 5–50 | 11.8* | 11.2* | 0.68 | 0.92 | SC | $\gamma$ = 22, $\beta$ = 12.8, $\Theta$ = 210, SdH, AMRO | [55,56] |
| III | 4 $\kappa$-$(ET)_2Cu[N(CN)_2]Cl$ | $N(CN)_2$, Cl | 2 | 12.8/0.03 GPa | 13.1 | 0.75 | 0.90 | AF | SdH, AMRO, $T_N$ = 27 K | [57–60] |
| | 5 $\kappa$-$(ET)_2Cu_2(CN)_3$ | CN, CN/NC | 2–7 | 6.8–7.3** | | 1.06 | 0.9 | SL | SdH, AMRO | [49,61–65] |
| | ET•TCNQ | | 10 | | | – | | AF | $T_N$ = 3 K | [66] |
| | $\beta'$-$(ET)_2ICl_2$ | | $3\times10^{-2}$ | 14.2/8.2 GPa | | – | | AF | $T_N$ = 22 K | [67,68] |
| | $\beta'$-$(ET)_2AuCl_2$ | | ~$10^{-1}$ | | | – | | AF | $T_N$ = 28 K | [68] |
| | $\beta'$-$(ET)_2BrICl$ | | ~$10^{-2}$ | 7.2/8.0 GPa | | | | AF | $T_N$ = 19.5 K | [69] |

a) *: mid-point. **on-set under uniaxial strain (see Fig. 9a). Others are the on-set values under hydrostatic pressure. b) SC: superconductor, SL: spin liquid.

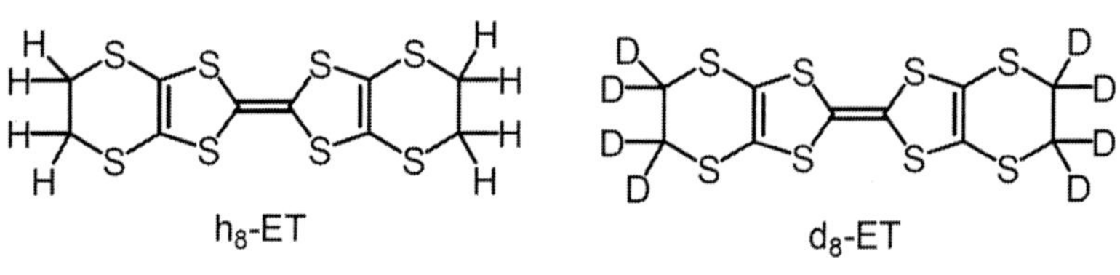


**Scheme 4.**

Next, we will focus mainly on the κ-type superconductors with polymerized anions.

### 3.1.2.3.κ-type ET conductors

The κ-type superconductors κ-$(ET)_2CuL_1L_2$ (1–5 in Table 2) share some common structural and physical properties [49–65]. Figure 7 shows the crystalstructureof theprototypeH-salt of 2, anion structures, donor packing pattern, and calculated Fermi surface [51–54].Table 2 summarizes the two kinds of ligand ($L_1$, $L_2$) in a salt and ratio $t'/t$ for triangle geometry of ET dimers. These salts have polymerized anions in which ligand $L_1$ forms infinite chain by coordinating to $Cu^{1+}$ and ligand $L_2$ coordinates to $Cu^{1+}$ as pendant. ET molecules form a dimer and the ET dimers are arranged nearly orthogonally to each other forming two-dimensional conducting ET layer in the *bc*-plane which is sandwiched by the insulating anion layers along the *a*-axis (Fig. 7a). $Cu^{1+}$ and SCN form Cu SCN Cu SCN zigzag infinite chain along the *b*-axis and other ligand SCN coordinates to $Cu^{1+}$ by N atom to make an open space (indicated by ellipsoid in Fig. 7b, also see Fig. 5) to which an ET dimer fits.An ET dimer has one spin, and the dimers form triangle (Fig. 7c) whose shape is represented by the ratio $t'/t$ (Fig. 7d).

H-salt of 4showed a complicated *T-P* phase diagram (Fig. 8a) [72–76]. Thoroughgoing studies under pressure showed a firm evidence of the coexistence of superconducting (I-SC-2 phase: I-SC = incomplete superconducting) and AF phases [72–78], where the radical electrons of ET molecules played both roles of localized and itinerant ones. Under a pressure of ca. 20–30 MPa another incomplete superconducting phase (I-SC-1) appeared and the complete superconducting (C-SC) phase neighbored to this phase at higher pressures. Below these superconducting phases, reentrant nonmetallic (RN) phase was observed. Similar *T-P* phase diagrams were obtained for κ-$(d_8\text{-}ET)_2X$ (X = $Cu[N(CN)_2]Cl$ and $Cu[N(CN)_2]Br$) with a parallel shift of pressure. They occur at the higher and lower pressure sides of the κ-$(h_8\text{-}ET)_2Cu[N(CN)_2]Cl$ for the Br and Cl salts, respectively.Contrary to the H-salt, κ-$(d_8\text{-}ET)_2Cu[N(CN)_2]Cl$ exhibitedno coexistence of the superconducting and AF phases.At AP, κ-$(ET)_2Cu[N(CN)_2]I$ is a semiconductive and becomes superconducting under hydrostatic pressure above 0.12 GPa with $T_c$ of 7.7 K (on-set $T_c$ = 8.2 K), though the magnetic ordering was not clarified at AP [79].

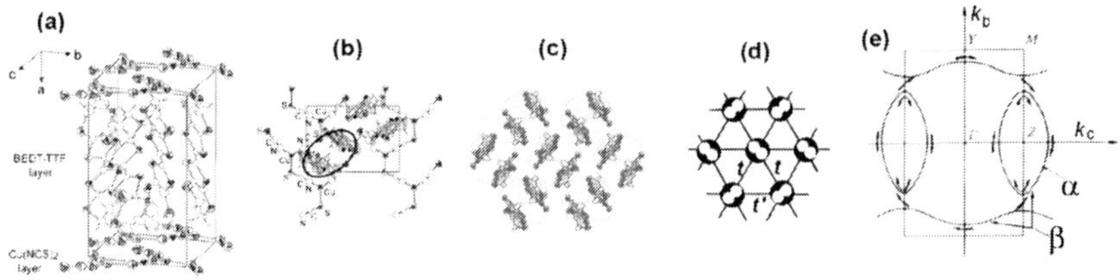


**Figure 7.** κ-$(ET)_2Cu(NCS)_2$: (a) Crystal structure. (b) Anion layer viewed along the *a*-axis has anion openings (indicated by ellipsoid) to which an ET dimer fits. Picture is the dextrorotatory form. (c) Packing pattern (κ-type) of ET dimers along the *a*-axis. (d) Schematic view of triangular lattice of ET dimers which has one spin. White and black circles represent ET molecule and ET dimer, respectively. The$t'/t$ represents the shape of the triangle. (e) Calculated Fermi surfaces of the $P2_1$salts (κ-$(ET)_2Cu(NCS)_2$, κ-$(ET)_2Cu(CN)[N(CN)_2]$) showed the certain energy gap between a one-dimensional electron like Fermi surface ($//k_c$) and a two-dimensional cylindrical hole-like one (α-orbit), while such a gap is absent in the *Pnma*salts (κ-$(ET)_2Cu[N(CN)_2]Br$, κ-$(ET)_2Cu[N(CN)_2]Cl$). For κ-$(ET)_2Cu(NCS)_2$, electrons move along the closed ellipsoid (α-orbit) to exhibit SdH oscillations[53], and at higher magnetic field (> 20 T) electrons hop from the ellipsoid to open Fermi surface to show circular trajectory (β-orbit, magnetic breakdown oscillations)[71].

Alternating mixed donor packing motifs were observed in two phases of α'-κ-$(ET)_2Ag(CF_3)_4(TCE)$, $T_c$(on-set) of which are 9.5 K and 11.0 K for the phase having two-layered (α'+κ) and four-layered (α'+$\kappa_1$+α'+ $\kappa_2$) phase, respectively[81,82]. Since α'-packing generally imparts semiconducting state, both systems have nano-scale hetero junction of semiconductive/superconductive interface, which is thought to give higher $T_c$ in these systems in comparison with $\kappa_l$-$(ET)_2Cu(CF_3)_4(TCE)$ ($T_c$ = 4.0 K). If this explanation is correct, this is an example of interface superconductivity [83,84].

With increasing the distance between the ET dimers in Fig. 7a–d, the transfer interactions between ET dimers decrease; this may correspond to the decrease of *W* and to increase of $D(\varepsilon_F)$, and consequently$T_c$ is expected to increase. According to this line of thought, higher $T_c$ is expected for the salt having alarger anion spacing. Such a κ-type salt may be found near the border between poor metals and Mott insulators. It is true not only for $\kappa$-type but also for other types, *e.g.*, $\beta'$-$(ET)_2X$ (X = $ICl_2$[67,68] and BrICl [69]) having high $T_c$ in the ET family are Mott insulators at AP.

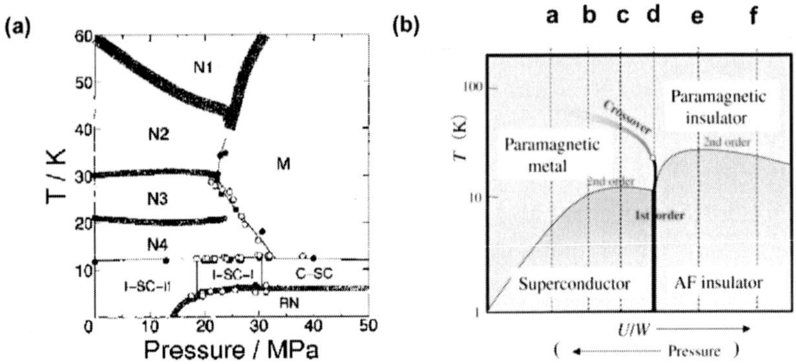


**Figure 8.** (a) Phase diagram of $\kappa$-$(h_8\text{-}ET)_2Cu[N(CN)_2]Cl$ determined from electrical conductivity and magnetic measurements[72–76]. N1–N4: non-metallic phases, M: metallic phase, RN: reentrant non-metallic phase, I-SC-I and I-SC-II: incomplete superconducting phases. N3 shows growth of three-dimensional AF ordered phase. N4 is a weak ferromagnetic phase. (b) Proposed simplified phase diagram[80].a:$\beta$-$(ET)_2I_3$, b: $\kappa$-$(ET)_2Cu(NCS)_2$, c: H-salt of $\kappa$-$(ET)_2Cu[N(CN)_2]Br$, d: D-salt of$\kappa$-$(ET)_2[N(CN)_2]Br$, e: $\kappa$-$(ET)_2[N(CN)_2]Cl$, f: $\beta'$-$(ET)_2ICl_2$.

Topological structures of their Fermi surfaces studied by SdH, dHvA, and AMRO (Table 2)[19,20,53,70,71,85–87] show that the area of the closed Fermi surface relative to the first Brillouin zone and cyclotron mass calculated from SdH oscillations are 15.7% (α-orbit in Fig. 7e, $3.5m_e$) and 105% (β-orbit in Fig. 7e, $6.5m_e$) for 2 at AP, 4.4% ($0.95m_e$) at 0.9 GPa and *ca.* 100% ($6.7m_e$) at AP for 3, and 15.5% ($1.7m_e$) and 102% ($3.5m_e$) at 0.6 GPa for 4. Fermi surface of 2 (Fig. 7e) calculated based on the crystal structure is in good agreement with these observed data.

The followings are the superconducting characteristicsof $\kappa$-type ET superconductors, some of which differ from those of the conventional BCS superconductors.

1) Upper critical magnetic field $Hc_2$: 2gave higher $H_{c2}$ values for the magnetic field parallel to the two-dimensional plane than $H_{Pauli}$based on a simple BCS model[88,89].

2) Coherent length $\xi$: The superconducting coherent lengths are 29 and 3.1 Å for 2at 0.5 K along the two-dimensional plane ($\xi_{//}$) and perpendicular to the plane ($\xi_{\wedge}$). The $\xi_{//}$is larger than the lattice constants, however, the $\xi_{\wedge}$is much smaller than the lattice constant indicating the conducting layers along this direction is Josephson coupled.

3) Symmetry of superconducting state: No Hebel-Slichter coherence peak was observed in both 2and 3 in $^1$H NMR measurements, ruling out the BCS *s*-wave state. The symmetry of the superconducting state of 2had been controversially described as normal BCS-type or non-BCS type, however, STM spectroscopy showed the *d*-wave symmetry with line nodes along the direction near π/4 from the $k_a$- and $k_c$-axes ($d_{x^2-y^2}$) [90], and thermal conductivity measurements were consistent with that [91]. STM on 3also showed the same symmetry [92].A recent specific heat measuremen on 2 and 3 was consistent with these results [93].

4) Inverse isotope effect: Inverse isotope effect has so far been observed for 1 [50],2[54], and 4[58], while normal isotope effect for3[56]. The reason of the observed isotope effects is not fully understood yet consistently.

5) A very simplified *T-P* phasediagram for κ-$(ET)_2X$ was proposed (Fig. 8b),where only the parameter $U/W$ is taken into account. Fig. 8b includes the salts 2,3, 4, β-$(ET)_2I_3$, and β'-$(ET)_2ICl_2$[80],however, the metallic behavior of 2above 270 K and that of 1, whole behavior of 5, and low-temperature reentrant behavior of3and 4 (Fig. 8a) cannot be allocated in this diagram. The β-$(ET)_2I_3$ in Fig. 8b should be$\beta_H$-phase ($T_c$ ~ 8 K, see Section 3-1-2-5) and other two β-$(ET)_2I_3$salts of $T_c$ ~ 1.5 K and ~2 K phases cannot be allocated though they should have the same$U/W$values. $T_c$ of 4 is higher than that of β'-$(ET)_2ICl_2$ in Fig. 8b contrary to the experimentally observed $T_c$ results. This phase diagram and "geometrical isotope effect"[94] point out that $T_c$'sofβ-$(ET)_2I_3$, 2, and 3decrease with increasing pressure if only the parameter $U/W$ or $D(\varepsilon_F)$ is taken into account. This tendency has been observed under hydrostatic pressure but not under uniaxial pressure (see Sections3-1-2-4, 3-1-2-5, 3-1-3). Thus the phase diagram in Fig. 8b remains incomplete,

despite it is frequently used to explain the general trends for these salts.

### 3.1.2.4. Quantum spin liquid state in κ-(ET)2Cu2 (CN)3 and neighboring superconductivity

As mentioned,κ-type packing is characterized by the triangular spin-lattice (Fig. 7c,7d)where an ET dimer is a unit with $S$ = 1/2 spin [95,96].The line shape of $^1$HNMR absorption of κ-$(ET)_2Cu[N(CN)_2]Cl$ [60] exhibited adrastic change below 27 K owing to the formation of three-dimensional AF ordering. On the other hand, the absorption band of κ-$(ET)_2Cu_2$ $(CN)_3$remains almost invariant down to 32 mK indicating non-spin-ordered state: quantum spin liquid state [63]. The appearance of spin liquid state inκ-$(ET)_2Cu_2$ $(CN)_3$is the consequence of significant spin frustration in this salt ($t'/t$ = 1.06) in comparison with the less frustrated AF state in κ-$(ET)_2Cu[N(CN)_2]Cl$($t'/t$ =0.75).

It has long been predicted that the geometrical spin frustration of antiferromagnets caused by the spin correlation in particular spin geometry (triangle, tetrahedral, Kagome (Scheme 5), etc.) prevents the permanent ordering of spins. So the spins of Ising system with AF interaction in the equilateral triangle spin lattice will not show any long-range order even at 0 K, and hence the phase, namely quantum spin liquid phase, has high degeneracy [97]. Such spin liquid state has only been predicted theoretically [98], and a variety of ideal materials have been designed and examined for long [99–102]. Since the discovery of the spin liquid state in κ-$(ET)_2Cu_2(CN)_3$, several materials have been reported to have such exotic spin state: $EtMe_3Sb[Pd(dmit)_2]_2$[103], $ZnCu_3(OH)_6Cl_2$[104,105], $Na_4Ir_3O_8$[106], and $BaCu_3V_2O_8(OH)_2$[107,108]. Some inorganic materials reported as spin liquid candidates were eliminated owing to the spin ordering at extremely low temperatures, etc [109–113].$Na_4Ir_3O_8$ and two organic compounds (κ-$(ET)_2Cu_2(CN)_3$,$EtMe_3Sb[Pd(dmit)_2]_2$)may be recognized as “soft” Mott insulators and have metallic state under pressure.Only κ-$(ET)_2Cu_2(CN)_3$ has the superconducting phase next to spin-liquid state so far as described below.

The phase diagrams of TMTSF (Fig. 4), ET (Fig. 8), $C_{60}$[114] families and also electron-correlated cuprate and iron pnictide high $T_c$ systems[115] indicate that a magnetic ordered state (SDW, AF) is allocated next to the superconducting state. Figure 9a shows the $T$-$P$ phase diagram of κ-$(ET)_2Cu_2(CN)_3$ at low temperature regionby applying uniaxial strain along *c*- ($t'/t$ decreases in this direction)and *b*- ($t'/t$ increases in this direction) axes with epoxy-method[62,65]. In both cases, a superconducting statereadily appeared nearly above 0.1 GPa since the $t'/t$ deviates from unity; *i.e.*, strong spin frustration was released. It is very noteworthy that the spin liquid phase is neighboring to the superconducting state and its $T_c$ is fairly anisotropic as shown in Fig. 9b.A plot of $T_c$ vs. $T_{IM}$, which is a Mott insulator-metal transition temperature, indicates thatin comparison with the hydrostatic pressure results, the uniaxial method afforded: 1) a much higher $T_c$ value, 2) an increase of $T_c$ at the initial pressure region, 3) an anisotropic pressure dependence, and 4) superconducting phase remains at higher pressure. The uniaxial strain experiments including other $\kappa$-type superconductors clearly revealed that the $T_c$ increased as the $U/W$ approaches unity and as the $t'/t$ departs from unity [116].

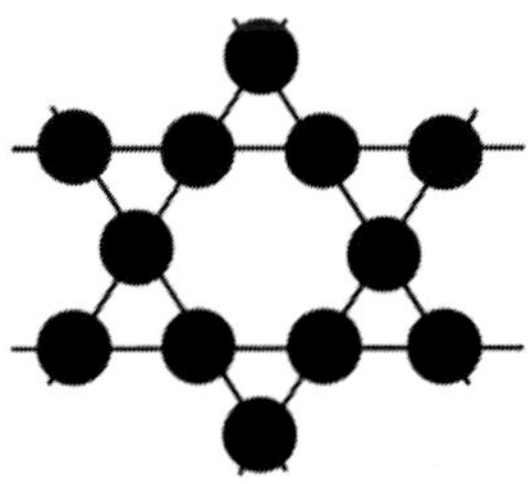


**Scheme 5.**

### *3.1.2.5. Other ET superconductors*

One of the most intriguing ET superconductors is the salt with $I_3$ anion, which afforded $\alpha$-, $\alpha_t$-, $\beta_L$-, $\beta_H$-,$\delta$-, $\varepsilon$-, $\gamma$-, $\theta$-, and $\kappa$-type salts with different crystal and electronic structures. Among them, $\alpha$-, $\alpha_t$-, $\beta_L$-,$\beta_H$-, $\gamma$-, $\theta$-, and $\kappa$-type salts are superconductors with $T_c$ = 7.2, ~8, 1.5, 8.1, 2.5, 3.6, and 3.6 K, respectively[36,41,42,44,45,117–120]. The $\beta_L$-saltwas converted to the $\beta_H$-salt by pressurizing (hydrostatic pressure)above 0.04 GPa and then by depressurizing while keeping the sample below 125 K[45,46]. The $\beta_H$-salt returned to $\beta_L$-salt when the salt was kept above 125 K at AP. The $\beta_L$-salt is characterized by having a superlattice appearing at 175 K with incommensurate modulations of ET and $I_3$ to each other [121]. The formation of the superlattice was suppressed by the pressure above 0.04 GPa. Then the two ethylene groups in an ET molecule were fixed in the eclipsed conformation to give rise to more than 5 times higher $T_c$ in $\beta_H$-salt. The $T_c$ of $\beta_H$-salt decreased with increasing hydrostatic pressure monotonously, however, under the uniaxial stress the further $T_c$ increase taking a maximum at a piston pressure of 0.3–0.4 GPa is observed for both directions parallel and perpendicular to the donor stack [120]. The superconducting coherent lengths are $\xi_{//}$ = 630 (//*a*) – 610 Å (//*b*') and $\xi_{\wedge}$ = 29 Å (//*c**) for the $\beta_L$-salt, and $\xi_{//}$ = 130 Å and$\xi_{\wedge}$ = 10 Å for the $\beta_H$-salt.

The $\alpha$-$(ET)_2I_3$ exhibited nearly temperature independent resistivity down to 135 K, at which charge-ordered MI transition occurred[122,123]. It has been claimed that $\alpha$-$(ET)_2I_3$ has a zero-gap state with a Dirac cone type energy dispersion like graphene[124,125]. Under hydrostatic pressure it became two-dimensional metal down to low temperatures (2 GPa), however, it became superconductor under the uniaxial pressure along the *a*-axis (0.2 GPa, on-set $T_c$ = 7.2 K), though along the *b*-axis it remained metal down to low temperature (0.3–0.5 GPa)[117]. $\alpha$-$(ET)_2I_3$ was able to be converted to mosaic polycrystal with $T_c \sim$ 8 K by tempering at 70–100 ºC for more than 3 days, giving $\alpha_t$-salt which exhibited similar NMR pattern to that of the $\beta_H$-salt[36]. Other $\alpha$-type superconductors ($\alpha$-$(ET)_2$ MHg $(SCN)_4$: M = K, $NH_4$, Rb, Tl) seem to have charge-ordered state near or next to the superconducting state with low $T_c$ (highest $T_c$ = 1.7 K for M = $NH_4$). Uniaxial strain increased $T_c$anisotropically (6 K at 0.5 GPa //*c*, 4.5 K at 1 GPa //*b** for M = $NH_4$) [126]. The salts $\alpha$-$(ET)_2$ $MHg(SCN)_4$ were confirmed to retain their donor packing patterns under pressure at low temperature by the SdH observation. However, for $\alpha$-$(ET)_2I_3$under pressure at low temperature, no exact information is reported for the donor packing in the superconducting state.

Only one $\theta$-type superconductor, $\theta$-$(ET)_2I_3$, is known, however, one third of the obtained crystals are superconducting and others remain metallic [127]. Several $\theta$-type salts are arranged by their inter-column transfer interactions, and the dihedral angle between columns in a phase diagram showing superconducting $\theta$-$(ET)_2I_3$is next to both the charge ordered state of $\theta$-$(ET)_2MM'(SCN)_4$(M = Tl, Rb, Cs, M' = Zn, Co) and metallic phase of $\theta$-$(ET)_2Ag(CN)_2$ [128]. It is not clear which point the non-superconducting $\theta$-$(ET)_2I_3$ occupies in this phase diagram. Tempering (70 ºC, 2 h) all crystals of $\theta$-$(ET)_2I_3$ induced superconductivity with higher $T_c$ (named $\theta_T$-$(ET)_2I_3$: sharp drop of resistivity at 7 K and dull drop at ~5 K) [129]. The tempering changed the ethylene conformation and position of $I_3$from disordered state in $\theta$-phase to ordered state in $\theta_T$-phase. Therefore, the phase diagram of the

$\theta$-type salts needs further parameters concerning with the structure of the salts (ethylene conformation and position or disorder of anions).

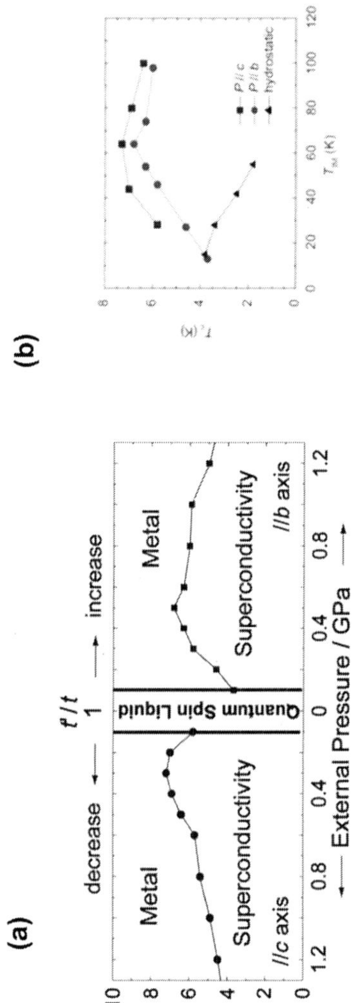


**Figure 9.** a) Temperature-uniaxial pressure phase diagram in the low temperature region of $\kappa$-$(ET)_2Cu_2(CN)_3$[62,65]. The strain along the *c*-axis corresponds to a decrease of $t'/t$ (left side), while the strain along the *b*-axis increases $t'/t$ (right side). b) Pressure dependence of on-set $T_c$ by the uniaxial strain and hydrostatic pressure methods. $T_{IM}$: a Mott insulator-metal transition temperature [65].

### 3.1.2.6. BETS superconductors

The most intriguing phenomenon among the 8 BETS (Scheme 6) superconductors (highest $T_c$ = 5.5 K) is the reentrant superconductor-insulator-superconductor transition under magnetic field for $(BETS)_2FeCl_4$. The λ- or κ-type BETS salts formed with tetrahedral anions $FeX_4$ (X: Cl and Br) were studied in terms of the competition of magnetic ordering and superconductivity[130–136]. The λ-$(BETS)_2FeCl_4$ exhibited coupled AF and MI transitions at 8.3 K. For the $FeCl_4$ salt, a relaxor ferroelectric behavior in the metallic state below 70 K[133] and a firm nonlinear electrical transport associated with the negative resistance effect in the magnetic ordered state have been observed[135]. Moreover it has been found that the $FeCl_4$ salt shows the field-induced superconducting transition under a magnetic field of 18–41 T applied exactly parallel to the conducting layers[132]. The λ-$(BETS)_2Fe_xGa_{1-x}X_4$ passes through a superconducting to insulating transition on cooling[135]. The κ-$(BETS)_2FeX_4$ (X = Cl, Br) are AF superconductors ($T_N$ = 2.5 K, $T_c$ = 1.1 K for X = Br: $T_N$ = 0.45 K, $T_c$ = 0.17 K for X = Cl)[136]. Similar phenomena, namely AF, ferromagnetic, or field-induced superconductivity, have been observed in several inorganic solids such as Chevrel phase[137] and heavy-fermion system[138].Recently it has been reported that λ-$(BETS)_2GaCl_4$ ($T_c$ = 8 K(on-set), 5.5 K(mid-point)) exhibited superconductivity in the minute size of four pairs of $(BETS)_2GaCl_4$ based on the STM study[139]. This salt has two-dimensional superconducting character ($\xi_{//}$ = 125 Å, $\xi_{\wedge}$ = 16 Å).

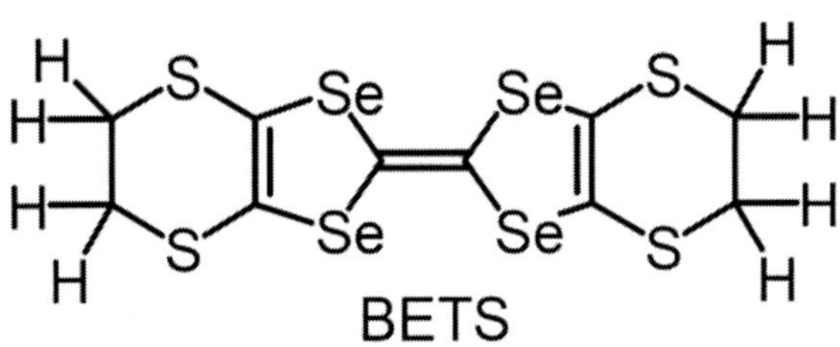


**Scheme 6.**

## 3.1.3. Superconductors Of Other Electrondonor Molecules

Besides TMTSF, TMTTF, BO, ET, and BETS superconductors, there are other superconductors (Scheme 7, numbers in bracket are the total members of each superconducting family and the highest $T_c$) of CT salts based on symmetric (BEDSe-TTF[140] and BDA-TTP [141–144]) and asymmetric donors (ESET-TTF[145], *S,S*-DMBEDT-TTF[146], *meso*-DMBEDT-TTF[147,148], DMET[149], DODHT[150], TMET-STF[151], DMET-TSF[152], DIETS[153], EDT-TTF[154], MDT-TTF[155,156], MDT-ST[157,158], MDT-TS[159], MDT-TSF[160–163], MDSe-TSF[164], DTEDT[165], and DMEDO-TSeF [166,167]).

Donor molecules for organic superconductors except TMTSF, TMTTF, BO, ET, and BETS systems. Numbers in bracket are the total members of each superconductor and the highest $T_c$. κ-$(MDT\text{-}TTF)_2AuI_2$ ($T_c$ = 3.5 K) exhibited a Hebel-Slichter coherent peak just below $T_c$, indicating a BCS-type gap with *s*-symmetry[156]. On the other hand, *d*-wave like superconductivity has been suggested forβ-$(BDA\text{-}TTP)_2SbF_6$[143,144].β-$(BDA\text{-}TTP)_2X$ (X = $SbF_6$, $AsF_6$) exhibited a slight $T_c$increase at the initial stage of uniaxial strain parallel to the donor stack and interlayer direction while$T_c$ decreased perpendicular to the donor stack[142]. θ-$(DIETS)_2[Au(CN)_4]$ exhibited superconductivity under uniaxial strain parallel to the *c*-axis ($T_c$ = 8.6 K at 1 GPa), though under hydrostatic pressure a sharp MI transition remained even at 1.8 GPa[153]. MDT-ST, MDT-TS, and MDT-TSF superconductors [157–163] have non-integer ratio of donor and anion molecules such as (MDT-TS)$(AuI_2)_{0.441}$ making the Fermi level different from the conventional 3/4 filled band for TMTSF and ET 2:1 salts. The Fermi surface topology of (MDT-TSF)X (X = $(AuI_2)_{0.436}$, $(I_3)_{0.422}$) and (MDT-ST)$(I_3)_{0.417}$has been studied by SdH and AMRO[158,161–163]. DMEDO-TSeF afforded eight superconductors. Six of them are κ-$(DMEDO\text{-}TSeF)_2[Au(CN)_2]$(solvent) and the $T_c$'s of them (1.7–5.3 K) are tuned by the use of cyclic ethers as solvent of crystallization[167]. The superconducting coherent lengths indicate that β-$(BDA\text{-}TTP)_2$ $SbF_6$ has two-dimensional character ($\xi_{//}$= 105 Å, $\xi_{\wedge}$= 26 Å) while $(DMET\text{-}TSeF)_2AuI_2$ has quasi-one-dimensional character (1000, 400, and 20 Å).

## 3.2. Superconductors Based On Electron Acceptors

Icosahedral $C_{60}$ molecule with $I_h$ symmetry has triply degenerate LUMO and LUMO+1 orbitals with$t_{1u}$ and $t_{1g}$ symmetries, respectively. In 1991, superconducting phase wasobservedbelow 19 K for the potassium-doped compounds prepared by a vapor-solid reaction [168],immediately after the isolation of macroscopic quantities of $C_{60}$ solid [169].Powder X-ray diffraction profile revealed that the composition of the superconducting phase is $K_3C_{60}$ and the diffraction pattern can be indexed to be a face-centered cubic (fcc) structure [170].

The lattice constant ($a$ = 14.24 Å) is apparently expanded relative to the undoped cubic $C_{60}$ ($a$ = 14.17 Å). The superconductivity has been observed for many $M_3C_{60}$ (M: alkali metal), *e.g.*, $Rb_3C_{60}$ ($T_c$ = 29 K [171]), $Rb_2CsC_{60}$ ($T_c$ = 31 K [172]), and $RbCs_2C_{60}$($T_c$ = 33 K [172]), and their structures determined to be analogous to that of $K_3C_{60}$ with varying lattice constants. The $T_c$ varies monotonously with lattice constant, independently of the type of the alkali dopant (Fig. 10) [172,173]. The observation of a Hebel-Slichter peak in relaxation rate just below $T_c$ in NMR [174] and μSR [175]indicatethe BCS-type isotopic gap. The decrease in $T_c$ due to the isotopic substitution [176] also supports the phonon-mediated pairing in $M_3C_{60}$, where the $\alpha$ value in $T_c \propto (\text{mass})^{-\alpha}$ (ideal value of $\alpha$ predicted by the BCS model is 0.5) is estimated to be 0.30(6) for $K_3{}^{13}C_{60}$and 0.30(5) for $Rb_3{}^{13}C_{60}$.

Keeping the $C_{60}$ valence invariant (−3), the intercalation of $NH_3$ molecules (*e.g.*, $(NH_3)K_3C_{60}$) results in the lattice distortion from cubic to orthorhombic accompanied by the appearance of AF ordering instead of superconductivity [177]. Changing the valence in cubic system also has a pronounced effect on $T_c$. For example, $T_c$ in $Rb_{3-x}Cs_xC_{60}$ prepared in liquid ammonia gradually increases as the mixing ratio approaches to $x$ = 2 [178]. Further increasing the nominal ratio of Cs leads to a sizable decrease of $T_c$, despite the lattice keeps the fcc structure for $x$< 2.65. Such a band-filling control has been realized for $Na_2Cs_xC_{60}$ ($0 \le x \le 1$) [179] and $Li_xCsC_{60}$ ($2 \le x \le 6$) [180], and shows that the $T_c$ decreases sharply as the valence state on $C_{60}$ deviates from −3.

In 2008, A15 or body-centered cubic (bcc) $Cs_3C_{60}$ phase, which shows the bulk superconductivity under an applied hydrostatic pressure, was obtained together with a small amount of by-products of body-centered orthorhombic (bco) and fcc phases, by a solution process in liquid methylamine [181]. Interestingly, the lattice contraction with respect to pressure results in the increase in $T_c$ up to around 0.7 GPa, above which $T_c$ gradually decreases. The trend in the initial pressure range is not explicable within the simple BCS theory. At AP, on the other hand, the A15 $Cs_3C_{60}$ shows an AF ordering below 46 K, verified by means of $^{133}Cs$ NMR and μSR [182]. Very recently, it has been found that the fcc phase also shows an AF ordering at 2.2 K at AP and superconducting transition at 35 K under an applied hydrostatic pressure of about 0.7 GPa [183]. We note that $T_c$ of the both phases follows the universal relationship for $M_3C_{60}$ superconductors in the vicinity of the Mott boundary, as seen in Fig. 10. So far about 40 superconductors were prepared with the highest $T_c$ of 33 K ($RbCs_2C_{60}$) at AP and 38 K (A15 $Cs_3C_{60}$) under pressure (0.7 GPa).

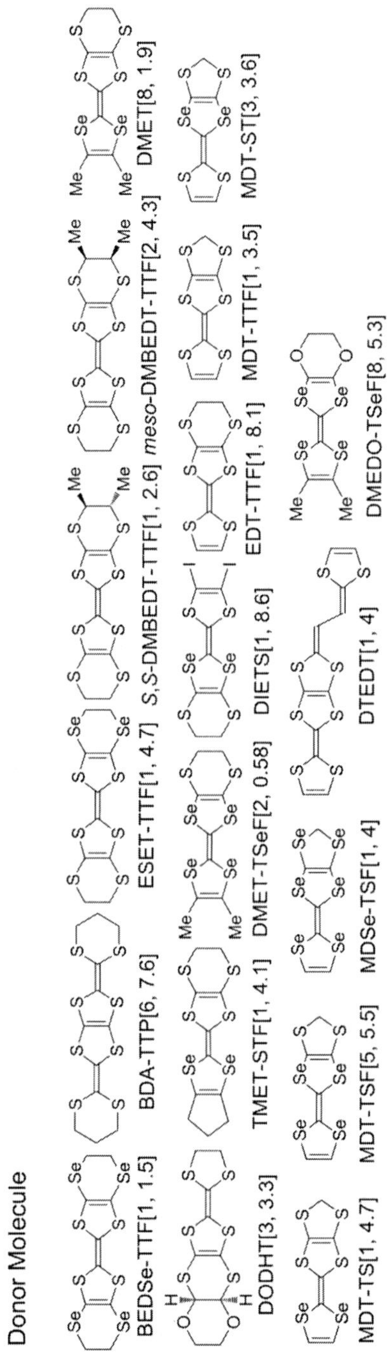
Donor Molecule
BEDSe-TTF[1, 1.5]
BDA-TTP[6, 7.6]
ESET-TTF[1, 4.7]
S,S-DMBEDT-TTF[1, 2.6]
meso-DMBEDT-TTF[2, 4.3]
DMET[8, 1.9]
DODHT[3, 3.3]
TMET-STF[1, 4.1]
DMET-TSeF[2, 0.58]
DIETS[1, 8.6]
EDT-TTF[1, 8.1]
MDT-TTF[1, 3.5]
MDT-ST[3, 3.6]
MDT-TS[1, 4.7]
MDT-TSF[5, 5.5]
MDSe-TSF[1, 4]
DTEDT[1, 4]
DMEDO-TSeF[8, 5.3]

**Scheme 7.**

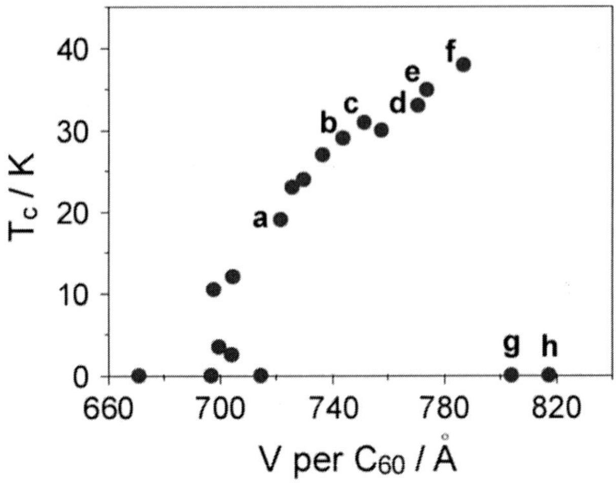


**Figure 10.** $T_c$ as a function of volume occupied per $C_{60}^{3-}$ in cubic $M_3C_{60}$ (M: alkali metal). a: $K_3C_{60}$, b: $Rb_3C_{60}$, c: $Rb_2CsC_{60}$, d: $RbCs_2C_{60}$, e: fcc $Cs_3C_{60}$ at 0.7 GPa, f: A15 $Cs_3C_{60}$ at 0.7 GPa, g: fcc $Cs_3C_{60}$ at AP, h: A15 $Cs_3C_{60}$ at AP.

# 4. POLYAROMATIC HYDROCARBON SUPERCONDUCTORS

Doping of alkalimetals in picene (Scheme 8), that has a wider band gap (3.3 eV) than 1.8 eV for pentacene, introduced superconductivity [184]. The bulk superconductingphase was observed below 7 K and 18 K for $K_3$picene, and 7 K for $Rb_3$picene, which are comparable to that of $K_3C_{60}$($T_c$= 19 K). Since the energy difference between LUMO and LUMO+1 is very small (< 0.1 eV), the three electrons reside in the almost two-fold degenerate LUMO. Considering the fact that $Ca_{1.5}$picene also shows a superconducting phase below 7 K [185], three-fold charge transfer from dopants to one picene molecule would be responsible for emergence of the superconductivity. At present, although the crystal structures of thedoped compounds are unclear, the refined lattice parametersare indicative of the deformation of the herringbone structureof pristine picene and the intercalation of dopants withinthe two-dimensional picene layers. After the discovery of the picene-based superconductors, several superconductors have been found for alkali metal ($T_c$ = 7 K) [186] and alkaline-earth metal ($T_c$ ~ 5.5 K) [187] doped phenanthrene, potassium-doped 1,2:8,9-dibenzopentacene ($T_c$ = 33 K; partially decomposed) [188], and potassium-doped coronene ($T_c$< 15 K) [185]. Among them, phenanthrene-based superconductors shows an enhancement of $T_c$ with increasing pressure, which is indicative of the non-BCS behavior.

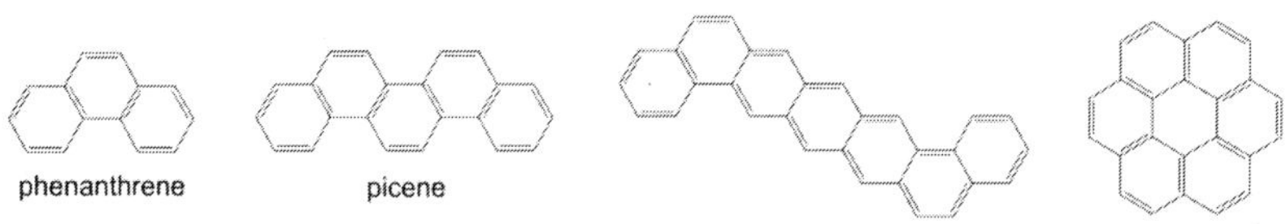

phenanthrene picene 1,2:8,9-dibenzopentacene coronene

**Scheme 8.**

Contrary to the electron-doped system above described, it has been found that a cation radical salt (perylene)$_2$Au(mnt)$_2$, in which each perylene has an average charge of +0.5 and form segregated columns, show a superconductivity with $T_c$ = 0.3 K when the hydrostatic pressure above 0.5 GPa was applied to suppress CDW phase [189].So far 8 aromatic hydrocarbon superconductors were prepared with the highest $T_c$ of 33 K at AP.

# 5. CARBON NANOTUBES, GRAPHITE, AND DIAMOND SUPERCONDUCTORS

In 2001, first superconducting carbon nanotubes were discovered for ropes of single-walled carbon nanotubes (SWNTs) with diameters of the order of 1.4 nm ($T_c$ = 0.4 K) [190], and immediately after that SWNT with diameters of 0.4 nm embedded in a zeolite matrix ($T_c$ = 15 K) [191,192]. The drop in magnetic susceptibility is more gradual than expected for three-dimensional superconductors, and superconducting gap estimated from the *I-V* plot shows the temperature dependency characteristic of one-dimensional fluctuations. It is apparent that the isolation of carbon nanotubes from each other is responsible for the realization of the almost ideal one-dimensional system. Multi-walled carbon nanotubes (MWNTs) also show the superconductivity; namely, MWNT with diameters of 10–17 nm that were grown in nanopores of alumina templates was found to show superconductivity with $T_c$ = 12 K [193]. We note that this superconducting system is classified into single component superconductors, contrary to the $C_{60}$ and graphite (*vide infra*) based superconductors.

Graphite has a layered structure composed of infinite benzene-fused $\pi$-planes (graphenes) with $sp^2$character. First-stage alkali metal doped graphite intercalation compounds (GICs)were known to superconduct with $T_c$= 0.15K for $KC_8$[194].In 1980s and 1990s, further efforts werepoured to synthesize GICs with higher $T_c$, such as $LiC_2$with $T_c$= 1.9 K [195].In 2005, these efforts culminated in the discovery of $CaC_6$with $T_c$as high as 11.5 K at AP [196],which goes up to 15.1 K under pressures up to 7.5 GPa [197]. In other alkaline-earth metal doped GICs, the superconducting phase was observed below 1.65 K for $SrC_6$and 6.5 K for $YbC_6$ [198]. The apparent reduction of $T_c$ strongly suggests that the interlayer states of graphite have an impact on the electronic state of GIC, which was supported by theoretical calculations [199]. The conventional phonon mechanism in the framework of conventional BCS theory isgenerally accepted, due mainly to the observation of theCa isotope effect with its exponent $\alpha$ =0.5[200].

A typical $sp^3$ covalent system, diamond, is an electrical insulator with a wide band gap of 5.5 eV, and is well known for its hardness as well as its unique electronic and thermal properties. Superconductivity in diamond was achieved through heavy p-type doping by boron in 2004 ($T_c$ = 4 K), which was performed under high pressure (8–9 GPa) and high temperatures (2500–2800 K) [201]. Enhanced $T_c$ in homoepitaxial CVD films has been achieved as high as 11 K [202]. Doped boron introduces an acceptor level with a hole bindingenergy of 0.37 eV and results in a metallic state above acritical boron concentration in the range of a few atoms perthousand. The $T_c$ varies between 1 and 10 K with the doping level[203]. Superconducting gap estimated from STM [204] and isotopic substitution of boron and carbon [205] follow the BCS picture, as $MgB_2$ ($T_c$ = 39 K).

Accordingly, all of the carbon polymorphs, namely zero-dimensional $C_{60}$ ($sp^2/sp^3$ character), one-dimensional carbon nanotube ($sp^2$character), two-dimensional graphite ($sp^2$character), and three-dimensional diamond ($sp^3$character) could provide superconductors, despite their covalent characterbeing different. The superconductors with $sp^2/sp^3$ or $sp^2$ carbons were realized either in themselves or by doping of metal atoms, while those with $sp^3$carbons were realized by substitution of boron for carbon.

## 6. SINGLE COMPONENT SUPERCONDUCTORS

Since the discovery of the first metallic CT solid, TTF•TCNQ, in 1973[8], much attention for organic (super) conductors has been devoted to plural component CT solids. Besides numerous studies on multi-component CT solids, several single-component organic conductors have been developed. Even though pentacene is known to be the first organic metal (semimetal) showing a decrease of resistivity down to ca. 200 K at 21.3 GPa [206], no superconductivity was reported so far on the solids composed of aromatic hydrocarbon solely. Electric conductivity increases by the enhancement of intermolecular interactions by appropriate use of hetero-atomic contacts.There are two single-component superconductors under extremely high pressure, p-iodanil ($\sigma_{RT}$ = 1 × $10^{-12}$ S $cm^{-1}$ at AP, $\sigma_{RT}$ = 2 × 10 S $cm^{-1}$ at 25 GPa, and superconductor at$T_c$~ 2 K at 52 GPa) [207,208] and hexaiodobenzene ($T_c$ = 0.6–0.7 K at around 33 GPa andca. 2.3 K at 58 GPa) [209]. Both have peripheral chalcogen atoms, iodine, which may cause the increased electronic dimensionality of the solid under pressure owing to intermolecular iodine···iodine contacts.

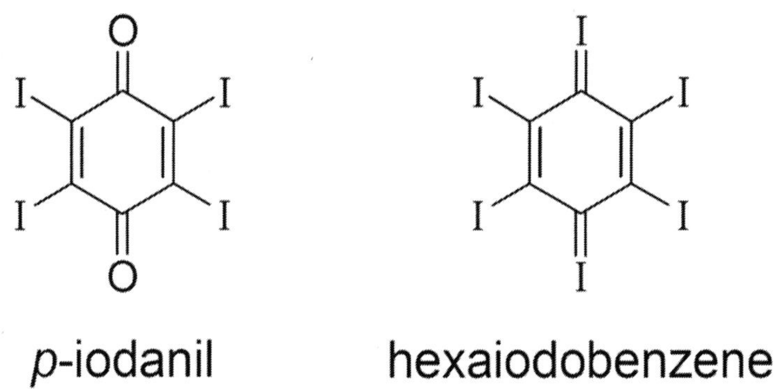


**Scheme 9.**

## REFERENCES

1. Little W A1964Phys. Rev.A13414161424

2. Greene R L, Street G B, Suter L J1975Phys. Rev. Lett. 34577579

3. H. Kawamura, I. Shirotani, K. Tachikawa, 1984Solid State Commun. 49879881

4. K. Lee, S. Cho, S. H. Park, A. J. Heeger, C. Lee-W, S. Lee-H, 2006Nature 4416568

5. a. review, for. see, Prassides. K. example, (200, 2000In: AndreoniW, editor. The Physics of Fullerene-Based and Fullerene-Related Materials. Boston: Kluwer Academic Publishers. 175202

6. H. Akamatu, H. Inokuchi, Y. Matsunaga, 1954Nature 173168169

7. a. review, for. see, Hertler. W. R. example, W. Mahler, L. R. Melby, J. S. Miller, R. E. Putscher, O. W. Webster, 1989Mol. Cryst. Liq. Cryst. 171205216

8. J. Ferraris, D. O. Cowan, V. Walatka, J. H. Perlstein, 1973J. Am. Chem. Soc. 95948949

9. D. Jerome, A. Mazaud, M. Ribault, K. Bechgaard, 1980J. Phys. Lett. 41: L95L98.

10. Lee I J, Naughton M J, Tanner G M, Chaikin P M1997Phys. Rev. Lett. 7835553558

11. F. Wudl, E. Aharon-Shalom, D. Nalewajek, J. V. Waszeczak, W. M. Walsh Jr, L. W. Rupp Jr, P. M. Chaikin, R. Lacoe, M. Burns, T. O. Poehler, J. M. Williams, M. A. J. Beno, Phys. Chem, 1982

12. M. Sakata, Y. Yoshida, M. Maesato, G. Saito, K. Matsumoto, R. Hagiwara, 2006Mol. Cryst. Liq. Cryst. 45299108

13. Jerome D, Schulz H J (1982)Adv.Phys.31:299-490.

14. D. Jerome, 2004Chem. Rev. 10455655591

15. S. Horiuchi, H. Yamochi, G. Saito, K. Sakaguchi, M. Kusunoki, 1996J. Am. Chem. Soc. 11886048622

16. G. Saito, Y. Yoshida, 2007Bull. Chem. Soc. Jpn. 801137and references cited therein.

17. T. Ishiguro, K. Yamaji, G. Saito, 1998Organic Superconductorsnd ed.: Springer-Verlag, Berlin.

18. J. M. Williams, J. R. Ferraro, R. J. Thorn, K. D. Carlson, U. Geiser, H. H. Wang, A. M. Kini, M. Whangbo, M-H (1992Organic Superconductors (IncludingFullerenes):PrenticeHall,EnglewoodCliffs,NJ.

19. J. Wosnitza, (199, 1996Fermi Surfaces of Low-Dimensional Organic Metals and Superconductors: Springer-Verlag, Berlin.

20. J. Singleton, 2000Rep. Prog. Phys. 6311111207

21. Rev. Chem, 200411Special Issue on Molecular Conductors.

22. L. Balicas, K. Behnia, W. Kang, E. Canadell, P. Auban-Senzier, D. Jerome, M. Ribault, J. M. Fabre, 1994J. Phys. I (France) 415391549

23. T. Adachi, E. Ojima, K. Kato, H. Kobayashi, T. Miyazaki, M. Tokumoto, A. Kobayashi, 2000J. Am. Chem. Soc. 12232383239

24. P. Auban-Senzier, C. Pasquier, D. Jerome, C. Carcel, J. M. Fabre, 2003Synth. Met. 133-134: 11-14.

25. M. Itoi, C. Araki, M. Hedo, Y. Uwatoko, T. Nakamura, 2008J. Phys. Soc. Jpn. 77: 023701/14

26. M. Takigawa, H. Yasuoka, G. Saito, 1987J. Phys. Soc. Jpn. 56873876

27. Y. Hasegawa, H. Fukuyama, 1987J. Phys. Soc. Jpn. 56877880

28. S. Belin, K. Behnia, 1997Phys. Rev. Lett. 7921252128

29. D. Jerome, 1991Science 25215091514A more detail diagram was then proposed; Dumm M, Loidl A, Fravel B W, Starkey K P, Montgomery L K, Dressel M (2000) Phys. Rev. B61: 511-521.

30. M. A. Beno, H. H. Wang, A. M. Kini, K. D. Carlson, U. Geiser, W. K. Kwok, J. E. Thompson, J. M. Williams, J. Ren, M. Whangbo-H, 1990Inorg. Chem. 2915991601

31. S. Kahlich, D. Schweitzer, I. Heinen, S. E. Lan, B. Nuber, H. J. Keller, K. Winzer, H. W. Helberg, 1991Solid State Commun. 80191195

32. G. Saito, T. Enoki, K. Toriumi, H. Inokuchi, 1982Solid State Commun. 42557560

33. P. C. W. Leung, T. J. Emge, M. A. Beno, H. H. Wang, J. M. Williams, V. Patrick, P. Coppens, 1985J. Am. Chem. Soc. 10761846191

34. G. Saito, A. Otsuka, A. A. Zakhidov, 1996Mol. Cryst. Liq. Cryst. 284314

35. O. Baram, L. I. Buravov, L. S. Degtyarev, M. E. Kozlov, V. N. Laukhin, E. E. Laukhina, V. G. Onishchenko, K. I. Pokhodnya, M. K. Sheinkman, R. P. Shibaeva, E. B. Yagubskii, 1986JETP Lett. 44376378

36. D. Schweitzer, P. Bele, H. Brunner, E. Gogu, U. Haeberlen, I. Hennig, I. Klutz, R. Sweitlik, H. J. Keller, 1987Z. Phys. B 67489495

37. M. Yoshimura, H. Shigekawa, H. Yamochi, G. Saito, A. Kawazu, 1991Phys. Rev. B4419701972

38. M. Yoshimura, H. Shigekawa, H. Nejoh, G. Saito, Y. Saito, A. Kawazu, 1991Phys. Rev. B431359013593

39. M. Yoshimura, N. Ara, M. Kageshima, R. Shiota, A. Kawazu, H. Shigekawa, Y. Saito, M. Oshima, H. Mori, H. Yamochi, G. Saito, 1991Surf. Sci. 2421822

40. H. Yamochi, T. Komatsu, N. Matsukawa, G. Saito, T. Mori, K. Kusunoki, K. Sakaguchi, 1993J. Am.Chem. Soc. 1151131911327

41. A. Kobayashi, R. Kato, H. Kobayashi, S. Moriyama, Y. Nishio, K. Kajita, W. Sasaki, 1987Chem. Lett. 459462

42. E. B. Yagubskii, I. F. Shchegolev, V. N. Laukhin, P. A. Kononovich, M. V. Karstovnik, A. V. Zvarykina, L. I. Buravov, 1984JETP Lett. 391215

43. Shibaeva R P, Kaminskii V F, Yagubskii E B1985Mol. Cryst. Liq. Cryst. 119361373

44. S. Kagoshima, H. Mori, Y. Nogami, N. Kinoshita, H. Anzai, M. Tokumoto, G. Saito, 1990In: Saito G, Kagoshima S, editors. The Physics and Chemistry of Organic Superconductors: Springer-Verlag, Berlin, 126129

45. V. N. Laukhin, E. E. Kostyuchenko, Yu. V. Sushko, I. F. Shchegolev, E. B. Yagubskii, 1985Pis'ma Zh. Eksp. Teor. Fiz. 416870

46. K. Murata, M. Tokumoto, H. Anzai, H. Bando, G. Saito, K. Kajimura, T. Ishiguro, 1985J. Phys. Soc.Jpn. 5412361239

47. Schirber J E, Azevedo L J, Kwak J F, Venturini E L, Leung P C W, Beno M A, Wang H H, Williams J M1986Phys. Rev. B3319871989

48. H. H. Wang, L. Nunez, G. W. Carlson, J. M. Williams, J. L. Azevedo, J. F. Kwak, J. E. Schirber, 1985Inorg. Chem. 2424652466

49. T. Komatsu, T. Nakamura, N. Matsukawa, H. Yamochi, G. Saito, 1991Solid State Commun. 80843847

50. Fig.54 inRef. 16.

51. H. Urayama, H. Yamochi, G. Saito, K. Nozawa, T. Sugano, M. Kinoshita, S. Sato, K. Oshima, A. Kawamoto, J. Tanaka, 1988Chem. Lett. 5558

52. H. Urayama, H. Yamochi, G. Saito, S. Sato, A. Kawamoto, J. Tanaka, T. Mori, Y. Maruyama, H. Inokuchi, 1988Chem. Lett. 463466

53. K. Oshima, T. Mori, H. Inokuchi, H. Urayama, H. Yamochi, G. Saito, 1988Phys. Rev. B38938941

54. Fig. 53 inRef. 16.

55. M. Kini, U. Geiser, H. H. Wang, K. D. Carlson, J. M. Williams, W. K. Kwok, K. G. Vandervoort, J. E. Thompson, D. L. Stupka, D. Jung, M. Wangbo-H, 1990Inorg. Chem. 2925552557

56. T. Komatsu, N. Matsukawa, T. Nakamura, H. Yamochi, G. Saito, 1992Phosphorus, Sulfur Silicon Relat. Elem. 67295300and Section 3-5-4-2 in Ref. 16.

57. J. M. Williams, A. M. Kini, H. H. Wang, K. D. Carlson, U. Geiser, L. K. Montgomery, G. J. Pyrka, D. M. Watkins, J. M. Kommers, S. J. Boryschuk, A. V. Crouch, W. K. Kwok, J. E. Schirber, D. L. Overmyer, D. Jung, M. Whangbo-H, (199, 1990Inorg. Chem. 2932723274

58. Schirber J E, Overmyer D L, Carlson K D, Williams J M, Kini A M, Wang H H, Charlier H A, Love B J, Watkins D M, Yaconi G A1991Phys. Rev. B4446664669

59. U. Welp, S. Fleshler, W. K. Kwok, G. W. Crabtree, K. D. Carlson, H. H. Wang, U. Geiser, J. M. Williams, V. M. Hitsman, 1992Phys. Rev. Lett.69840843This paper mistakenly described that the AF transition occurred at 45 K and weak ferromagnetic transition at 23 K. Themagnetic susceptibility measurements could not identify the transition either AF or SDW.

60. K. Miyagawa, K. Kawamoto, Y. Nakazawa, K. Kanoda, 1995Phys. Rev. Lett. 7511741177

61. U. Geiser, H. H. Wang, K. D. Carlson, J. M. Williams, H. A. Charlier Jr, J. E. Heindl, G. A. Yaconi, B. H. Love, M. W. Lathrop, J. E. Schirber, D. L. Overmyer, J. Ren, M. Whangbo-H, 1991Inorg. Chem. 3025862588

62. Y. Shimizu, M. Maesato, G. Saito, O. Drozdova, L. Ouahab, 2003Synth. Met. 133-134: 225-226.

63. Y. Shimizu, K. Miyagawa, K. Kanoda, M. Maesato, G. Saito, 2003Phys. Rev. Lett. 91: 107001/14

64. F. L. Pratt, P. J. Baker, S. J. Blundell, T. Lancaster, S. Ohira-Kawamura, C. Baines, Y. Shimizu, K. Kanoda, I. Watanabe, G. Saito, 2011Nature 471612616

65. Y. Shimizu, M. Maesato, G. Saito, 2011J. Phys. Soc. Jpn. 80: 074702/17

66. Y. Iwasa, K. Mizuhashi, T. Koda, Y. Tokura, G. Saito, 1994Phys. Rev. B4935803583

67. N. Yoneyama, A. Miyazaki, T. Enoki, G. Saito, 1999Bull. Chem. Soc. Jpn. 72639651

68. H. Taniguchi, M. Miyashita, K. Uchiyama, K. Satoh, N. Mori, H. Okamoto, K. Miyagawa, K. Kanoda, M. Hedo, Y. Uwatoko, 2003J. Phys. Soc. Jpn. 72468471

69. K. Uchiyama, M. Miyashita, H. Taniguchi, K. Satoh, N. Mori, K. Miyagawa, K. Kanoda, M. Hedo, Y. Uwatoko, 2004J. Phys. IV 114387389

70. Appendix (pp. 455-458) inRef. 17.

71. T. Sasaki, H. Sato, N. Toyota, 1990Solid State Commun. 76507510

72. H. Ito, M. Watanabe, Y. Nogami, T. Ishiguro, T. Komatsu, G. Saito, N. Hosoito, 1991J. Phys. Soc. Jpn. 6032303233

73. Y. V. Sushko, H. Ito, T. Ishiguro, S. Horiuchi, G. Saito, 1993J. Phys. Soc. Jpn. 6233723375

74. H. Ito, T. Ishiguro, M. Kubota, G. Saito, 1996J. Phys. Soc. Jpn. 6529872993

75. M. Kubota, G. Saito, H. Ito, T. Ishiguro, N. Kojima, 1996Mol. Cryst. Liq. Cryst. 284367377

76. H. Ito, T. Ishiguro, T. Kondo, G. Saito, 2000J. Phys. Soc. Jpn. 69290291

77. H. Posselt, H. Muller, K. Andres, G. Saito, 1994Phys. Rev. B491584915852

78. S. Lefebvre, P. Wzietek, S. Brown, C. Bourbonnais, D. Jérome, C. Mézière, M. Fourmigué, P. Batial, 2000Phys. Rev. Lett. 8554205423

79. M. A. Tanatar, T. Ishiguro, S. Kagoshima, N. D. Kushch, E. B. Yagubskii, 2002Phys. Rev. B65: 064516/15

80. K. Kanoda, 2006J. Phys. Soc. Jpn. 75: 051007/116

81. J. A. Schulueter, L. Wiehl, H. Park, M. de Souza, M. Lang, H. Koo-J, M. Whangbo-H, 2010J. Am. Chem. Soc. 1321630816310

82. T. Kawamoto, T. Mori, A. Nakao, Y. Murakami, J. A. Schlueter, 2012J. Phys. Soc. Jpn. 81: 023705/14

83. J. Pereiro, A. Petrovic, C. Panagopoulos, I. Božović, 2011Phys. Express 1208241

84. S. Gariglio, J. Triscone-M, (201, 2011C. R. Phys. 12591599

85. M. V. Kartsovnik, G. Logvenov, Ito. H. Yu, T. Ishiguro, G. Saito, 1995Phys. Rev. B52: R15715R15718.

86. H. Weiss, M. V. Kartsovnik, W. Biberacher, E. Steep, E. Balthes, A. G. M. Jansen, K. Andres, N. D. Kushch, 1999Phys. Rev. B591237012378

87. M. V. Kartsovnik, W. Biberacher, K. Andres, N. D. Kushch, 1995JETP Lett. 62905909

88. K. Oshima, H. Urayama, H. Yamochi, G. Saito, 1988J. Phys. Soc. Jpn. 57730733

89. M. Nam-S, J. A. Symington, J. Singleton, S. J. Blundell, A. Ardavan, J. A. A. J. Perenboom, M. Kurmoo, P. Day, 1999J. Phys.: Condens. Matter11: L477484

90. T. Arai, K. Ichimura, K. Nomura, 2001Phys. Rev. B63: 104518/15

91. K. Izawa, H. Yamaguchi, T. Sasaki, Y. Matsuda, 2002Phys. Rev. Lett. 88: 27002/14

92. K. Ichimura, M. Takami, K. Nomura, 2008J. Phys. Soc. Jpn. 77: 114707/16

93. L. Malone, O. J. Taylor, J. A. Schlueter, A. Carrington, 2010Phys. Rev. B82: 014522/15

94. N. Toyota, 1996In: Physical Phenomena at High Magnetic Fields-II.World Sci., 282293

95. McKenzie R H1998Comments Condens. Matter Phys. 18309337

96. H. Kino, H. Fukuyama, 1995J. Phys. Soc. Jpn. 6427262729

97. Wannier G H1950Phys. Rev. 79357364

98. Anderson P W1973Mater. Res. Bull. 8153164

99. L. Balents, 2010Nature 464199208

100. Norman M R2011Science 332196200

101. Y. Furukawa, Y. Sumida, K. Kumagai, F. Borsa, H. Nojiri, Y. Shimizu, H. Amitsuka, K. Tenya, P. Kögerler, L. Cronin, 2011J. Phys.: Conf. Ser. 320: 012047/16

102. G. Seeber, P. Kögerler, B. M. Kariuki, L. Cronin, 2004Chem. Commun. 15801581

103. T. Itou, A. Oyamada, S. Maegawa, M. Tamura, R. Kato, (200, 2007J. Phys.: Condens. Matter 19: 145247/15

104. Shores M P, Nytko E A, Bartlett B M, Nocera D G2005J. Am. Chem. Soc. 1271346213463

105. P. Mendels, F. Bert, 2010J. Phys. Soc. Jpn. 79: 011001/110

106. Y. Okamoto, M. Nohara, H. Aruga-Katori, H. Takagi, 2007Phys. Rev. Lett. 99: 137207/14

107. Y. Okamoto, H. Yoshida, Z. Hiroi, 2009J. Phys. Soc. Jpn. 78: 033701/14

108. R. H. Colman, F. Bert, D. Boldrin, A. D. Hiller, P. Manuel, P. Mendels, A. S. Wills, 2011Phys. Rev. B83: 180416/14

109. R. Coldea, D. A. Tennante, Z. Tylczynski, 2003Phys. Rev. B68: 134424/116

110. S. Nakatsuji, Y. Nambu, H. Tonomura, O. Sakai, C. Broholm, H. Tsunetsugu, Y. Qiu, Y. Maeno, 2005Science 30916971700

111. A. Olariu, P. Mendels, F. Bert, B. G. Ueland, P. Schiffer, R. F. Berger, R. J. Cava, 2006Phys. Rev. Lett. 97: 167203/14

112. A. Kimmel, M. Mucksch, V. Tsurkan, M. M. Koza, H. Mutka, A. Loidl, 2005Phys. Rev. Lett. 94: 237402/14

113. H. Yoshida, Y. Okamoto, T. Tayama, T. Sakakibara, M. Tikunaga, A. Matsuo, Y. Narumi, K. Kindo, M. Yoshida, M. Takigawa, Z. Hiroi, 2009J. Phys. Soc. Jpn. 78: 043704/14

114. Y. Iwasa, T. Takenobu, 2003J. Phys.: Condens. Matter 15: R495R519.

115. Chu C W2009Nat. Phys. 5787789

116. M. Maesato, Y. Shimizu, T. Ishikawa, G. Saito, K. Miyagawa, K. Kanoda, (200, 2004J. Phys. IV 114227232

117. N. Tajima, A. Ebina-Tajima, M. Tamura, Y. Nishio, K. Kajita, 2002J. Phys. Soc. Jpn. 7118321835

118. Shibaeva R P, Kaminskii V F, Yagubskii E B1985Mol. Cryst. Liq. Cryst. 119361373

119. H. Kobayashi, R. Kato, A. Kobayashi, Y. Nishio, K. Kajita, W. Sasaki, 1986Chem. Lett. 789792

120. H. Ito, T. Ishihara, M. Niwa, T. Suzuki, S. Onari, Y. Tanaka, J. Yamada, H. Yamochi, G. Saito, 2010Physica B 405: S262264

121. Emge T J, Leung P C W, Beno M A, Schultz A J, Wang H H, Sowa L M, Williams J M1984Phys. Rev. B3067806782

122. K. Bender, K. Dietz, H. Endres, H. V. Helberg, I. Hennig, H. J. Keller, H. V. Schafer, D. Schweitzer, 1984Mol. Cryst. Liq. Cryst. 1074553

123. Y. Takano, K. Hiraki, H. M. Yamamoto, T. Nakamura, T. Takahashi, 2001J. Phys. Chem. Solids 62393395

124. N. Tajima, S. Sugawara, M. Tamura, Y. Nishio, K. Kajita, 2006J. Phys. Soc. Jpn. 75: 051010/110

125. S. Katayama, A. Kobayashi, Y. Suzumura, 2006J. Phys. Soc. Jpn. 75: 054705/16

126. R. Kondo, S. Kagoshima, M. Maesato, 2003Phys. Rev. B67: 134519/16

127. H. Kobayashi, R. Kato, A. Kobayashi, Y. Nishio, K. Kajita, W. Sasaki, 1986

128. H. Mori, S. Tanaka, T. Mori, 1998Phys. Rev. B571202312029

129. B. Salameh, A. Nothardt, E. Balthes, W. Schmidt, D. Schweitzer, J. Strempfer, B. Hinrichsen, M. Jansen, D. K. Maude, 2007Phys. Rev. B75: 054509/113

130. H. Tanaka, E. Ojima, H. Fujiwara, Y. Nakazawa, H. Kobayashi, A. Kobayashi, 2000J. Mater. Chem.10245247

131. H. Kobayashi, H. Tomita, T. Naito, A. Kobayashi, F. Sakai, T. Watanabe, P. Cassoux, (199, 1996J. Am. Chem. Soc.118368377

132. S. Uji, H. Shinagawa, T. Terashima, T. Yakabe, Y. Terai, M. Tokumoto, A. Kobayashi, H. Tanaka, H. Kobayashi, 2001Nature 410908910

133. H. Matsui, H. Tsuchiya, T. Suzuki, E. Negishi, N. Toyota, 2003Phys. Rev. B68: 155105/110

134. N. Toyota, Y. Abe, H. Matsui, E. Negishi, Y. Ishizaki, H. Tsuchiya, H. Uozaki, S. Endo, 2002Phys. Rev. B66: 033201/14

135. H. Kobayashi, A. Sato, E. Arai, H. Akutsu, A. Kobayashi, P. Cassoux, 1997J. Am. Chem. Soc. 1191239212393

136. E. Fujiwara, H. Fujiwara, H. Kobayashi, T. Otsuka, A. Kobayashi, 2002Adv. Mater. 1413761379

137. H. W. Meul, C. Rossel, M. Decroux, Ø. Fischer, G. Remenyi, A. Briggs, 1984Phys. Rev. Lett. 53497500

138. C. L. Lin, J. Teter, J. E. Crow, T. Mihalisin, J. Brooks, A. I. Abou-Aly, G. R. Stewart, 1985Phys. Rev. Lett. 5425412544

139. K. Clark, A. Hassanien, S. Khan, K. Braun-F, H. Tanaka, S. Hla-W, 2010Nature Nanotech. 5261265

140. J. Sakata, H. Sato, A. Miyazaki, T. Enoki, Y. Okano, R. Kato, 1998Solid State Commun. 108377381

141. J. Yamada, M. Watanabe, H. Akutsu, S. Nakatsuji, H. Nishikawa, I. Ikemoto, K. Kikuchi, 2001J. Am. Chem. Soc.12341744180

142. H. Ito, T. Ishihara, H. Tanaka, S. Kuroda, T. Suzuki, S. Onari, Y. Tanaka, J. Yamada, K. Kikuchi, 2008Phys. Rev. B78: 172506/14

143. Y. Shimojo, T. Ishiguro, T. Toita, J. Yamada, 2002J. Phys. Soc. Jpn. 71717720

144. K. Nomura, R. Muraoka, N. Matsunaga, K. Ichimura, J. Yamada, 2009Physica B 404562564

145. Y. Okano, M. Iso, Y. Kashimura, J. Yamaura, R. Kato, 1999Synth. Met. 10217031704

146. J. S. Zambounis, C. W. Mayer, K. Hauenstein, B. Hilti, W. Hofherr, J. Pfeiffer, M. Buerkle, G. Rihs, 1992Adv. Mater. 43335

147. S. Kimura, T. Maejima, H. Suzuki, R. Chiba, H. Mori, T. Kawamoto, T. Mori, H. Moriyama, Y. Nishio, K. Kajita, (200, 2004Chem. Commun. 24542455

148. S. Kimura, H. Suzuki, T. Maejima, H. Mori, J. Yamaura, T. Kakiuchi, H. Sawa, H. Moriyama, 2006J. Am. Chem. Soc. 12814561457

149. K. Kikuchi, K. Murata, Y. Honda, T. Namiki, K. Saito, T. Ishiguro, K. Kobayashi, I. Ikemoto, 1987J. Phys. Soc. Jpn. 5634363439

150. H. Nishikawa, T. Morimoto, T. Kodama, I. Ikemoto, K. Kikuchi, J. Yamada, H. Yoshino, K. Murata, 2002J. Am. Chem. Soc. 124730731

151. R. Kato, K. Yamamoto, Y. Okano, H. Tajima, H. Sawa, 1997Chem. Commun. 947948

152. R. Kato, S. Aonuma, Y. Okano, H. Sawa, M. Tamura, M. Kinoshita, K. Oshima, A. Kobayashi, K. Bun, H. Kobayashi, 1993Synth. Met. 61199206

153. T. Imakubo, N. Tajima, M. Tamura, R. Kato, Y. Nishio, K. Kajita, 2002J. Mater. Chem. 12159161

154. Lyubovskaya R N, Zhilyaeva E I, Torunova S A, Mousdis G A, Papavassiliou G C, Perenboom J A A J,Pesotskii S I, Lyubovskii R B2004J. Phys. IV114463466

155. G. C. Papavassiliou, G. A. Mousdis, J. S. Zambounis, A. Terzis, A. Hountas, B. Hilti, C. W. Mayer, J. Pfeiffer, Met. Synth, 1988B27379383

156. T. Takahashi, Y. Kobayashi, T. Nakamura, K. Kanoda, B. Hilti, J. S. Zambounis, (199, 1994Physica C235-240: 2461-24562.

157. K. Takimiya, A. Takamori, Y. Aso, T. Otsubo, T. Kawamoto, T. Mori, 2003Chem. Mater. 1512251227

158. T. Kawamoto, T. Mori, K. Enomoto, T. Koike, T. Terashima, S. Uji, A. Takamori, K. Takimiya, T. Otsubo, (200, 2006Phys. Rev. B73: 024503/15

159. K. Takimiya, M. Kodani, N. Niihara, Y. Aso, T. Otsubo, Y. Bando, T. Kawamoto, T. Mori, 2004Chem. Mater. 1651205123

160. K. Takimiya, Y. Kataoka, Y. Aso, T. Otsubo, H. Fukuoka, S. Yamanaka, 2001Angew. Chem., Int. Ed. 4011221125

161. T. Kawamoto, T. Mori, T. Konoike, K. Enomoto, T. Terashima, S. Uji, H. Kitagawa, K. Takimiya, T. Otsubo, 2006Phys. Rev. B73: 094513/18

162. T. Kawamoto, T. Mori, C. Terakura, T. Terashima, S. Uji, K. Takimiya, Y. Aso, T. Otsubo, 2003Phys. Rev. B67: 020508/14

163. T. Kawamoto, T. Mori, C. Terakura, T. Terashima, S. Uji, H. Tajima, K. Takimiya, Y. Aso, T. Otsubo, 2003Eur. Phys. J. B 36161167

164. M. Kodani, A. Takamori, K. Takimiya, Y. Aso, T. Otsubo, 2002J. Solid State Chem. 168582589

165. Y. Misaki, N. Higuchi, H. Fujiwara, T. Yamabe, T. Mori, H. Mori, S. Tanaka, 1995Angew. Chem., Int. Ed. Engl. 3412221225

166. T. Shirahata, M. Kibune, T. Imakubo, 2006Chem. Commun. 15921594

167. T. Shirahata, M. Kibune, H. Yoshino, T. Imakubo, 2007Chem. Eur. J. 1376197630

168. Hebard A F, Rosseinsky M J, Haddon R C, Murphy D W, Glarum S H, Palstra T T M, Ramirez A P,Kortan A R1991Nature350600601

169. W. Krätschmer, L. D. Lamb, K. Fostiropoulos, D. R. Huffman, 1990Nature 347354358

170. P. W. Stephens, L. Mihaly, P. L. Lee, R. L. Whetten, S. Huang-M, R. Kaner, F. Diederich, K. Holczer, 1991Nature 351632634

171. Rosseinsky M J, Ramirez A P, Glarum S H, Murphy D W, Haddon R C, Hebard A F, Palstra T T M,Kortan A R, Zahurak S M, Makhija A V1991Phys. Rev. Lett. 6628302832

172. K. Tanigaki, T. W. Ebbesen, S. Saito, J. Mizuki, J. Tsai-S, Y. Kubo, S. Kuroshima, 1991Nature 352222223

173. Fleming R M, Ramirez A P, Rosseinsky M J, Murphy D W, Haddon R C, Zahurak S M, Makhija A V1991Nature352787788

174. S. Sasaki, A. Matsuda, C. W. Chu, 1994J. Phys. Soc. Jpn. 6316701673

175. R. F. Kiefl, Farlane. W. A. Mac, K. H. Chow, S. Dunsiger, T. L. Duty, T. M. S. Johnston, J. W. Schneider, J. Sonier, L. Brard, R. M. Strongin, J. E. Fischer, I. I. I. A. B. Smith, 1993Phys. Rev. Lett. 7039873990

176. C. Chen-C, C. M. Lieber, 1993Science 259655658

177. T. Takenobu, T. Muro, Y. Iwasa, T. Mitani, (200, 2000Phys. Rev. Lett. 85381384

178. P. Dahlke, M. S. Denning, P. F. Henry, M. J. Rosseinsky, 2000J. Am. Chem. Soc. 1221235212361

179. T. Yildirim, L. Barbedette, J. E. Fischer, C. L. Lin, J. Robert, P. Petit, T. T. M. Palstra, 1996Phys. Rev. Lett. 77167170

180. M. Kosaka, K. Tanigaki, K. Prassides, S. Margadonna, A. Lappas, C. M. Brown, A. N. Fitch, 1999Phys. Rev. B59: R6628R6630.

181. A. Y. Ganin, Y. Takabayashi, Y. Z. Khimyak, S. Margadonna, A. Tamai, M. J. Rosseinsky, K. Prassides, 2008Nat. Mat. 7367371

182. Y. Takabayashi, A. Y. Ganin, P. Jeglič, D. Arčon, T. Takano, Y. Iwasa, Y. Ohishi, M. Takata, N. Takeshita, K. Prassides, M. J. Rosseinsky, 2009Science 32315851590

183. A. Y. Ganin, Y. Takabayashi, P. Jeglič, D. Arčon, A. Potočnik, P. J. Baker, Y. Ohishi, M. T. Mc Donald, M. D. Tzirakis, A. Mc Lennan, G. R. Darling, M. Takata, M. J. Rosseinsky, K. Prassides, 2010Nature 466221225

184. R. Mitsuhashi, Y. Suzuki, Y. Yamanari, H. Mitamura, T. Kambe, N. Ikeda, H. Okamoto, A. Fujiwara, M. Yamaji, N. Kawasaki, Y. Maniwa, Y. Kubozono, (201, 2010Nature4647679

185. Y. Kubozono, H. Mitamura, X. Lee, X. He, Y. Yamanari, Y. Takahashi, Y. Suzuki, Y. Kaji, R. Eguchi, K. Akaike, T. Kambe, H. Okamoto, A. Fujiwara, T. Kato, T. Kosugi, H. Aoki, 2011Phys. Chem. Chem. Phys. 131647616493

186. X. F. Wang, R. H. Liu, Z. Gui, Y. L. Xie, Y. J. Yan, J. J. Ying, X. G. Luo, X. H. Chen, 2011Nat. Commun. 2507513

187. X. F. Wang, Y. J. Yan, Z. Gui, R. H. Liu, J. J. Ying, X. G. Luo, X. H. Chen, 2011Phys. Rev. B84: 214523/14

188. M. Xue, T. Cao, D. Wang, Y. Wu, H. Yang, X. Dong, J. He, F. Li, G. F. Chen, 2012Sci. Rep. 2: 389/14

189. D. Graf, J. S. Brooks, M. Almeida, J. C. Dias, S. Uji, T. Terashima, M. Kimata, 2009Europhys. Lett. 85: 27009/15

190. M. Kociak, A. Kasumov, Guéron. S. Yu, B. Reulet, I. I. Khodos, Yu. B. Gorbatov, V. T. Volkov, L. Vaccarini, H. Bouchiat, 2001Phys. Rev. Lett. 8624162419

191. Z. K. Tang, L. Zhang, N. Wang, X. X. Zhang, G. H. Wen, G. D. Li, J. N. Wang, C. T. Chan, P. Sheng, 2001Science 29224622465

192. R. Lortz, Q. Zhang, W. Shi, J. T. Ye, C. Qiu, Z. Wang, H. He, P. Sheng, T. Qian, Z. Tang, N. Wang, X. Zhang, J. Wang, C. T. Chan, 2009Proc. Natl. Acad. Sci. USA 10672997303

193. I. Takesue, J. Haruyama, N. Kobayashi, S. Chiashi, S. Maruyama, T. Sugai, H. Shinohara, (200, 2006Phys. Rev. Lett. 96: 057001/14

194. N. B. Hannay, T. H. Geballe, B. T. Matthias, K. Andres, P. Schmidt, Nair. D. Mac, (196, 1965Phys. Rev. Lett. 14225226

195. Belash I T, Bronnikov A D, Zharikov O V, Pal'nichenko A V1989Solid State Commun. 69921923

196. T. E. Weller, M. Ellerby, S. S. Saxena, R. P. Smith, N. T. Skipper, 2005Nat. Phys. 13941

197. A. Gauzzi, S. Takashima, N. Takeshita, C. Terakura, H. Takagi, N. Emery, C. Hérold, P. Lagrange, G. Loupias, 2007Phys. Rev. Lett. 98: 067002/14

198. J. S. Kim, L. Boeri, J. R. O'Brien, F. S. Razavi, R. K. Kremer, 2007Phys. Rev. Lett. 99: 027001/14

199. G. Csányi, P. B. Littlewood, A. H. Nevidomskyy, C. J. Pikard, B. D. Simons, 2005Nat. Phys. 14245

200. D. G. Hinks, D. Rosenmann, H. Claus, M. S. Bailey, J. D. Jorgensen, 2007Phys. Rev. B75: 014509/16

201. E. A. Ekimov, V. A. Sidorov, E. D. Bauer, N. N. Mel'nik, N. J. Curro, J. D. Thompson, S. M. Stishov, (200, 2004Nature 428542545

202. Y. Takano, T. Takenouchi, S. Ishii, S. Ueda, T. Okutsu, I. Sakaguchi, H. Umezawa, H. Kawarada, M. Tachiki, 2007Diamond Relat. Mater. 16911914

203. E. Bustarret, J. Kačmarčik, C. Marcenat, E. Gheeraert, C. Cytermann, J. Marcus, T. Klein, (200, 2004Phys. Rev. Lett. 93: 237005/14

204. B. Sacépé, C. Chapelier, C. Marcenat, J. Kačmarčik, T. Klein, M. Bernard, E. Bustarret, 2006Phys. Rev. Lett. 96: 097006/14

205. E. A. Ekimov, V. A. Sidorov, A. V. Zoteev, J. B. Lebed, J. D. Thompson, S. M. Stishov, (200, 2008Sci. Technol. Adv. Mater. 9:044210/16

206. Aust R B, Bentley W H, Drickamer H G1964J. Chem. Phys.4118561864

207. T. Yokota, N. Takeshita, K. Shimizu, K. Amaya, A. Onodera, I. Shirotani, S. Endo, (199, 1996Czech. J. Phys. 46817818

208. K. Amaya, K. Shimizu, N. Takeshita, M. I. Eremets, T. C. Kobayashi, S. Endo, 1998J. Phys.: Condens. Matter 101117911190

209. E. Iwasaki, K. Shimizu, K. Amaya, A. Nakayama, K. Aoki, R. P. Carlon, 2011Synth. Met. 12010031004

9

# STRUCTURAL ASPECTS OF THE BECHGAARD AND FABRE SALTS: AN UPDATE

**Jean-Paul Pouget**

Laboratoire de Physique des Solides, Université Paris-sud, CNRS, UMR 8502, Bâtiment 510, Orsay Cedex F-91405, France

## ABSTRACT

We review structural aspects of the Bechgaard and Fabre salts in relationship with their electronic, magnetic and superconducting properties. We emphasize the role of bond and charge modulations of the quarter filled organic stack in the various instabilities and ground states exhibited by these salts. A special consideration is also devoted to the influence of anions and methyl groups in these processes. In particular we point out the importance of the anions in achieving the inter-stack coupling by either direct or indirect (via the polarization of the methyl group cavities) interactions with the donors. In this framework we discuss the role of anions and methyl group disorders in the inhibition of the divergence of the high temperature bond order wave instability of the Bechgaard salts. We analyze the modulation in the magnetic ground states by considering explicitly the coupling of the magnetization with structural degrees of freedom. We consider the role of the anions and methyl groups in stabilizing the charge ordering pattern in the Fabre salts. We also discuss the spin-Peierls transition of the Fabre salts in relation with the charge ordering transition and the adiabaticity of the phonon field. We review the anion ordering transitions by considering more particularly the influence of the ordering process on the electronic structure and on the ground states which results. In this framework we show that the texture of the anion ordered structure has direct consequences on the superconducting properties of $(TMTSF)_2ClO_4$. Finally we conclude on the essential implication of the structural degrees of freedom on the generic phase diagram of the Bechgaard and Fabre salts.

## KEYWORDS

one-dimensional metals; strongly correlated fermions; organic conductors; organic superconductors; charge order; spin-Peierls transitions; anion ordering transitions; charge density wave; spin density wave; X-ray diffuse scattering

## 1. INTRODUCTION

For several decades the synthesis of new classes of materials has triggered original fields of research where the discovery of new electronic states of matter has increasingly enriched the number of fundamental concepts in solid state physics while opening new routes for potential applications. This has been the case, to name a few examples, of oxides and carbon based compounds including conducting polymers. In this respect, the study of organic conductors since the beginning of the 1970s has largely contributed to this revival [1,2,3].

Organic conductors presenting a regular array of quasi-planar organic molecules, developing a preferential overlap of their $p_\pi$ HOMO in one or two crystallographic directions, provide very anisotropic materials exhibiting a relatively simple electronic structure. The originality of these materials relies on the following facts:

1) Their $p_\pi$ electron bandwidth is comparable to the electron–electron Coulomb repulsions which, being weakly screened, decrease slowly with the intermolecular separation.

2) Their lattice is soft and quite compressible, which leads to rapid change of the electronic properties under pressure or uniaxial stress, while providing a sizeable coupling between the electronic and structural degrees of freedom.

3) Their crystal array is quite perfect, which allows a fine control of the perturbation of physical properties by defects intentionally introduced in the structure.

If one restricts to quasi one-dimensional (1D) systems, the organic conductors provide classes of materials which realize a very large panel of different physical situations. The charge transfer salts of the TTF-TCNQ family are composed of segregated stacks of donors (D: TTF) and acceptors (A: TCNQ) with generally an incommensurate charge transfer between the two metallic 1D subsystems. These salts are subject to a charge density wave (CDW) instability which triggers a Peierls insulating ground state [1]. The 2:1 $A_2Y$ or $D_2X$ salts, where Y or X is a monovalent cationic or anionic entity, form quasi-1D quarter filled electronic systems built exclusively on acceptors, A, or donors, D. These "less screened" metallic systems generally experience quite strong Coulomb repulsions driving high temperature charge localization. This localization, accompanied by a spin charge decoupling, leaves the spin degrees of freedom active for a low temperature anti-ferromagnetic (AF) or spin-Peierls (SP) instability. Among these 2:1 series considerable studies have been devoted to the so-called Fabre and Bechgaard salts built with TMTTF and TMTSF donors respectively. The Fabre salts show charge localization, spin-charge decoupling and prototypal examples of AF and SP ground states while the Bechgaard salts exhibit a Fermi surface (FS) nesting driven spin density wave (SDW) ground state. The suppression of the SDW ground state under pressure restores the metallic state which is thus subject to superconductivity [1,2]. To complete this survey of different classes of organic 1D system, let us briefly mention very original 2:1 series of quasi-1D organic systems based on acceptors incorporating a metal transition element M. The corresponding salts, in presence of perylene or of TTF based donor, achieve segregated metallic and magnetic stacks [$Per_2$-$M(mnt)_2$ series [4]] or spin ladders [$(DT\text{-}TTF)_2$-$M(mnt)_2$ series [5]]. Finally, let us mention conducting and even superconducting 2:1 $(M(dmit)_2)_2Y$ or charge transfer $TTF(M(dmit)_2)_2$ systems [2], where the electronic structure of the $M(dmit)_2$ acceptor, composed of dmit side groups weakly coupled through a central M atom, involves both HOMO and LUMO electronic states [6]. These highly original conducting salts, which are not considered here, deserve a special review in their own right.

This present review focuses on structural aspects of Bechgaard and Fabre salts and their relationship with electronic, magnetic and superconducting properties. It will cover the most recent findings concerning the structural instabilities exhibited by these salts, and will present quantitative analyses of earlier data. In this respect this paper updates our 15 year old reviews [7,8]. The review is organized as follows: The structure of the Bechgaard and Fabre salts is presented in Section 2 with special consideration of entities achieving a subtle coupling to the electronic degrees of freedom. The underlying bond order wave (BOW) instability, always present in these salts, is quantitatively analyzed in Section 3. The AF and SDW magnetic ground states are described inSection 4, which also includes the often ignored coupling between magnetic and structural order parameters.Section 5 considers charge ordering (CO) effects in the Fabre salts with a special investigation of the role of the anions and methyl groups in the stabilization of the CO ground state. Section 6 is devoted to the SP transition and its coupling with the CO. Section 7 will review the most recent findings concerning the influence of the anion ordering (AO) transitions on the electronic structures. Finally we conclude this review by summarizing the influence of the structural degrees of freedom on the generic phase diagram of the Fabre and Bechgaard salts.

# 2. BASIC ASPECTS OF THE CRYSTALLOGRAPHIC STRUCTURE IN RELATIONSHIP WITH ELECTRONIC PROPERTIES

## 2.1. Structural Arrangement

As shown in Figure 1, Bechgaard and Fabre salts are made of a zig-zag stacking of TMTSF and TMTTF molecules arranged nearly perpendicularly to the 1D direction, a. In the short direction of the molecule, stacks, shifted from each other along b, form (a, b) donor layers. In the long direction of the molecule, out of phase first neighboring zig-zag stacks delimit methyl group cavities which are filled by the anions X. This packing leads to an alternation of layers of well coupled donors with anions along c. Two aspects of the structure at ambient conditions are noticeable: (i) the slight dimerization of the zig-zag stack; and (ii) the presence of a sizeable thermal motion of the anion X in its cavity [9]. These salts crystallize in the triclinic space group P-1 with the inversion centers located in between the molecules and in the center of the methyl group cavities. If one ignores the slight stack dimerization and the location of the anions close to the shortest zig-zag bond, the structure has an underlying monoclinic C2/m symmetry (a:

binary axis, b' = 2b-a, c) tending to promote a $2_1$ screw axis symmetry in the stack direction [8]. In this respect it is interesting to remark that both Bechgaard and Fabre salts, with X = $PF_6$, undergo, at 5.5GPa and 8.5GPa respectively, a first order phase transition to a monoclinic (a: binary axis, b', c) structure [10]. In the (a, b) donor layer, this discontinuous structural change could be achieved by the relative shift, along a, of first neighboring stacks.

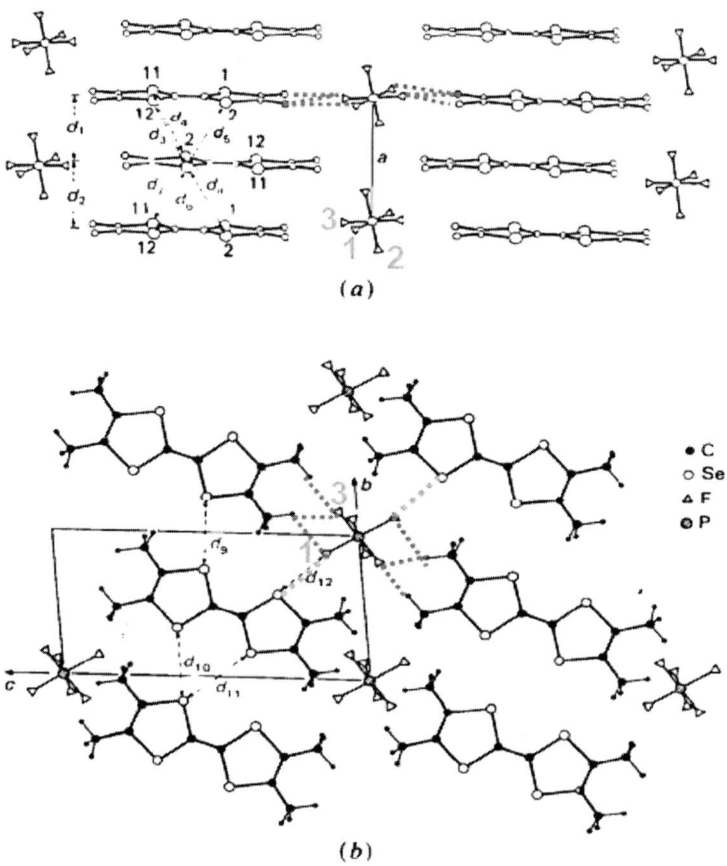


**Figure 1.** Structure of $(TMTSF)_2PF_6$: (**a**) Projection in the (a, [011]) plane; (**b**) Projection along a. The F atoms are labeled. In (b) the methyl groups located in the low temperature neutron scattering structural refinement [11] are shown. The shortest F(1)-Se distances, F(1)-H and F(3)-H bonds are indicated by the blue and red dotted lines respectively.

## 2.2. Dimerization of the Stack

In a 1D description of the quarter filled interacting electron gas, the dimerization of the zig-zag stack activates an umklapp $4k_F = a^*$ electron–electron scattering term which, in presence of sizeable Coulomb repulsions, facilitates the charge localization ($k_F$ is the Fermi wave vector of the quarter filled 1D electron gas) [12]. The strength of this umklapp process depends of the amplitude of the dimerization gap $2\Delta_D$ which opens at $\pm a^*/2$ in the band structure well below the Fermi level. $\Delta_D$ amounts to the difference of transfer integrals (in presence of the Hartree anion potential) on the two inequivalent bonds of the zig-zag. $\Delta_D$ calculated by the DFT differs from $\Delta_D$ calculated by the Extended Hűckel Theory (EHT) which does not take into account the anion potential. For example in $(TMTSF)_2\,ClO_4$ at room temperature (RT) DFT gives $\Delta_D = 26$ meV [13], while EHT gives $\Delta_D = 37$–$54$ meV (see Table 4.3 in [2]). Another complication arises if one considers the interstack coupling: the inter-chain transfer integrals lead to a significant variation of $\Delta_D$ along b*. In $(TMTSF)_2\,ClO_4\Delta_D$ varies from ~50meV in (a*/2, 0) to ~0meV in (a*/2, b*/2) [13] at RT. $\Delta_D$ decreases upon cooling and under pressure, while $\Delta_D$ increases from the TMTSF'salts to the TMTTF'salts. $\Delta_D$ is controlled by the shape and volume of the anion which deforms the donor packing via its interaction with the terminal methyl groups.

## 2.3. Anion Cavities

Structural effects associated to anions require a special attention because the anions are located in soft cavities delimited by the methyl groups of the TMTSF or the TMTTF molecules. The anions fit these cavities in a more or less compact manner depending upon the volume and shape of the anion [14,15,16]. The anions were considered in earlier studies as spacers acting on the inter-layer distance as does the pressure [17]. In fact recent works summarized in this review show that their role is more subtle than it was previously believed. First, the anions are located in centro-symmetric cavities. This location causes no problem for centro-symmetric anions such as octahedron $MF_6$ (M = P, As, Sb) or sphere (Br), but it implies at RT the existence of an orientation disorder for tetrahedral ($ReO_4$, $ClO_4$, $BF_4$), triangular ($NO_3$) or asymmetric linear (SCN) anions. Second, each anion experiences quite a symmetric environment from its six first neighbor donors [18] due to the six closest methyl groups. Among them, four groups are located in the (b', c*) plane perpendicular to the stacking axis, while the two remaining groups, located in the (a, c*) plane, are significantly tilted with respect to a. These six methyl groups delimit three sets of three-fold symmetry axis close to the symmetry axis of the octahedral anions. The four planar methyl groups delimit two sets of two-fold symmetry axis close to the symmetry axis of the tetrahedral anions [18]. Third, the F or O outer atoms of the anion form weak H-bonds with the closest methyl groups, especially at low temperatures when thermal disorder is reduced [19]. The 4K neutron scattering structural refinement [11], which allows precisely locating the H of the methyl groups of $(TMTSF)_2PF_6$, shows the formation of 6 H-bonds with F(1) and F(3) of the $PF_6$ (red dotted lines in Figure 1). All these H-bonds are located in the plane perpendicular to the stacking axis a. In addition, the anion develops short contact distances with the Se of the closest donors. Figure 1 shows (blue dotted lines) the two short Se...F(1) contacts established in $(TMTSF)_2PF_6$. The anion and thermal dependences of the geometry of the S...F interactions have been analyzed in $(TMTTF)_2PF_6$, $AsF_6$ and $SbF_6$ [20]. Note that F(2), which presents the strongest disorder [11], is not involved in any interactions. Any change of the shape of the methyl group cavity, caused by an elastic deformation of the donor stack for example, should react on the location of the anions and *vice versa*.

These two kinds of short contacts provide respectively indirect (via the polarization of the H-bond network and of the intra-molecular $\sigma$ bonds) [21] and direct (via the Se or S atoms) attractive interactions between the anion and the $\pi$ holes cloud located in the inner part of the donor. In this respect infrared measurements provided evidence of a considerable electron-molecular-vibration coupling of the methyl groups of both TMTSF and TMTTF with the charges on the molecules [22]. These interactions are involved in the inter-chain coupling mechanism (see Section 3.1.2) and for this reason play a key role in the stabilization of various ground states exhibited by Bechgaard and Fabre salts such as the AO transition involving non centro-symmetric anions (seeSection 7), the CO transition breaking the inversion symmetry (see Section 5) and the SP transition subject to an important magneto-elastic coupling (see Section 6).

## 2.4. Anion and Methyl Group Disorders

The location of anions in methyl group cavities provide flexibility in the structure due to the soft interface provided by the methyl groups between the anions and the core of the donors and the incomplete fit of the anion in the volume delimited by these methyl groups [14,15,16]. This soft environment tolerates the presence of disordered anions in the structure even for centro-symmetrical anions [9]. At ambient conditions, the non centro-symmetrical tetrahedral anions are disordered in Bechgaard and Fabre salts while they are ordered in others structurally related 2:1 salts, such as $(BEDT\text{-}TTF)_2ReO_4$, $(t\text{-}TTF)_2X$ with X = $ClO_4$ and $BF_4$ and $(DMtTTF)_2ClO_4$, built with donors which do not provide a methyl group environment at the anion.

The lattice softness is revealed by thermal expansion measurements. First, linear expansion coefficients, $\alpha_i$, are one order of magnitude larger in the $PF_6$ salt than in the Br salt where the anion better fits the volume of its methyl group cavity [23]. Second, in salts incorporating $PF_6$ or $AsF_6$ anions, the rate of variation of $\alpha_i$ upon heating becomes negative at high temperature [24]. This finding points out peculiar lattice dynamics, possibly due to "free" rotation and/or translation of the anions in their cavities. This effect, which is observed both for TMTTF and TMTSF salts, is mostly pronounced along the c* direction where layers of donors and anions alternate [23,24,25].

The enhanced values of Debye Waller factors obtained in structural refinements show the presence of an atomic disorder for both the methyl groups and the outer atoms of the anions. An orientation disorder is expected for non centro-symmetrical anions located in a centro-symmetric cavity. This disorder is removed at a well defined AO transition which will be the object of Section 7. Below we shall consider the case of centro-symmetrical octahedral anions such as $PF_6$. At RT, F atoms exhibit a sizeable thermal motion but the finding of well defined maxima in electron-density map, obtained in the structural refinement of $(TMTSF)_2PF_6$ [9], show that there is no free rotation of the centro-symmetrical anion. At 20K, the anion disorder is almost completely suppressed especially for F(1) and F(3) atoms forming H bonds and a linkage with the Se (see Figure 1) [11]. However neutron refinements show that methyl groups sustain a certain disorder even at 4K [11], probably due to quantum-mechanical tunneling. NMR studies show that $PF_6$ disorder is established upon heating above ~70K in $(TMTSF)_2PF_6$ [26]. More recent NMR studies performed in $(TMTTF)_2SbF_6$ [27,28] show, by the measurement of different activation energies in the molecular motion, that in fact anion disorder is progressively set upon heating. A detailed analysis [29] of the Debye Waller factor of $(TMTSF)_2AsF_6$ at 125K suggests that the disorder could take place between two well defined orientations of the $AsF_6$ anion in its cavity.

Several activation energies also govern the motion of the methyl groups in TMTSF salts [30,31,32,33].

Thermal variation of the principal directions of the dilatation tensor of $(TMTSF)_2PF_6$ [11] suggests a lattice modification around 60–50K which could be related to the "freezing out" of the $PF_6$ disorder previously considered. More precisely, the thermal variation of the lattice expansion in the direction i divided by *T*, $\alpha_i/T$, or more likely the volume expansion divided by *T*, $(\Sigma_i\alpha_i)/T$, is proportional to the entropy derivative $\partial S/\partial T$ [34]. $\alpha_\perp/T$ taken from the data of [24] exhibits broad maxima around 30–40 K in $(TMTSF)_2PF_6$ before dropping to zero at lower temperature [34]. This finding implies that below these maxima a net decrease of the "lattice" entropy, probably due to the vanishing of the structural disorder, should occur. The best way to reduce simultaneously the $PF_6$ and methyl group disorders is to link these two entities, as suggested from methyl protons NMR [32]. These findings have important consequences on the density wave instabilities which will be considered in Section 3.

The $PF_6$ anion disorder could be even more significantly reduced by a well defined thermodynamic transformation recalling the AO transition (see Section 7). This could be achieved by an orientation ordering transition achieving a well defined orientation at the anion which oscillates between several positions in its methyl group cavity at high temperature [29]. Such a transition has been observed in $(EDO\text{-}TTF)_2PF_6$ [35] where the orientation ordering of $PF_6$, accompanied by its shift, stabilizes a 3D pattern of CO and BOW distortions of the EDO-TTF stack. All these effects lead to a unit cell doubling which causes the $2k_F$ MI transition of $(EDO\text{-}TTF)_2PF_6$. In P-1 triclinic $(TMTSF)_2PF_6$, where there is no cell doubling (above $T_{SDW}$), the only possible symmetry breaking operation while keeping the translation symmetry will be the loss of inversion centers as for the CO transition of the $(TMTTF)_2X$'s (see Section 5). However neutron scattering structural refinements of $(TMTSF)_2PF_6$performed at 4K and 20K [11] do not provide any evidence of the loss of the inversion centers (Note that this also implies that $PF_6$ does not shift from the centre of its methyl group cavity).

# 3. THE BOND ORDER WAVE INSTABILITY IN THE BECHGAARD AND FABRE SALTS

## 3.1. The Interplay between $2k_F$ BOW and SDW Instabilities in $(TMTSF)_2PF_6$

### 3.1.1. Quantitative Analysis of the 1D $2k_F$ BOW Instability

All charge transfer salts of the TTF-TCNQ family exhibit a divergent $2k_F$ CDW instability which, when coupled to the lattice degrees of freedom, drives a low temperature $T_P$ (~30–50 K) metal-insulator (MI) transition [1,36]. In TTF-TCNQ the electron-phonon coupling of the 1D electron gas with the lattice modes leads to the formation above $T_P$ of a Kohn anomaly in a transverse acoustic (TA) branch [37] which softening at $T_P$ leads to a static modulation of the intra-stack bond distances. This stabilizes a $2k_F$ BOW ground state below $T_P$. Such instability of a 1D metal with respect to a $2k_F$ periodic lattice distortion (PLD) was first proposed by Peierls in 1955.

Above $T_P$ the 1D BOW structural instability gives also rise to critical fluctuations consisting, in reciprocal space, in a planar X-ray diffuse scattering (or diffuse lines on X-ray patterns) at the wave vector Q = G + q, where G is a reciprocal lattice wave vector and q ~ $2k_F$. The intensity, I(Q), of this diffuse scattering can be expressed in the form [36,38]:

$$I(Q) = |F(Q)|^2 < |u_q|^2 > \tag{1}$$

In expression (1) $u_q$ is the atomic displacement (of the TA critical phonon mode at q in TTF-TCNQ for example) and F(Q) is the structure factor of the modulation. In the classical limit when the thermal energy $k_BT$ is larger than the critical phonon energy, ħΩ, one has:

$$< |u_q|^2 > \approx k_B T \chi(q) \tag{2}$$

In expression (2) χ(q) is the CDW or BOW response function of the electron-phonon coupled system. For a standard Peierls chain χ(q) is peaked at the $2k_F$ wave vector. In the regime of 1D fluctuations χ(q) has a Lorentzian dependence in q:

$$\chi(q) = \chi(2k_F)/[1 + \xi^2(q - 2k_F)^2] \tag{3}$$

In expression (3) the correlation length of the intra-chain structural fluctuations, $\xi$, is given by the inverse half width at half maximum (HWHM) of $\chi(q)$. In the case of a Peierls instability (case of TMTSF-DMTCNQ considered below in Figure 2) both $\chi(2k_F)$ and $\xi$ diverge at $T_P$.

The surprise of the investigation of the Bechgaard salts was the observation in $(TMTSF)_2AsF_6$ [39] and $PF_6$ [40] of X-ray diffuse lines, corresponding to the 1D $2k_F$ BOW instability previously described, whose intensity I(Q) progressively vanishes upon cooling below ~50K. Below ~30K these structural fluctuations are no longer observable. The MI transition, which occurs at $T_{MI}$ = 12K in these salts, is thus not due to the setting of a Peierls ground state. The MI is caused by the establishment of a $2k_F$ SDW order [1,2] by a mechanism similar to the one proposed by Slater in 1951 [41]. In this mechanism the divergence of $2k_F$ SDW (AF in the model of Slater) correlations give rise to a $2k_F$ periodic exchange potential which opens a gap at the Fermi level in the 1D band structure and thus drives the electronic system to an insulating ground state (here the $2k_F$ periodic exchange potential plays the role of the $2k_F$ PLD of the Peierls transition).

The reason for the low temperature vanishing of the $2k_F$ BOW instability in the Bechgaard salts has never been elucidated. To this end, we compare below the $2k_F$ BOW instability of $(TMTSF)_2PF_6$ with the $2k_F$ BOW instability of TMTSF-DMTCNQ which diverges at the $T_P$ = 42 K Peierls transition [36]. This comparison is all the more justified because, both in $(TMTSF)_2PF_6$ and in TMTSF-DMTCNQ, the $2k_F$ BOW instability develops on the same TMTSF stack which presents similar electronic characteristics in the two salts (same bandwidth, same band filling and comparable dimensionless Luttinger parameter $K\rho$~0.25 [8]).

Figure 2 quantitatively compares the thermal dependence of the 1D $2k_F$ BOW fluctuations of $(TMTSF)_2PF_6$ [40] and of TMTSF-DMTCNQ [36,42]. Part (a) shows the $2k_F$ susceptibility, $\chi(2k_F)$, given by $I(Q)/T$ in the expressions (1) + (2), while part (b) shows the inverse intra-chain correlation length $\xi_a^{-1}$, given by expression (3). Above ~100K $\chi(2k_F)$ behaves similarly in the two salts. Below 100K $\chi(2k_F)$ of TMTSF-DMTCNQ grows rapidly and diverges at $T_P$ while $\chi(2k_F)$ of $(TMTSF)_2PF_6$ saturates before decreasing below ~50K. Above ~150K $\xi_a^{-1}$ of TMTSF-DMTCNQ and of $(TMTSF)_2PF_6$ are comparable. Below this temperature $\xi_a^{-1}$ of TMTSF-DMTCNQ drops rapidly to reach zero at $T_P$ while $\xi_a^{-1}$ of $(TMTSF)_2 PF_6$ decreases very slowly.

The thermal dependence of $\xi_a$ of TMTSF-DMTCNQ (named $\xi_{/T}$ below) has been quantitatively analyzed in [42] using exact calculations of the fluctuations of the Peierls chain (black continuous line of Figure 2b). Upon cooling, $\xi_{/T}$ is governed by the diverging growth of intra-chain electron-hole pair correlations associated with the increase of the thermal life time of the individual electron-hole pairs. In $(TMTSF)_2PF_6$ the saturation in the growth of $\xi_a$upon cooling means that the low temperature divergence of $\xi_{/T}$ expected for a standard Peierls chain is capped by a cut off length. We associate at this cut off length a "mean free path" $l = v^*\tau$, where $\tau$ corresponds to a non thermal lifetime of the electron-hole pair, and where v* is the "charge velocity" depending of the microscopic parameters of the TMTSF electron gas (v* amounts to $\sim v_F/2$ in TMTSF-DMTCNQ [42]). By adding these two independent contributions in quadrature:

$$\xi_a^{-2} = \xi_{/T}^{-2} + l^{-2} \tag{4}$$

$l^{-1}$ can be obtained if one assumes that $\xi_{/T}^{-1}$ is given by $\xi_a^{-1}$ measured in TMTSF-DMTCNQ. Figure 2b gives the thermal dependence of $l^{-1}$, deduced from expression (4), for $(TMTSF)_2PF_6$. $l^{-1}$ increases upon cooling and saturates below 100K, as does $\chi(2k_F)$. The life time $\tau$ smoothes the q dependence of the CDW response function near $2k_F$, with the result to kill the low temperature divergence of both $\xi_a$ and $\chi(2k_F)$. $l$ saturates at about 14 Å, which corresponds to one wave length 2a of the $2k_F$ BOW. From this length one deduces an inverse electron-hole life time $\hbar/\tau = \hbar v^* l^{-1}$ of 43meV. This value is one order of magnitude larger than the phonon frequency, $\hbar\Omega$ = 5 meV, of the transverse acoustic branch which bears the Kohn anomaly in TTF-TCNQ [37]. As $\Omega\tau \ll 1$ the $2k_F$ BOW dynamics of $(TMTSF)_2PF_6$ should occur in the anti-adiabatic limit and thus a Kohn anomaly should not form in the phonon spectrum. In this regime the dynamics is achieved by the critical growth of quasi-static $2k_F$ fluctuations which are visualized in reciprocal space by the presence of $2k_F$ X-ray diffuse lines above 30 K.

The life time $\tau$ of the electron-hole pairs could be caused by the orientation disorder of the anion. The change of orientation of the anion in its cavity should break both the short F...Se contact distance and the weak F...H-$CH_2$bonds (see Section 2.4). This will respectively modulate directly and indirectly (via the polarization of the H bonds and of the $\sigma$ intra-molecular electrons [21]) the Hartree potential experienced by the $\pi$ holes. These fluctuations of the Hartree potential should provide new scattering mechanisms of holes and electrons at the Fermi level, resulting in either an incoherent inter-stack charge transfer and/or intra-stack backward and forward scatterings. The inter-stack transfer and the backward scattering change the direction of the $k_F$ wave vector of the electron or of the hole of a pre-formed $2k_F$ electron-hole pair. This change breaks the pairing, and thus reduces the lifetime τ of the electron-hole pairs.

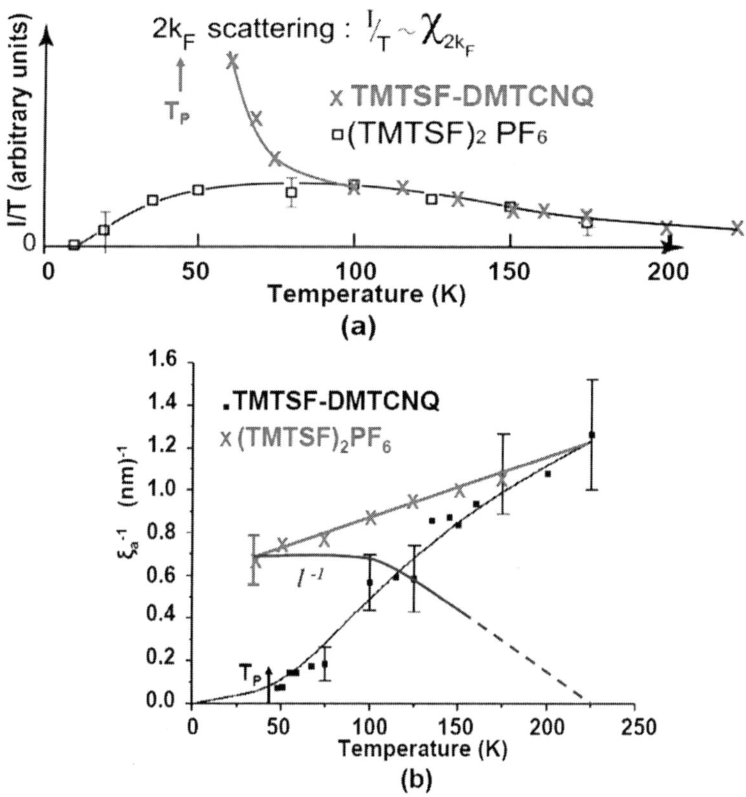


**Figure 2.** Thermal dependence of (**a**) the $2k_F$ BOW susceptibility, $\chi(2k_F)$; and (**b**) the inverse intra-chain correlation length $\xi_a^{-1}$ in $(TMTSF)_2PF_6$ and in TMTSF-DMTCNQ. In (b) the thermal dependence of the inverse mean free path $l^{-1}$ of the electron-hole pairs of $(TMTSF)_2PF_6$, given by (4), is given by the blue line.

Using the value at saturation of $l$, the FS should be broadened by $\hbar/\tau{\sim}43$ meV. This lifetime broadening amounts to the warping of the Fermi surface due to coherent interchain hopping ($2t'_b$ = 44meV from the transverse plasma edge measurement at 25 K [22]). Thus, the warping of the FS should be smeared out by this lifetime effect on all the temperature range of observation of the $2k_F$ BOW diffuse lines. In this framework it has been suggested from the absence of Drude absorption in intra-chain optical reflectance of $(TMTSF)_2ClO_4$ that the single particle transport is diffusive in all directions and temperatures and that the high dc conductivity in chain direction is due to collective sliding CDW [43]. Similar features are observed in $(TMTSF)_2PF_6$ [44], but the optical spectrum has been interpreted differently [45].

Note that this lifetime effect should similarly affect the $2k_F$ SDW fluctuations also driven by the divergence of the electron-hole response function of the correlated electron gas. In particular a reduced SDW correlation length, similarly to $\xi(T)$ of the $2k_F$ BOW fluctuations reported in Figure 2b, should control the spatial dependence of the $2k_F$ SDW fluctuations. Such SDW fluctuations have been probed by NMR [46] which however does not give any information on their spatial extent.

The $T_N$ = 12 K SDW transition of $(TMTSF)_2PF_6$ is achieved by a FS nesting process (see next section). However FS nesting should provide an efficient inter-chain coupling between density waves only if the FS is warped. Thus the setting of the inter-chain coupling requires a net sharpening of the FS upon cooling which can be achieved only if the extra scattering processes due to disorder vanish.

Evidence of a vanishing of the structural disorder comes from the thermal decrease of lattice entropy below ~30–40 K ascribed to the anchorage of the anions to the methyl groups (see Section 2.4). The cooperative locking of $PF_6$ due to the formation of H-bonds with the methyl groups could explain the stiffening of the lattice revealed by the hardening of sound velocity below ~ 45 K in $(TMTSF)_2PF_6$ [47]. Interestingly, a similar hardening of sound velocity is observed at the 24 K AO transition of $(TMTSF)_2ClO_4$ when the orientation ordering of the $ClO_4$ is accompanied by its linkage to the methyl groups (see section 7.3) [48]. This scenario is sustained by microwave transport measurements showing a dimension crossover in the charge transport at about the same temperature, $T_X$ ~ 35 K [49,50], at which the disorder vanishes.

Note that if the transverse coupling $t'_b$ is only considered, the warping of the FS and thus the 1D-2D spatial deconfinement of the electron gas should be effective at a larger $T_X \sim t'_b/\pi \sim 80$ K. However if the confinement due to electron-electron repulsions is considered the effective $t'_b$ is reduced and it is estimated [51] that $T_X$ can be depressed below 50K. On the basis of available experimental arguments the generic phase diagram of [52] was built with $T_X \sim 80$ K. However the expected development of inter-chain correlations below $T_X \sim 80$ K are neither detected in the $2k_F$ BOW X-ray diffuse scattering experiments nor in the thermal dependence of the $2k_F$ SDW fluctuations measured by NMR [46]. All these features show that the nature of the dimension crossover and of the deconfinement transition is unclear in the Bechgaard salts and that the value of the crossover temperature $T_X$remains debated in the literature [45,51,53]. Our finding of additional effects due to anion disorder adds new elements at the discussion.

$(TMTSF)_2PF_6$ exhibits below 200-150K critical 1D SDW fluctuations, probed by $T_1^{-1}$ NMR measurements [46]. These fluctuations occur in the same temperature range where the 1D $2k_F$ BOW fluctuations are also detected, which means that at high temperature $(TMTSF)_2PF_6$ exhibits both $2k_F$ SDW and $2k_F$ BOW instabilities. Below $T^* \sim 30$ K inter-chain $2k_F$ SDW critical fluctuations, due to the FS nesting process, develop [45] and the $2k_F$ BOW fluctuations are no longer detected. However the precise reason of the vanishing of the BOW fluctuations below$T^*$ requires clarifications which will be the object of the next section.

### 3.1.2. Interchain Coupling between the $2k_F$ Density Waves and the Vanishing of the Anion Disorder

It is now well established that the SDW transition of $(TMTSF)_2PF_6$ is due to a FS nesting because the interchain component $q_b \sim 0.20 \pm 0.05b^*$ [54] - $0.24 \pm 0.03b^*$ [55] of the SDW order, obtained by NMR, is close to the best nesting wave vector $\sim 0.3b^*$ of its FS determined from the 4 K structure [56]. In addition to this FS nesting process, inter-chain Coulomb interaction, $g_{1\perp}$, provides also an efficient coupling mechanism between CDW [1]. The inter-chain coupling $g_{1\perp}$ achieves the 3D ordering of $2k_F$ CDW/BOW in charge transfer Peierls systems like TTF-TCNQ and TMTSF-DMTCNQ [36,38].

However there is the possibility for the Bechgaard salts, and as previously suggested for the Fabre salts [57], of conflicting interchain coupling mechanisms if Coulomb interactions, $g_{1\perp}$, and FS nesting tend to promote different b* transverse periodicities: $g_{1\perp}$ (possibly mediated by the anion shift–see below) favoring $q_b = 1/2b^*$, as observed for the AO and SP modulations (see Section 7 and Section 6), while the nesting of the FS favoring $q_b \sim 1/4b^*$ as discussed above.

In all the superstructures achieving a transverse order between BOW, such as those formed at the (1/2, 1/2, 1/2) AO transition [58,59] (see Section 7.2), the shift of the anion from the inversion center seems necessary to stabilize the 3D pattern of $2k_F$ BOW in the Bechgaard and Fabre salts. In these superstructures:

- The $(2k_F)^{-1}$ periodicity is achieved in stack direction by repeating periodically two neighboring maxima of hole concentration with two neighboring minima of hole concentration.
- The anion shift allows to achieve a 3D coupling between neighboring $2k_F$ CDW by locking a maxima of the $2k_F\pi$ hole CDW on the molecule towards which the anion approaches and concomitantly a minimum of π hole density on the molecule that the anion leaves.
- By combining these two features, the 3D CDW pattern requires a synchronous staggered shift of near-neighbor anions along a, along the $\mathbf{d_1} = \mathbf{a} - \mathbf{b} + \mathbf{c}$ direction (which achieves the coupling of anions with the closest Se atom of the donor) and along the $\mathbf{d_2} = \mathbf{a} + \mathbf{b} + \mathbf{c}$ direction (which achieves the coupling of the anions with the methyl groups and the polarization of σ bonds of the donor).

Figure 3 resumes this synchronous CDW – anion shift pattern.

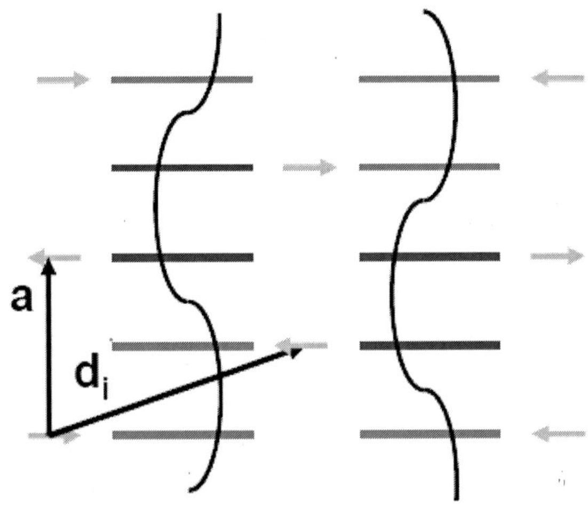


**Figure 3.** Schematic representation in the (**a**, $\mathbf{d_i}$) plane of the array of $2k_F$CDW (in black) coupled to anion shifts (in green). Donors with an excess (defect) of hole are shown in red (blue).

In this model the CDW wave vector **q** allowing such π phase shifts between the anion displacements is given by:

$\mathbf{qa} = \pi$ and $\mathbf{qd_i} = \mathbf{q}(\mathbf{a} \pm \mathbf{b} + \mathbf{c}) = \pi$ (mod. $2\pi$), which leads to:

$$q_a = 1/2 \text{ and } q_a \pm q_b + q_c = 1/2 \text{ (mod.1)} \tag{5}$$

The relationships (5) lead to $\pm\, q_b + q_c = 0$ (mod.1) which has two solutions: $q_b = qc = 1/2$ and $q_b = qc = 0$. The first solution (1/2, 1/2, 1/2) corresponds to the $2k_F$ BOW/AO modulation, $q_{AO}$, (see Section 7.2) and the SP modulation, $q_{SP}$ (see Section 6). The second solution (1/2, 0, 0) corresponds to the $2k_F$ BOW/AO modulation in the $NO_3$ salts (see Section 7.5). As usual the $2k_F$ BOW has a phase shift of π/2 with the $2k_F$ CDW. In [60] this model is used to derive the anion shift pattern setting the $4k_F$ CDW or CO.

In this picture the suppression of the $2k_F$ BOW at low temperature could be explained by the impossibility to establish an inter-chain coupling between CDW via the anion shift. The absence of a low temperature anion shift, ascribed the lock-in of the $PF_6$ to the methyl groups via the formation of H-bonds, was previously discussed inSection 2.4 and Section 3.1.1.

## 3.2. Evolution of the $2k_F$ BOW Instability along the TMTSF-TMTTF Series

The 1D $2k_F$ BOW instability of $(TMTSF)_2ClO_4$ [40] behaves as the one of $(TMTSF)_2PF_6$ and $AsF_6$. Figure 4reports the thermal dependence of the $2k_F$ peak intensity, $I(2k_F)$ given by expression (1), of the X-ray diffuse scattering from $(TMTSF)_2ClO_4$. In this salt the low temperature vanishing of $I(2k_F)$ has to be associated with the 3D critical growth below ~40 K of the (0, 1/2, 0) AO structural instability promoting the uniform $ClO_4$ ordering and anion shift in stack direction (see Section 7.3) [61]. Figure 4 shows surprisingly that the thermal dependence of the $2k_F$ BOW fluctuations of $(TMTSF)_2ClO_4$ is strongly modified upon substitution by the TMTTF [62].

A low temperature divergence of $I(2k_F)$ appears for very small amount of TMTTF substituent. For $x$ = 0.5% this additional divergence pushes down at lower temperatures the drop of $I(2k_F)$ which remains due to the competing AO instability as in pure $(TMTSF)_2ClO_4$. The thermal behavior of I(2k$_F$) in the $x$ = 0.5% salt resembles the one observed (see Figure 2 in [63]) in the $(TMDTDSF)_2PF_6$ iso-structural salt built on a donor hybrid between the TMTSF and TMTTF molecules. However in $(TMDTDSF)_2PF_6$ the drop of I(2k$_F$) below 20 K is not due to an AO transition but, as for $(TMTSF)_2PF_6$, to the stabilization of a SDW ground state below 7 K. Note that because of the orientation disorder of the TMDTDSF molecule in the structure [64], I(2k$_F$) does not vanish completely at low temperature in $(TMDTDSF)_2PF_6$ so that a 1D SP short range order (on $\xi_a$~25 Å) still coexists with the 2k$_F$ SDW modulation.

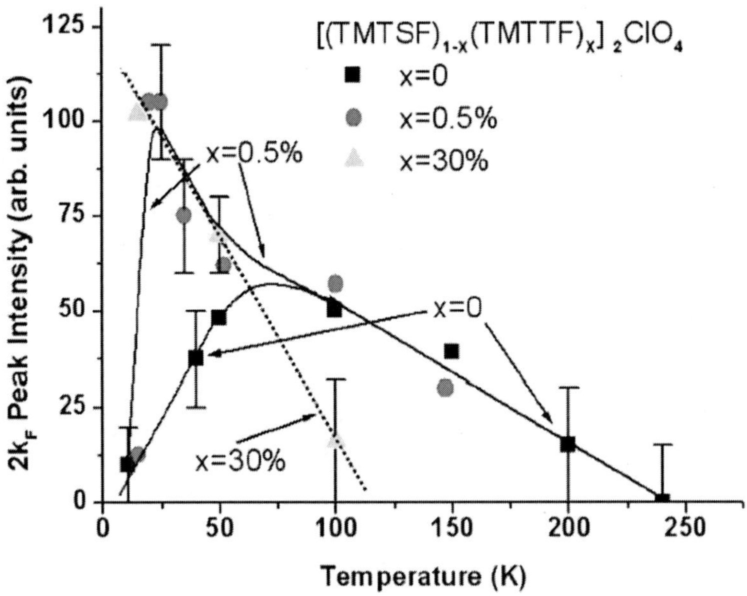


**Figure 4.** Thermal dependence of the 2k$_F$ peak intensity, I(2$k_F$), in the solid solution $[(TMTSF)_{1-x}(TMTTF)_x]_2ClO_4$ for $x$= 0 (black squares), 0.5% (red circles) and 30% (green triangles) of TMTTF deduced from the data reported in [40] and [62].

Only the low temperature divergence of I(2k$_F$) remains in the $x$ = 30% salt below 100K. This low temperature divergence of the 2k$_F$ diffuse scattering intensity recalls the divergence of the SP fluctuations observed in $(TMTTF)_2Br$ and $PF_6$ below 80 K and 60 K respectively [65].

In TMTTF salts the occurrence of a low temperature divergence of $I(2k_F)$ coincides with the presence of a high temperature charge localization [51,52] which leaves the spin degrees of freedom available for a low temperature SP or AF instability. This charge localization manifests below a temperature, $T_\rho$ [66], higher than the onset temperature of the critical growth of the SP fluctuations. In the $x$ = 30% salt the $2k_F$ fluctuations detected below the spin-charge separation temperature $T_\rho$ [67] are thus the structural fingerprint of an incipient SP instability which fully diverges in $(TMTTF)_2PF_6$ .

In conclusion the $[(TMTSF)_{1-x}(TMTTF)_x]_2ClO_4$ solid solution illustrates quite well when $x$ increases that the evolution from a $2k_F$ BOW instability, which vanishes at low temperature, to a divergent SP instability follows the growth of the $4k_F$ charge localization

## 3.3. THE SPIN-PEIERLS INSTABILITY

$(TMTTF)_2PF_6$ develops $2k_F$ SP critical fluctuations below about 60 K (taken as the mean field transition temperature of the SP transition, $T_{SP}^{MF}$). The critical nature of the SP fluctuations is assessed by the divergent growth of the SP susceptibility, $\chi(2k_F)$ [7,8,65], and of the intra-chain correlation length, $\xi_{SP}$; both quantities being defined by expression (3). The divergence is achieved at $T_{SP} \sim 17$ K, temperature at which a long range (1/2, 1/2, 1/2) SP stack tetramerization occurs [68].

Figure 5 compares the thermal dependence of $\xi_{SP}^{-1}$ of $(TMTTF)_2PF_6$ and of $(TMTTF)_2Br$. It clearly shows that the SP fluctuations of $(TMTTF)_2Br$ do not diverge. Thus Figure 5 resembles for the SP fluctuations of $(TMTTF)_2Br$ at Figure 2b for the $2k_F$ BOW fluctuations of $(TMTSF)_2PF_6$. We shall consider at the end of this section this aspect of the SP fluctuations of $(TMTTF)_2$ Br.

The thermal dependence of $\xi_{SP}^{-1}$ of $(TMTTF)_2$ $PF_6$ can be well accounted for by the calculation of the fluctuations of the weakly localized SP chain [69]. From this calculation one deduces that the SP coherence length $\xi_0$ amounts to 12 Å. This value is very close to $\xi_0 \sim 10$ Å found for the SP instability of $(BCPTTF)_2$ $PF_6$ [7]. From the relationship linking $\xi_0$ to $T_{SP}^{MF}$:

$$\xi_0 = \hbar v_\sigma / \pi k_B T_{SP}^{MF} \tag{6}$$

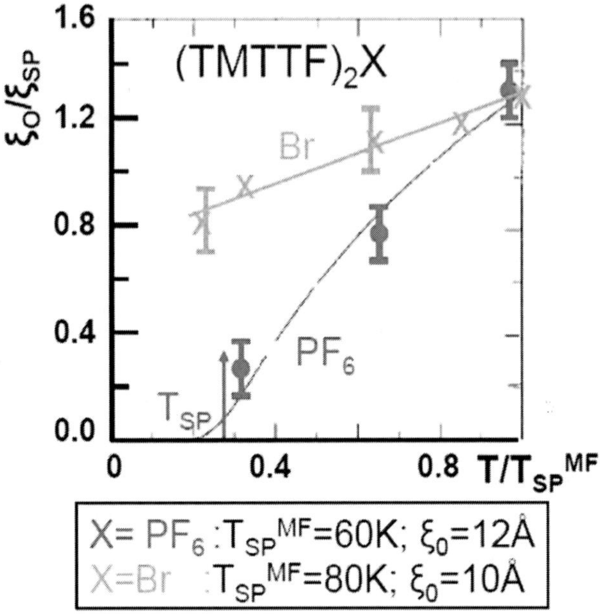


**Figure 5.** Thermal dependence of the inverse intra-chain correlation length $\xi_{SP}^{-1}$ of the SP instability of $(TMTTF)_2PF_6$ (red dots) and of $(TMTTF)_2Br$ (blue crosses). $T_{SP}^{MF}$ and $\xi_0$ are indicated in the lower panel for each salt. The continuous line for the $PF_6$ salt is $\xi_{SP}^{-1}$ calculated [69] for the weakly localized SP chain.

one gets a "spin velocity" $\hbar v_\sigma \sim 0.2$ eVÅ in $(TMTTF)_2PF_6$. A similar value $\hbar v_\sigma = 0.27$ eVÅ is obtained in $(BCPTTF)_2PF_6$ where $T_{SP}^{MF} \sim 100$ K. For this last compound which can be described by the Heisenberg Hamiltonian coupling well localized spins one gets, with $\hbar v_\sigma = \pi Ja/2$, an AF exchange coupling J = 280K which is close to J = 330 K deduced from the fit of the thermal dependence of the spin susceptibility of $(BCPTTF)_2PF_6$ [70]. If the same analysis is performed for $(TMTTF)_2PF_6$, where the holes are less localized than in $(BCPTTF)_2PF_6$, one gets, using the above quoted $\hbar v_\sigma$ value, J = 210 K, which is twice smaller than J = 420K obtained from the fit of the thermal dependence of the spin susceptibility [71]. A similar discrepancy is obtained for $(TMTTF)_2Br$ where expression (6) leads, with $\xi_0 \sim 10$ Å and $T_{SP}^{MF} \sim 80$ K, to a J value of 230 K twice smaller than J = 500 K obtained from the fit of the thermal dependence of the spin susceptibility [71]. It thus appears that the analysis using the Heisenberg Hamiltonian, strictly valid for localized spins, does not hold for the TMTTF salts where the holes bearing the spins are too delocalized.

Let us remark that in the TMTTF salts $k_B T_{SP}^{MF}$ amounts to the energy of the transverse acoustic phonon mode, $\hbar\Omega \sim 5$meV, which controls the dynamics of the SP instability. With $k_B T_{SP}^{MF} \sim \hbar\Omega$, the SP transition should occur in the adiabatic limit with however a strongly damped soft mode (see Figure 4b in [72]). This is consistent with the adiabatic limit used in [69] for the calculation of $\xi_{SP}^{-1}$ shown in Figure 5. However with $k_B T_{SP}^{MF} \sim \hbar\Omega$, $(TMTTF)_2$ $PF_6$ should not be too far from the adiabatic-antiadiabatic crossover. As $T_{SP}^{MF}$ decreases when the size of the anion increases, the $AsF_6$ salts could be located in the anti-adiabatic regime and the $SbF_6$ salt in the gapless regime where the SP ground state cannot be stabilized (see Section 6.2 and the analysis performed in [73]). The location of the $SbF_6$ salt in the gapless regime agrees with the non detection of SP X-ray diffuse scattering fluctuations in this salt. Due to the absence of a SP instability the $SbF_6$ salt undergoes an AF transition. The SCN salt behaves similarly. However the mutual exclusion between AF and SP instabilities is not general among the TMTTF salts because the Br salt exhibits below 80 K quasi-1D SP fluctuations which coexist with AF fluctuations whose divergence [74] stabilizes below $T_N$ = 13 K the AF ground state.

As previously mentioned, the thermal dependence of the SP fluctuations of $(TMTTF)_2Br$ resembles those of the $2k_F$ BOW fluctuations of $(TMTSF)_2PF_6$. Figure 5 shows that $\xi_{SP}$ reaches ~13 Å before the vanishing of the SP fluctuations, a similar value of $\xi_{BOW}$ (14 Å) was found in $(TMTSF)_2PF_6$ (Figure 2b). The vanishing of the SP fluctuations in the Br salt could have the same origin as for the BOW fluctuations of $(TMTSF)_2PF_6$. However there is no low temperature structural refinement in the Br salt allowing sustaining this assertion. Such a structure would be all the more desirable than thermal expansion measurements [23] show that the lattice (as well as electronic properties, see Section 4.1) exhibits unusual features just above $T_N$. Upon cooling below 18 K $\alpha_{c^*}$abruptly drops, exhibits a negative maximum around 17 K, then gently increases towards zero. These low temperature thermal expansion anomalies, which recall glassy phenomena involving ethylene groups in $(BEDT\text{-}TTF)_2X$ [3], could be also associated to the freezing of the disorder of methyl groups of the TMTTF.

# 4. THE MAGNETIC GROUND STATES OF THE BECHGAARD AND FABRE SALTS

At ambient pressure $(TMTTF)_2Br$ undergoes a commensurate AF order at $T_N$ = 13 K while $(TMTSF)_2PF_6$undergoes an incommensurate SDW order at $T_{SDW}$ = 12 K. Pressure studies show that the AF ground state evolves into a SDW ground state when the electron localization (*i.e.*, $T_\rho$), occurring on the TMTTF side of the generic phase diagram drops [51,52]. This section will show that, because of the presence of a sizeable magneto-elastic coupling, the magnetic ground states of the Fabre and Bechgaard salts, which involve also structural degrees of freedom [65], are quite a bit more complex than those expected for a simple magnetic transition. This magneto-elastic coupling is also responsible of the occurrence of SP transitions which will be considered inSection 6.

## 4.1. The AF Phase of $(TMTTF)_2Br$ and the Pre-Transitional Anomalies

AF order which develops below a 2nd order phase transition in $(TMTTF)_2Br$ stabilizes a $q_{AF}$ = (1/2, 1/4, ?) commensurate magnetic modulation [75]. The magnetic order is accompanied by a structural modulation at the $q_S$ = (1, 1/2, ?) reduced wave vector (the component 1a*, corresponding to the $4k_F$ wave vector, means that the superstructure reflections are detected in H odd layers of Bragg reflections [65]). As $q_S = 2q_{AF}$ the structural modulation should result from a "magneto-elastic" coupling which will be considered below.

The coupling between the structural and AF order parameters is analyzed in Annex A.1 using the simplest Landau development of the free energy. Its minimization gives two solutions shown in Figure 6.

Solution (a) corresponds to an AF order with one site out of two bearing the magnetization, while solution (b) gives an AF order where all the sites bear the magnetization. In the Landau theory of Annex A.1 only solution (a) gains energy by setting a $4k_F$ structural modulation. NMR lineshape analysis [75] better agrees with the spin configuration shown in (a). In the schematic representation of Figure 5a the presence of a spin on one site out of two is accompanied by an excess of hole on the magnetic site and a defect of hole on the non magnetic site. This leads to the formation of a $4k_F$ CDW or CO [76]. CO should be accompanied by a

$4k_F$ internal deformation wave of the TMTTF molecules in phase with the charge density per molecule. The CO pattern could be also stabilized by a displacement wave of the Br towards the molecule bearing the excess of hole (if there is no H-bonds blocking the anion shift), recalling the one found in the structural refinement of δ-(EDT-TTF-CONMe$_2$)$_2$Br in its CO ground state [77]. As previously mentioned in Section 3.3, (TMTTF)$_2$Br exhibits unusual features few degrees above $T_N$. In addition to the lattice thermal expansion anomaly previously reported, (TMTTF)$_2$Br exhibits below$T^*$ ~ 18–22 K anomalous electronic properties:

- The thermal dependence of the electrical conductivity exhibits, depending on the measurements, either a kink at $T^*$ or a slope anomaly at $T^*$, or even a plateau below $T^*$ (to my knowledge the $T^*$ anomaly was first reported in Figure 1 of [78]).

- The spin susceptibility abruptly decreases below $T^*$ [79,80]; this effect is apparently not due to the formation of a pseudo gap because SP fluctuations vanish also below $T^*$.

- A broadening of the NMR spectra, interpreted as being due to non-homogeneities in the electronic states, is observed below $T^*$ [80].

(TMTTF)$_2$ Br bears some resemblance with (TMTTF)$_2$SCN which also stabilizes the same modulations. (TMTTF)$_2$ SCN (see Section 7.4) sets at $T_{CO}$ = 160K an anti-ferroelectric (1, 1/2, 1/2) CO superstructure followed at$T_N$ = 7 K by a $q_{AF}$ = (1/2, 1/4, ?) magnetic order while in (TMTTF)$_2$Br both modulations are established at the same 13K transition. This analogy is better revealed by dielectric permittivity measurements showing in both salts upon cooling an increase of dielectric constant, probably due to the growth of CO ferroelectric segments on individual stacks, followed by an abrupt drop of the dielectric constant ~20K above $T_{CO}$ in (TMTTF)$_2$SCN [81] and below ~$T^*$ in (TMTTF)$_2$Br [82], probably when the inter-chain anti-ferroelectric coupling develops. 3D CO local order should induce the charge non homogeneities revealed by NMR [80] below $T^*$ in (TMTTF)$_2$Br. These charge non homogeneities could trigger local random Br displacements, possibly inducing methyl group disorder, which could cause the lattice expansion anomaly observed along the c* direction [23].

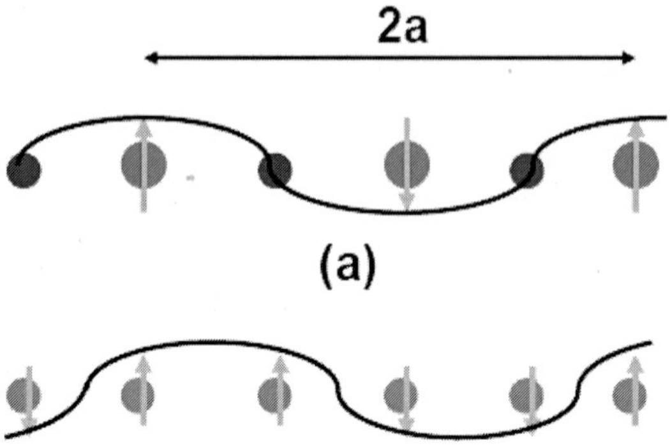


**Figure 6.** (**a**) Mixed AF–CO order where the hole rich molecule bears the spin and the hole poor molecule is a node of the magnetization wave; (**b**) AF modulation where each site bears magnetization and the same charge. In (b) the nodes of the magnetization wave are located on one bond out of two. The spin direction is given by the sign of η(q) defined in Annex A.1. In (a), the charge rich/poor (+ρ/−ρ) sites are represented by the red/blue circles of a different size. The dimerization of the stack is ignored in this figure.

## 4.2. The SDW Phase of $(TMTSF)_2PF_6$

$(TMTSF)_2PF_6$ stabilizes below $T_{SDW}$ = 12 K an incommensurate $q_{SDW}$ = (1/2, ~1/4, ?) magnetic modulation [54,55] stabilized by the nesting of its FS [56] and where the exchange field due to the intra-site coulomb repulsion U opens a full gap at the Fermi energy. However the phase transition of $(TMTSF)_2PF_6$ exhibits additional features. First, the $2k_F$ SDW order is stabilized by a first order transition [83], while the Peierls or Slater transition caused by the simple divergence of CDW/BOW or SDW electron-hole response functions is of second order [1,2]. It follows that the order parameter, the magnetization [71] or the square root of the $2k_F$ satellite intensity [65] (see below), does not continuously vanish to zero at $T_{SDW}$. There is also a jump of electrical conductivity at the MI transition. Second, a "specific heat" anomaly is observed at $T_{SDW}$ in thermal expansion measurements [25]; a feature not expected for a pure electronic SDW transition. Third, very weak X-ray satellite reflections (but too intense to be due to magnetic scattering) are observed below $T_{SDW}$ at the reduced $q_{SDW}$ ($2k_F$CDW reflections) and $2q_{SDW}$ ($4k_F$ CDW reflections) wave vectors [7,65,84].

Thus the ground state of $(TMTSF)_2PF_6$ must be described with 3 order parameters related to the $2k_F$ SDW, $2k_F$CDW and $4k_F$ CDW modulations. A simplified analysis of the phase transition is given Annex A.2 in the frame work of the Landau development of the free energy. This simple analysis shows, in agreement with experimental observations, that for an attractive enough coupling between the $2k_F$ SDW and $2k_F$ CDW order parameters a mixed SDW-CDW modulated ground state can be directly obtained through a first order transition.

The stabilization of a first order phase transition particularly requires that the Landau coefficients:

- $a_1$, associated to the $2k_F$ SDW order parameter, changes of sign near $T_{SDW}$, which means that the $2k_F$ SDW susceptibility, $a_1^{-1}$, should diverge. Such a divergence is detected by NMR [46].

- $a_2$, associated to the $2k_F$ CDW order parameter, should be very small, which means that the system should present incipient $2k_F$ CDW instability. This is the case above ~30K for the $2k_F$ BOW previously discussed inSection 3.1.

These conditions together with additional requirements given in Annex A.2, lead to phase diagrams shown inFigure 7.

The phase diagrams shown in Figure 7 leave the possibility (interrupted lines) of an additional transition at lower temperature towards a pure $2k_F$ SDW phase. This transition could occur in $(TMTSF)_2$ $PF_6$ at ~3.5 K, temperature at which earlier NMR measurements [86,87] exhibit anomalies. Consistently with the occurrence of a pure SDW sub-phase, a vanishing of the $2k_F$ CDW satellite reflection intensity is observed below ~3.5 K [84].

The finding of a mixed $2k_F$ SDW/CDW ground state implies that both $2k_F$ SDW and $2k_F$ CDW response functions should be strong. Let us first consider the electronic phase diagram. Recent calculation [88,89] shows that, in addition to the FS nesting process tending to promote, in presence of sizeable intra-chain coulomb repulsions, the $2k_F$ SDW ground state, the inter-chain backward repulsive coulomb interaction $g_{1\perp}$ previously considered in Section 3.1.2 can in parallel stabilize a $2k_F$ CDW ground state for strong enough $g_{1\perp}$. These calculations show that with both kinds of inter-chain coupling there is a proximity between the

$2k_F$ SDW and $2k_F$CDW phases, and that, even if the SDW ground state is stabilized, sizeable $2k_F$ CDW electronic correlations remain above the transition temperature. Interestingly when the FS nesting breaking effects destabilize the insulating density wave ground state, strong $g_{1\perp}$ tends to stabilize triplet f superconductivity instead of the singlet d superconductivity occurring for small $g_{1\perp}$ (for a recent review of the interplay between superconductivity and $2k_F$ density waves see [90]).

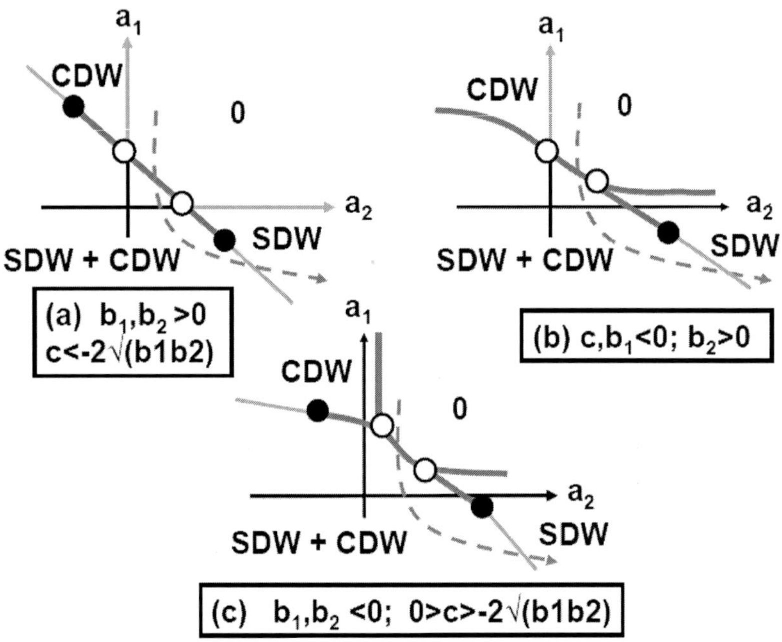


**Figure 7.** $2k_F$ SDW/CDW phase diagrams in function of the Landau coefficients $a_1$ and $a_2$ for an attractive interaction, c, between the $2k_F$ SDW and $2k_F$ CDW order parameters (adapted from Figure 3 in [85]). (**a**) Corresponds to $c < -2\sqrt{(b_1 b_2)}$ and $b_1, b_2 > 0$ (case also displayed in Figure A.1); (**b**) Corresponds to c, $b_1 < 0$ and $b_2 > 0$; (**c**) Corresponds to $0 > c > -2\sqrt{(b_1 b_2)}$ and $b_1$, $b_2 < 0$. The notations are defined in Annex A.2. Thin (thick) lines represent 2nd (1st) order transition lines. White dots indicate merging points between transition lines and black dots represent tricritical points. In each phase diagram the grey interrupted lines show a trajectory starting from the uniform phase "0", crossing, after a 1st order transition, the SDW + CDW phase to finally reach, after a 2nd order transition, a pure SDW phase.

However in the present case, already discussed in Section 3.1.2, $g_{1\perp}$ stabilizes a different periodicity as does the FS nesting, a feature not considered in the theory performed in [88,89]. This competition together with the lock-in of the anion destabilizes the $2k_F$ BOW as seen in Section 3.1. However the CDW counterpart of $2k_F$ instability could remain active and contribute to the ground state modulation if the 3D ordering of CDW is achieved by the same FS nesting mechanism as for the $2k_F$ SDW. As $g_{1\perp}$ is not activated below 30K, the structural modulation observed below $T_{SDW}$ must not be of the BOW type (The absence of BOW structural features explains why the $2k_F$ satellite reflections are of very weak intensity). For this reason it was proposed in [7] and [65] that the weak satellite reflections detected below $T_{SDW}$ could correspond to the X-ray scattering by the electronic $2k_F$ CDW. Eventually a molecular distortion could follow the setting of the $2k_F$ CDW (see below).

The separation between pure $2k_F$ SDW and $2k_F$ CDW ground states implies that there is a natural repulsion between the associated density waves, which means that the bare coupling term ν introduced in the Landau free energy of Annex A.2 is repulsive. However if these two density waves are simultaneously stabilized in the same phase, the free energy is minimized if the $2k_F$ SDW and $2k_F$ CDW are in phase quadrature ($\omega = \pi/2$ phase shift). This corresponds also to a $\pi/2$ phase shift between the spin ↑ and spin ↓ components of the SDW and of the electronic CDW [8,65].

The "specific heat" anomaly observed at $T_{SDW}$ in thermal expansion measurements [25] shows that the electronic CDW should have a structural counterpart. This counterpart is not a BOW modulation (see Section 3.1). However if the CDW consists primarily in a modulation of the hole occupancy on the molecules, one expects that molecular deformations should follow the electronic density wave by elongating the TMTSF hole rich molecules and contracting the TMTSF electron rich molecules. This should lead to a modification of intra-molecular vibration modes in Raman and/or infrared spectra of $(TMTSF)_2PF_6$ below $T_{SDW}$. The electronic CDW could also trigger "phase phonons" which should be observable in reflectance measurements. These latter effects apparently have been detected below $T_{SDW}$ in the parent $(TMTSF)_2SbF_6$ compound [91]. However there are very few optical studies performed below $T_{SDW}$ in $(TMTSF)_2X$'s to assess this statement.

The coupling between the $2k_F$ SDW and the $4k_F$ CDW could be achieved by a "magneto-elastic" effect similar to the one considered in $(TMTTF)_2Br$. By analogy with the findings of Annex A.1, this coupling should lead to an increase of hole concentration on the site having the larger magnetization. However as the intensity of the $4k_F$CDW reflections of $(TMTSF)_2PF_6$ is much weaker than the one of $(TMTTF)_2Br$, the lattice distortion which eventually accompanies the $4k_F$ CDW of $(TMTSF)_2PF_6$ must be much smaller than the $4k_F$ distortion of $(TMTTF)_2Br$. This difference could be explained by the incommensurate nature of the modulation (leading to the loss of umklapp effects) and the smaller amplitude of modulation [a reduction of the magnetization by a factor 1.75 ($0.08\mu_B$ for $(TMTSF)_2PF_6$ *versus* $0.14\mu_B$ for $(TMTTF)_2Br$ [71]) reduces, according to (A.2), the amplitude of the distortion by a factor 3 and the $4k_F$ reflection intensity by a factor 9]. Whatever its microscopic origin, the coupling between the $2k_F$ SDW and the lattice remains appreciable in the density wave ground state because optical studies [92] show that the SDW condensate presents a large dynamical mass enhancement, which is however several times smaller than the mass enhancement of a CDW condensate.

# 5. THE CHARGE ORDERING TRANSITION IN THE FABRE SALTS

## 5.1. Basic Features: The Historical Scenario

The first indication of the now well established symmetry breaking CO phase transition was revealed by anomalies in conductivity measurements performed in the $(TMTTF)_2X$ salts with X = $PF_6$, $AsF_6$ and $SbF_6$ and their solid solutions [93]. The most striking finding was the observation of a MI transition at $T_{CO}$ = 154 K in the $SbF_6$salt, recalling the one previously observed at 160 K in the SCN salt [66]. Although this transition was correctly interpreted as due to a $4k_F$ electronic localization phenomenon, no lattice symmetry breaking could be detected at$T_{CO}$ in this preliminary study, in the difference of the SCN salt [94]. For this reason this transition was labeled "structureless" in later studies [95] which also revealed an anomaly at $T_{CO}$ in the thermal dependence of the thermo-power (possibly due to a gap opening) of these salts and similar anomalies in the $ReO_4$ salt well above its AO transition. During the following year two important, but often ignored studies, revealed a dielectric divergence [81] and a few percent Young modulus softening [96] at $T_{CO}$. The symmetry breaking consisting of a loss of all the inversion centers at $T_{CO}$ was only revealed 15 years later by the NMR observation of a charge differentiation between the two

TMTTF molecules of the unit cell [97] providing the first evidence of the occurrence of a charge disproportion below $T_{CO}$. Charge disproportion between molecules was soon confirmed by the observation of split intra-molecular vibration modes below $T_{CO}$ [98,99]. This charge disproportion together with the incipient stack dimerization leads to a stack dielectric polarization and, as there is no cell doubling (at the difference of the SCN salt), to the establishment of electronic ferroelectricity. This was sustained by accurate measurements of the dielectric divergence at $T_{CO}$ [100].

Although it was suspected in these first works that the anions should control the CO transition [93] or even should be displaced with respect to donors at $T_{CO}$[38], several attempts to find a structural modification at $T_{CO}$were unsuccessful [7,101,102,103]. The reason, found recently [104], is that minute irradiation defects created by X-ray beams in laboratory diffraction conditions kill the CO. It was only recently that the finding [24] of a lattice thermal expansion anomaly at $T_{CO}$ provides convincing evidences of a structural counterpart at the electronic ferroelectricity. Then evidence of a tiny structural modification was found from the detection of weak (<15%) variations at $T_{CO}$ of the intensity of several main Bragg reflections using neutron diffraction on $(TMTTF)_2PF_6$deuterated powders [21,105]. These quite small structural modifications explain why earlier attempts to solve the $(TMTTF)_2PF_6$ and $AsF_6$ structures on hydrogenated crystals in the P1 space group from conventional neutron scattering data collection at 4K were unsuccessful [106]. Recently the low temperature non centro-symmetric P1 structure of $(TMTTF)_2PF_6$ has been assessed using high energy X-ray beam from synchrotron radiation [107]. The main result of this study was the finding of intra-molecular deformations associated to charge disproportion. However no sizeable change of the F-S contact distance could be detected at $T_{CO}$.

## 5.2. Structural Ingredients of the CO Transition

From a general point of view, three main structural features are expected at the CO transition of the Fabre salts [108]:

- an internal deformation of the TMTTF molecules following their charge occupancy,
- a deformation of the methyl group cavities associated to the loss of inversion symmetry,

- a shift of the anions from the inversion centers.

The first feature, expected from the splitting of intra-molecular mode frequencies at $T_{CO}$ [98,99,109], has been recently detected [107]. Although the differentiation of the two molecules of the unit cell already suppresses the inversions centers, a deformation of the methyl group cavities and/or a shift of the anions should also contribute to stabilize the non centro-symmetric structure. In this case both the polarization of the methyl groups and/or the shift of the anions provide the efficient couplings with the π cloud [21,110] necessary to stabilize the CO ground state [111].

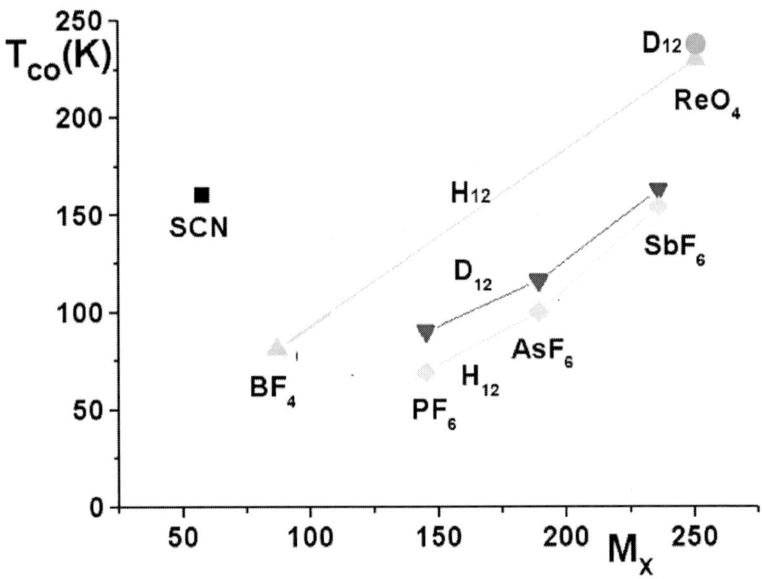


**Figure 8.** $T_{CO}$ in function of the anion (labeled by its mass $M_X$) in hydrogenated ($H_{12}$) and deuterated ($D_{12}$) $(TMTTF)_2X$ (adapted from [112]). The continuous lines connect $T_{CO}$ of $H_{12}$ and $D_{12}$ salts with anions of same symmetry.

The influence of the anions in the CO process is assessed by the observation that:

- $T_{CO}$ strongly varies with the volume, shape and polarizability of the anion (see Figure 8) [112],
- $T_{CO}$ increases when the anion –S contact distance is shortened [93,113].

Figure 9 shows two different processes where the shift of the anion could lead to a disproportion of charge on the TMTTF. In Process 2 of Figure 9, the anion moves towards the S of a TMTTF where the shortening of the anion –S contact distance below $T_{CO}$ enhances directly the π hole density of the TMTTF in the vicinity of the anion. In Process 1 of Figure 9, the anion moves inside its methyl group cavity and deforms it. This deformation polarizes the H-bond network. The H-bond polarization induces a displacement of charge in the σ bonds connected to the H-bonds. This shift of σ electrons towards the center of the TMTTF stabilizes the excess of π holes [21].

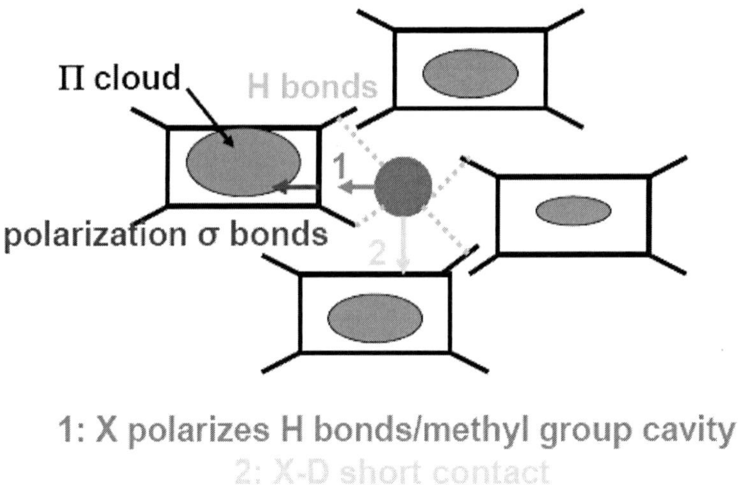


**Figure 9.** Schematic illustration of the two types of anion shifts. Process 1 (in red): displacement of the anion inside the methyl group cavity which polarizes the H-bond network and the σ electron skeleton of the TMTTF. Process 2 (in green): shift of the anion X towards the Se/S of a donor D establishing a short X-D contact distance. The modulation of the density of π holes is indicated for Process 1.

Both direct process 2 and indirect process1 stabilize an excess of π hole on the TMTTF towards which the anion moves. However the excess of hole is not localized on the same molecule because the direction of displacement of the anion is different in the two processes. However as the recent structural refinement [107] does not provide evidence of a sizeable change of the anion -S short contact distance below $T_{CO}$ in the $PF_6$ salt, the stabilization of the CO pattern probably occurs via the deformation of the methyl group cavity in Process 1. Such an assertion is sustained by the following observations:

- $T_{CO}$ increases when, with the larger size (and polarizability) of octahedral and tetrahedral anions, the contact between the anion and the periphery of the cavity delimited by the methyl groups is strengthened,

- $T_{CO}$ is enhanced by deuteration of the methyl groups of TMTTF [112,114] (the relative increase of $T_{CO}$reaches 30% in the $PF_6$ salt),

- CO is rapidly destroyed by X-ray irradiation damages [104] which particularly affect terminal methyl groups of the donor (in a general manner irradiation effects are especially strong in salts built with the TMTTF and TMTSF methyl-based donors [115]).

- Far infrared measurements show there is a modification of the coupling between the methyl groups and the anions at the CO transition [109].

Interactions via the methyl groups could also explain why strong $T_{CO}$ lattice parameter anomalies are observed in the (b', c*) plane containing the H-bond network (see Figure 1); the strongest anomaly being along c* [24]. In addition the easy squeezing of the soft methyl group cavity under pressure explains simply, via the blockade of the anion in its cavity, the rapid decrease of $T_{CO}$ under pressure [116]. Until now there is no direct evidence of the deformation of the methyl group cavity at the CO transition. This deformation should be better revealed by neutron diffraction more sensitive to H or D positions.

An eventual shift of the anion in the CO phase does not mean that the anion orientation disorder is completely removed at $T_{CO}$ in the TMTTF salts. $^{19}F$ NMR studies of $(TMTTF)_2SbF_6$ show the contrary [27,28]. Evidence of the decoupling between CO and orientation order of the anion is provided by the observation of successive CO and AO transitions in the $ReO_4$ ($T_{CO}$ = 230 K and $T_{AO}$ = 154 K) and $BF_4$ ($T_{CO}$ = 83 K and $T_{AO}$ = 40 K) salts [117].

Process 1 in Figure 9 which achieves an anion shift inside the methyl group cavity is structurally different from Process 2 which achieves an anion shift towards the S of the donor as found at the (1/2, 1/2, 1/2) AO transition (seeSection 7.2). It is thus possible that these two types of nearly perpendicular anion displacements could be successively active in the CO and (1/2, 1/2, 1/2) AO transitions of the TMTTF salts with the $ReO_4$ and $BF_4$tetrahedral anions. This decoupling allows to simply understand that the (1/2, 1/2, 1/2) AO transition, which does not substantially activate the H-bond network, is less sensitive to deuteration [114] and irradiation defects [118].

In this picture TMTTF salts incorporating small anions, such as $ClO_4$ and $NO_3$ which do not establish important interactions with the methyl groups [14,110] should not exhibit the CO transition. This seems to be the case for the $ClO_4$ salt [113]. In the opposite situation where an anion perfectly fits the methyl group cavity there is no room for an anion shift inside this cavity and thus no CO transition is expected. This could be the case of the Br salt. This better fit, which also occurs under pressure with the squeezing of the methyl group cavities, can be taken as responsible of the rapid drop of $T_{CO}$, as observed for example in pressurized $AsF_6$ salt [116].

In this framework the (0, 1/2, 1/2) anti-ferroelectric CO transition of $(TMTTF)_2SCN$, which coincides with the AO transition of the SCN [94], is singular. In this salt the small SCN anion, which has no real linkage with the methyl groups [110], can only strengthen its contacts with one donor out of two via the AO shift [103]. In this respect the wave vector of the AO superstructure of $(TMTTF)_2SCN$, which simultaneously stabilizes the CO, is different from the wave vectors stabilized at the others AO and CO transitions (see Section 7). The AO/CO transition of the SCN salt which does not directly involve the methyl groups is consistently weakly sensitive to X-ray irradiation damage.

Because of the involvement of the fragile methyl groups in the stabilization of the CO pattern, the ferroelectric ground state is subject to defects. Defects break the long range ferroelectric order into domains forming local clusters of polarization. This leads to a frequency dependant dielectric permittivity which does not really diverge at a well defined $T_{CO}$. Such features, recalling those of dielectric relaxors, are found in $(TMTTF)_2PF_6$ at ambient pressure and in $(TMTTF)_2SbF_6$ under pressure near its transformation to a (local) SP ground state [117]. The domain walls nucleated by these defects are described by charge soliton excitations of the CO ground state [119].

Electronic ferroelectricity [100,119] occurs in others organic systems [120]. Among them let us mention the 2D organic salt α–(BEDT-TTF)$_2$I$_3$ where ferroelectricity appears at the 135 K MI transition [121]. The interesting aspect of this phase transition is that, analogously to TMTTF salts, the charge rearrangement on type A BEDT-TTF molecules (called CO for this reason) is triggered by the deformation of the I$_3$ sublattice which thus modifies of the H-bond network between I$_3$ and the ethylene groups of the BEDT-TTF [122]. This mechanism, similarly to Process 1 in Figure 9, modulates the density of π holes on the BEDT-TTF.

## 5.3. The Underlying CO Instability

The CO transition observed in the Fabre salts is just a manifestation of the general tendency of organic conductors to form a Wigner lattice of localized charges because of the presence of long range intermolecular Coulomb repulsions (for a recent review see [73]). As the lattice is soft such an electronic instability is coupled to the lattice where it drives a 4k$_F$ CDW which also, in presence of an electron-phonon coupling with the acoustic modes, leads to the formation of a 4k$_F$ BOW [123,36]. However in quarter filled band systems, one has to distinguish between two different types of charge localization phenomena [73,76]:

- charge localization on the bonds, stabilizes a 4k$_F$ BOW or Dimer Mott (DM) ground state,
- charge localization on the sites, stabilizes a 4k$_F$ CDW or CO ground state.

For uniform quarter filled band systems, these two ground states have different inversion symmetry. However the TMTTF salts are more complex because with a stack being already dimerized (see Section 2.2) the divergence of the 4k$_F$ BOW instability is killed (the static dimerization itself gives rise to a charge localization [12] "visualized" by the development of an activated conductivity below $T\rho \sim 200$K higher than $T_{CO}$). Thus the only symmetry breaking instability remaining in the dimerized salts is the CO if the stacks are not too strongly dimerized [73,119].

Ferroelectric CO is not announced by easily detectable pre-transitional structural fluctuations (the pretransitional structural fluctuations at the anti-ferroelectic CO transition of (TMTTF)$_2$SCN are also barely detectable). Also the ferroelectric instability associated to CO appears to be mean-field because the dielectric susceptibility divergence follows a Curie Weiss behavior on a large (~30 K) temperature range above $T_{CO}$[100,117]. This means that CO in the TMTTF's is mostly announced by a regime of 3D fluctuations which isotropy reflects the importance of the inter-chain Coulomb coupling. In contrast, regular (*i.e.*, non dimerized) quarter filled systems such as $(DIDCNQI)_2Ag$ [124] or (o-DMTTF)$_2$X [125] exhibit a sizeable regime of 1D $4k_F$ BOW fluctuations above the "CO" transition which in fact presents a mixed CO/DM character.

# 6. THE SPIN-PEIERLS TRANSITION IN THE FABRE SALTS

## 6.1. Basic Features

The SP transition of $(TMTTF)_2PF_6$ is characterized by the appearance below $T_{SP} \sim 17$ K of superlattice reflections of very weak intensity at the (1/2, 1/2, ?) reduced reciprocal position according to earlier X-ray investigations [40] (see Figure 10). The third component 1/2c* has been determined by a recent neutron scattering investigation [68].

At ambient pressure the SP transition of the TMTTF's occurs in the CO phase. The SP order is rapidly destroyed by X-ray irradiation as for the CO transition. This irradiation sensitivity explains that recent structural studies of the SP order have been performed with neutron scattering. The structure of the SP ground state has not yet been determined, but it should present some analogies with the (1/2, 1/2, 1/2) AO superstructure [58,59]. Thus it is expected that the stack tetramerization, corresponding to a dimerization of stack of dimers, each dimer bearing a spin ½, will be accompanied by synchronized shift of the anions towards donors [68]. The involvement of the anions in the SP instability is assessed by the observation of a $T_{SP}$ critical divergence of the $^{75}$As NMR $T_1^{-1}$ in $(TMTTF)_2$ $AsF_6$ [126,127]. In this process the anion shift should tune the interchain coupling between the individual stack tetramerization, as discussed in Section 3.1.2. The 3D inter-chain coupling regime should occur in the near vicinity of $T_{SP}$ because the observation of precursor X-ray diffuse scattering lines on a large temperature range above $T_{SP}$ shows that the SP instability is basically 1D (see Section 3.3).

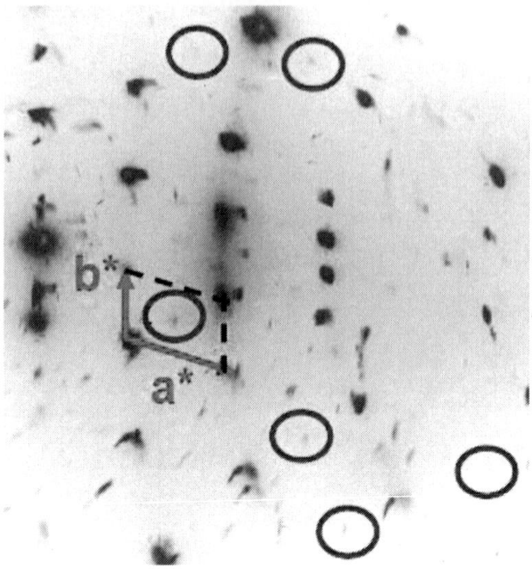


**Figure 10.** X-ray diffuse scattering pattern from $(TMTTF)_2PF_6$ at 10 K in its SP ground state showing inside the blue circles very weak (1/2, 1/2, ?) superlattice spots. The (a*, b*) unit cell is shown (original data of the study reported in [40]).

The temperature $T_{SP}$ at which the (1/2, 1/2, 1/2) superlattice reflections are observed coincides with the temperature at which a singlet gap develops in the spin susceptibility [112]. At $T_{SP}$ lattice expansion measurements exhibit a lambda type anomaly which coincides with the specific heat anomaly [128]. The value of the singlet-triplet gap in controversial in the literature which mixes calculated values, using approximate theories [129,130], and measured quantities [74,112]. These last measurements have been confirmed by a direct determination, using inelastic neutron scattering, of the SP gap in the spin excitation spectrum [131]. The SP wave vector is the same in the TMTTF salts and in the isostructural BCPTTF salts [7,70]. BCPTTF)$_2$PF$_6$ and AsF$_6$, which does not undergo a CO transition, have a twice higher $T_{SP}$ than $(TMTTF)_2PF_6$ and AsF$_6$.

The SP transition is very sensitive to the magnetic field, H, as in others SP compounds such as MEM-$(TCNQ)_2$ and $CuGeO_3$. In $(TMTTF)_2$ $PF_6T_{SP}$ decreases as $H^2$ [132] and above 19T this salt undergoes a phase transition to an incommensurate soliton-like structure [133,134].

## 6.2. Competition with the Charge Ordering

NMR measurements show that there is coexistence between CO and SP pairing in $(TMTTF)_2$ $PF_6$ and $AsF_6$. However the observation of an opposite variation of $T_{CO}$ and $T_{SP}$ under pressure shows that there is repulsion between the order parameters associated at the CO and SP transitions [116]. In this context, the real part of the microwave dielectric function, ε', which increases by ~10% in the SP phase, shows that the gap of charge (*i.e.*, CO order parameter) decreases at the SP transition [132]. This competition is analyzed in Annex A.3 in the framework of the Landau theory. In particular Landau development predicts the behavior of $T_{SP}$ in function of $T_{CO}$. This dependence, obtained with the restrictions outlined in Annex A3 and shown in Figure 11, accounts for the experimental results.

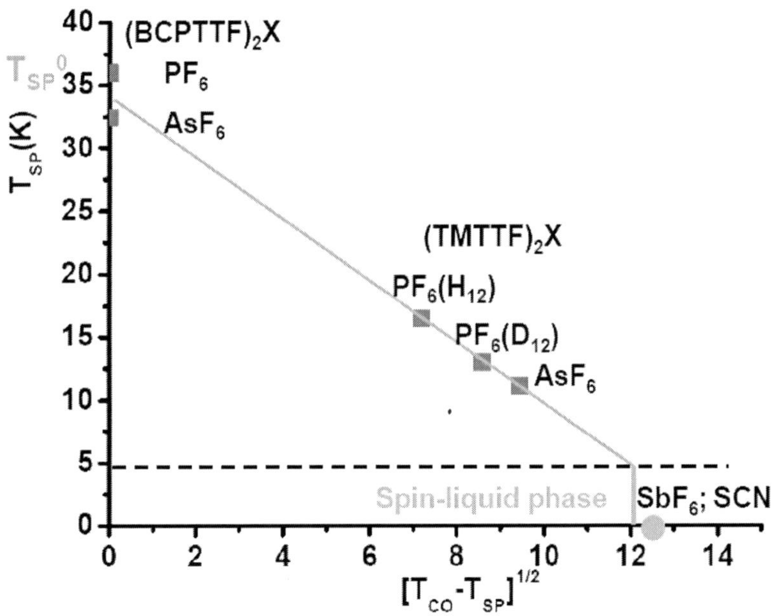


**Figure 11.** Spin-Peierls critical temperature, $T_{SP}$, of $(TMTTF)_2$ X *versus* $\sqrt{(T_{CO}-T_{SP})}$, as given by the expression (A12). $T_{SP}^0$ of $(BCPTTF)_2$ X, which does not undergo a CO, is also given, as well as $\sqrt{(T_{CO})}$ for the $SbF_6$ and SCN salts having no SP ground state. The spin-liquid phase occurring for $T_{SP} < 5$ K (see text) is indicated.

Figure 11 shows that $T_{SP}$ of $(BCPTTF)_2$ $PF_6$ and $AsF_6$ is consistent with the absence of CO (in that case the SP transition occurs at $T_{SP}{}^0$ defined Annex A3). Figure 11 shows also that the SP transition vanishes for $T_{CO}$ ~ 190 K or more likely for ~ 150 K if one assumes that for $T_{SP}$ < 5 K the spin-liquid phase is stabilized (see below). 150 K is close to $T_{CO}$ of $(TMTTF)_2SbF_6$ and SCN which do not exhibit a SP ground state.

To describe completely the SP phase diagram of the TMTTF salts, one has to consider explicitly the non-adiabatic effect of the phonon field (see Section 3.3). These effects have already been discussed in [73] with the main conclusions being:

- SP transition does not occur in $(TMTTF)_2SbF_6$ and SCN because the spin-phonon coupling is smaller than the zero-point lattice fluctuations (this condition corresponds to $T_{SP}{}^{MF} < \Omega/4 \sim 15$ K, or $T_{SP} < 5$ K because in the $PF_6$ and $AsF_6$ salts $T_{SP} \sim T_{SP}{}^{MF}/3$). These salts are thus located in the spin-liquid phase outlined in Figure 11phase the inter-chain exchange coupling stabilizes a low temperature ($T_N$ ~ 7K) AF ground state.

- SP transitions of $(BCPTTF)_2PF_6$ and $AsF_6$ occur in the adiabatic limit.

- SP transitions of $(TMTTF)_2PF_6$ and $AsF_6$ are located between the adiabatic limit and the spin-liquid phase, close to the crossover line between the adiabatic and anti-adiabatic regimes.

The microscopic origin of the competition between CO and SP instabilities relies on the fact that CO, which localizes the holes every second neighbor molecules, establishes also a reduced AF exchange integral, J, between the spin of the holes. With a spin-phonon coupling following the decrease of J, $T_{SP}{}^{MF}$ (*i.e.* the SP instability) decreases [73].

The weakening of the SP instability by the CO is just the manifestation of the repulsion existing between $4k_F$CDW (or CO) and $2k_F$ BOW (or SP) instabilities in quarter filled band systems: the next neighbour Coulomb repulsion $V_1$ which promotes CO in quarter filled band systems kills in the same time the divergence of the $2k_F$BOW response function [135,136]. In other words CO induces hetero-polar charge configurations which destabilize the modulation of bond distances in the $2k_F$ Peierls mechanism. This was clearly shown in the study of the $[(TMTSF)_{1-x}\ (TMTTF)_x]_2ReO_4$ solid solution [137,138]. For $x = 0.5$ the alternate stacking of TMTSF and TMTTF induces a chemical CO. In its presence it is observed a sizeable decrease of $T_{AO}$ of the (1/2, 1/2, 1/2)

AO transition which involves the formation of $2k_F$ BOW on the organic stacks (see Section 7.2).

## 6.3. The Vanishing of the SP Phase under Pressure

$T_{SP}$ increases when $T_{CO}$ decreases under pressure [116]. At higher pressure when the CO vanishes instead of reaching $T_{SP}^0$ at a value comparable to $T_{SP}$ measured in the BCPTTF's, $T_{SP}$ further decreases. In the $PF_6$ salts $T_{SP}$even exhibits a singular depression around Pc ~ 9 kbar before entering the AF phase [139]. This behavior suggests the existence of a quantum critical point where the normal phase is further stabilized around Pc.

A possible explanation could be, that with TMTTF salts being close to the crossover boundary between the adiabatic and anti-adiabatic regime, the hardening of the phonon mode under pressure will shift the SP instability to the anti-adiabatic regime followed by the spin-liquid regime around Pc. In the anti-adiabatic SP regime the enhancement of quantum fluctuations should strongly reduce the amplitude of 1D SP gap. In this scenario the suppression of the SP phase should occur through a critical point due to the enhancement of spin-liquid fluctuations. Then above $P_c$ the 3D AF ground state will be stabilized by the inter-chain exchange coupling. If the succession of ground states is described by the generic phase diagram [52], AF pressurized $(TMTTF)_2$ $PF_6$should be identical to AF ambient pressure $(TMTTF)_2Br$.

In order to assess the development of a pressure induced non-adiabaticity one must prove that the amplitude of the 1D SP gap decreases under pressure as a result of the enhanced quantum fluctuations. However, as it is observed that $T_{SP}^{MF}$ (linearly related to the mean field 1D SP gap when the phonon fluctuations are neglected) slightly increases from $(TMTTF)_2PF_6$ to $(TMTTF)_2Br$ (see Figure 5), the opposite situation seems to occur. In addition in order to sustain the non-adiabatic scenario one should have a considerable phonon frequency hardening (by a factor ~4) to stabilize the spin-liquid phase in pressurized $(TMTTF)_2PF_6$. This requires a phonon frequency as high as $\Omega > 4T_{SP}^{MF}$, which seems unrealistic for an acoustic mode. The difficulty with this scenario is that SP fluctuations are still detected in AF $(TMTTF)_2Br$, at the difference of $(TMTTF)_2SbF_6$ where it was stated in Section 6.2 that the AF ground state is caused by the enhanced non-adiabaticity of the phonon field which in the same time kills the SP fluctuations.

Another scenario could be that the squeezing of the methyl group cavities under pressure blocks the shift of the $PF_6$, the necessary ingredient setting the interchain coupling between the individual SP stack tetramerization. This is corroborated by a recent determination of the structure of pressurized $(TMTTF)_2PF_6$ [140] showing a sizeable contraction of the methyl group cavity. The mechanism of blockade of the anion in its methyl group cavity explains also that only a short range SP order is stabilized in pressurized $(TMTTF)_2SbF_6$ [27] when both the CO and AF phases are suppressed above 0.5 GPa.

# 7. THE ANION ORDERING TRANSITIONS

## 7.1. Basic Features

In Fabre and Bechgaard salts incorporating non-centrosymmetric anions, the anion orientation is disordered at RT. However for entropy reasons, the anions order upon cooling. The ordering is achieved at a well defined phase transition showing that the AO leads to a symmetry breaking. The AO transition stabilizes a superstructure which generally doubles the RT lattice periodicity in one or several directions [7]. The AO transition can be of 2$^{nd}$[(0, 1/2, 0), (1/2, 0, 0) and (0, 1/2, 1/2) AO] or 1st order [(1/2, 1/2, 1/2) AO] depending on the strength of the coupling of the AO process at the elastic lattice degrees of freedom The transition not only consists of an orientation ordering of the anion inside its methyl group cavity, which in itself breaks its inversion symmetry, but also of a concomitant deformation of this cavity associated with an anion shift. This shift can be either towards selected methyl groups as in the (0, 1/2, 0) AO transition of $(TMTSF)_2ClO_4$ (direction 1 in Figure 9) [141] or towards the S or Se atom of a given donor as in the (1/2, 1/2, 1/2) AO transition of $(TMTSF)_2ReO_4$ (direction 2 inFigure 9) [58,59]. The deformation of the methyl group cavities can be achieved by a cooperative PLD of the organic stacks. In the case of the (1/2, 1/2, 1/2) AO transition this corresponds to a $2k_F$ BOW distortion of the stacks. In $(TMTSF)_2ReO_4$ the $2k_F$ PLD leads to a MI transition [142]. Even if the coupling of the PLD with the 1D electron gas provides the necessary energy gain to open a gap at the Fermi level, the (1/2, 1/2, 1/2) AO transition is not achieved by a standard Peierls instability because the associated pretransitional fluctuations are 3D and not 1D [143]. This means that the anions located in between the stacks trigger by their ordering the $2k_F$ stack deformation while providing the necessary inter-chain coupling to stabilize the AO insulating ground state. This differs of a conventional Peierls transition where the driving force is due to the individual stack instability towards the formation of a $2k_F$ PLD. Indeed the electron-anion coupling displayed in Figure 9 plays in the AO transition the role played by the intra-stack electron-phonon coupling in the Peierls transition.

These few examples, as well as the case of $(TMTTF)_2SCN$ already considered Section 5.2, show that important changes of the electronic properties generally accompany the AO transition. AO transitions lead to substantial modifications of the electronic structure [144,145] and of the electronic ground states [146,147] of the Bechgaard and Fabre salts. These structural-electronic coupled properties will be revisited below in light of the most recent findings.

## 7.2. The (1/2, 1/2, 1/2) AO Transition of Tetrahedral Anions

Among salts undergoing an (1/2, 1/2, 1/2) AO transition, several cases must be distinguished:

- TMTSF salts where the AO transition coincides with a MI transition.
- TMTTF salts where $4k_F$ charge localization takes place above the AO transition.
- Salts with dipolar anions which incorporate additional degrees of freedom.

### 7.2.1. TMTSF Salts

$ReO_4$ and $BF_4$ salts undergo 1st order AO/MI transitions, at $T_{AO}$ = 176 K and 36 K respectively, which open a sizeable energy gap of ~0.2eV [142]. These MI transitions cannot be classified as standard Peierls transitions because these latter are 2nd order transitions and open much smaller energy gaps. For these reasons it has been suggested [148] that these AO transitions should be understood in the framework of a strong-coupling theory where the anion shift towards the Se of the closest TMTSF [58,59,149] is the essential ingredient. In this picture the anion shift, associated to its orientation ordering, localizes a $p_\pi$ hole on the molecule towards which the anion establishes a short contact. In this mechanism the magnitude of the energy gap should reflect the staggered Hartree anion potential experienced by the $p_\pi$ holes. However structural refinements of (TMTSF)2ReO$_4$ [59] and BF$_4$ [149] superstructures do not exhibit an appreciable internal deformation of the TMTSF revealing the formation of a sizeable 2k$_F$ CDW induced by the direct Coulomb coupling with the anions. Structural refinements [58,59] show that the 2k$_F$ BOW stack distortion consists mainly in a displacement wave of an amplitude ~0.07 Å (essentially directed along c) of dimers, composed of the two inequivalent donors below $T_{AO}$, which alternates in stack direction by inversion symmetry in order to set the 2k$_F$ periodicity. However tight binding band calculations [149,150] considering only the 2k$_F$ stack distortion cannot account for the entire value of the energy gap. Thus it should be interesting to perform a DFT calculation of the electronic structure which fully incorporates the shifted and ordered anions.

The AO transition of the ReO$_4$ salt occurs in the temperature range above $T_X$ where the warping of the FS is not relevant, but this is probably not the case for the BF$_4$ salt. Under pressure it is observed that the MI transition of the ReO$_4$ salt disappears [151] when the superstructure periodicity changes from (1/2, 1/2, 1/2), to (0, 1/2, 1/2) through a 1st order phase transition [152]. When, with the setting of the (0, 1/2, 1/2) superstructure, the metallic state is restored pressurized $(TMTSF)_2ReO_4$ becomes a superconductor with a $T_S$ as high as 1.5K [149]–1.7 K [153] in the coexistence region of the two AO superstructures. The (0, 1/2, 1/2) superstructure, occurring at a $T_{AO}$ of 240 K at 1.7 GPa, could be stabilized at ambient temperature around 2.5 GPa. The squeezing of the methyl group cavities under pressure is certainly responsible for the locking of the orientation of the anions in the (0, 1/2, 1/2) superstructure. A refinement of this superstructure would be desirable.

### 7.2.2. TMTTF Salts

Although presenting structural distortions of same nature as in TMTSF salts, the (1/2, 1/2, 1/2) AO ground state of the TMTTF salts exhibit different electronic properties. At atmospheric pressure, $T_{AO}$ occurs in the 4k$_F$ charge localization regime (*i.e.* charge localization in the TMTTF dimers below $T_\rho$). $T_{AO}$ occurs even below the CO transition ($T_{CO} < T_\rho$) in the ReO$_4$ and BF$_4$ salts. $T_{CO}$ and $T_{AO}$ are quite decoupled in these salts, which means, as already discussed in Section 5.2, that the anions, remaining disordered below $T_{CO}$, undergo a new shift towards the S of the TMTTF when they order at $T_{AO}$.

Structural refinement below $T_{AO}$ [103] shows that $(TMTTF)_2ReO_4$ exhibits both:

- a sizeable $2k_F$ CDW with an amplitude of 0.25 electrons estimated from the bond lengths of the two inequivalent TMTTF of the dimer, and which must add to a CO amplitude of 0.17 electrons estimated by NMR between $T_{CO}$ and $T_{AO}$ [154],

- a sizeable $2k_F$ BOW achieving a quite large dimerization of the stack of dimers with substantially different inter-dimer distances of 3.9 Å and 3.75 Å; the intra-dimer distance being of 3.7 Å.

Indeed a significant increase of the activation energy of the conductivity [113,117] and a rapid drop of spin susceptibility [118,130] follow the $2k_F$ AO structural distortion.

The AO transition of $(TMTTF)_2BF_4$ is quantitatively different from the AO transition of the $ReO_4$ salt because there is no appreciable change of the activation energy of the conductivity [66,113] and only a weak drop of spin susceptibility [118,130] at $T_{AO}$. This should be due to the presence of a very weak $2k_F$ PLD of the stack below$T_{AO}$. In this picture the slight increase of conductivity observed at $T_{AO}$ [113] can be explained by an increase of carrier mobility due to the reduction of the electron scattering by the AO. This finding is very surprising because a large gap of charge opens at the AO transition of $(TMTSF)_2BF_4$. A possible explanation could be that the high temperature CO of $(TMTTF)_2BF_4$ kills substantially the $2k_F$ BOW/SP response of the stack at the staggered AO potential (see the end of Section 6.2), leaving the stack nearly unaffected by the AO transition. In this case the AO transition should consist in a simple ordering of the $BF_4$.

Similar features [66,113,118,130] are observed at the AO transition of $(TMTTF)_2ClO_4$ where the structural refinement performed below $T_{AO}$ [155] shows that only very weak anion shift and stack tetramerization accompany the AO process. The AO transition of $(TMTTF)_2ClO_4$ thus corresponds to the simple ordering of well decoupled $ClO_4$, a finding in agreement with a measured entropy of transition of $k_B \ln 2$ per formula unit [66].

### 7.2.3. Salts with Dipolar Tetrahedral Anions

Bechgaard and Fabre salts incorporate also non symmetrical tetrahedral anions such as $PF_2O_2$ and $FSO_3$ which possess an electric dipolar momentum. These anions provide an additional degree of freedom in the structure related to the orientation of their electric dipoles.

$(TMTSF)_2PF_2O_2$ exhibits two competitive instabilities at ambient pressure which stabilize after a 2nd order transition the (1/2, ±1/4, 0) superstructure followed a few degrees below by a 1$^{st}$ order transition towards the more "conventional" (1/2, 1/2, 1/2) superstructure [156]. It is not known if the dipole of the anion is ordered in these superstructures. Note that the (1/2, ±1/4, 0) superstructure does not fulfill all the conditions (5) minimizing the interchain coupling via the anion shift.

$(TMTTF)_2FSO_3$ undergoes a 1st order transition at 58K which stabilizes the (1/2, 1/2, 1/2) AO superstructure. However at the difference of the (F-1) AO superstructure of $(TMTTF)_2ReO_4$, which contains two identical tetramerized stacks per unit cell related by the F centering (and inversion) symmetry, the superstructure of $(TMTTF)_2FSO_3$ contains two different stacks which breaks the F symmetry [155]. The ordered $FSO_3$ anions located between the stacks form ferroelectric chains. This removes also the inversion symmetry which relates the two donor stacks in $(TMTTF)_2ReO_4$ and lowers the space group symmetry to P1.

$(TMTSF)_2FSO_3$ exhibits also at 87.5K and atmospheric pressure a strongly 1st order MI transition which stabilizes the (1/2, 1/2, 1/2) AO superstructure [157], which, because of the ferroelectric order of the $FSO_3$ [158], is probably non centro-symmetric as the TMTTF analogue. Under pressure the MI is removed and above ~0.5GPa $(TMTSF)_2FSO_3$ becomes a superconductor with a $T_S$ as high as 3 K [159,160]. Preliminary X-ray diffraction measurements [158] show that in the pressure range of observation of superconductivity the (1/2, 1/2, 1/2) superstructure is transformed into the (0, 0, 1/2) superstructure, which itself transforms into the (0, 1/2, 0) superstructure at higher pressure. Recent transport measurements provide evidence that the (1/2, 1/2, 1/2) AO transition splits into several phases under pressure [160]. NMR studies [161] suggest that this splitting reveals the succession upon cooling of the ordering of the $FSO_3$ without ordering of the F-sites then of the ordering of the F-sites (*i.e.* ordering of electric dipoles).

## 7.3. The (0, 1/2, 0) AO Transition of $(TMTSF)_2ClO_4$

### 7.3.1. Basic Aspects of the AO Superstructure and of Its Electronic Structure

The AO transition of the $(TMTSF)_2ClO_4$ is significantly different from the AO transition of other salts of tetrahedral anions. For example in $(TMTTF)_2ClO_4$, the $ClO_4$ adopts a staggered order in all directions without important interaction of the $ClO_4$ with the organic stacks (see Section 7.2.2). In $(TMTSF)_2ClO_4$, the $ClO_4$ adopts a staggered order in the b direction and a uniform order in a and c directions [61]. Surprisingly, the ordering of the $ClO_4$ is accompanied by its shift which increases the closest Se-O distance [115] and a de-centering of the $ClO_4$inside its methyl group cavity [141]. This de-centering reinforces H-bonds of the $ClO_4$ with the methyl groups of the closest TMTSF (TMTSF A) and form loose contacts with the methyl groups of the other TMTSF (TMTSF B) of the cavity. The uniform shift of anions along a, thus stabilizes two types A and B of TMTSF stacks per unit cell.

The polarization of the methyl group cavities by the anion shift causes a charge transfer from stack A to stack B via Process 1 shown in Figure 9. TMTSF A towards which the anion moves becomes hole rich while TMTSF B which has looser contacts becomes hole poor [21]. The charge transfer from A to B molecules drives an incommensurate band filling [147]. EHT [141] and more recent DFT [162] band structure calculations estimate this charge transfer at ~0.025 electrons per TMTSF. In a 1D picture, this deviation from the quarter band filling should kill the $4k_F = a^*$ umklapp electron-electron scattering terms which tend to promote the Mott-Hubbard gap.

Another important consequence of the AO transition is that TMTSF A towards which the $ClO_4$ move is deformed (chair-like conformation), probably because of the constraints exercised by the H-bonds, while unconstrained TMTSF B remains planar [141]. These different molecular deformations achieve different HOMO energies for A and B molecules (separated by ~0.1 eV [141,162]). Also the non equivalent stacks A and B become differently dimerized.

The band structure and the FS of the AO superstructure have been calculated using EHT [141] and DFT [162,163]. All these calculations give an open FS composed of four quasi-1D sheets. The sheets originate from the "bonding" and "anti-bonding" combinations of the dimeric electronic structure of A and B stacks. Both the A-B inter-stack transfer integrals and the difference of site energies (including the HOMO energy separation plus the staggered anion potential) contribute to the warping and splitting of the FS. DFT calculation [162,163] shows that the anion gap has no significant effect on the FS shape and that the main contribution at the FS splitting comes from the difference of HOMO energies between the A and B donors. However the warping of the FS differs with the method of calculation [141,163]. In spite of this problem the assumption frequently done in the literature of a high temperature band structure only split by the staggered anion potential [144,145] is not valid to interpret the data obtained below $T_{AO}$.

In addition, as revealed by the most recent DFT calculations [162] and in contradiction with the earlier EHT calculation [141], the "bonding" and "anti-bonding" combinations of the dimeric electronic structure of A and B stacks lead to differently warped quasi-1D sheets that do not nest "perfectly" on each other. This could explain why the SDW ground state is not stabilized in the AO superstructure. This means also that the placement of $(TMTSF)_2$ $ClO_4$ in the generic phase diagram built on octahedral anions [52] is ambiguous because the vanishing of the SDW ground state in AO $ClO_4$ salt is not due to the enhanced warping of the FS, as it occurs under pressure in salts with octahedral anions such as $PF_6$, but to the reconstruction of the band structure. A proof is that in quenched $ClO_4$ salt (see below), where the band structure is not reconstructed by the AO process, the SDW ground state is kept. Another ambiguity with the placement of relaxed $(TMTSF)_2ClO_4$ in the generic phase diagram is that superconductivity occurs for an electronic structure substantially different of the pressurized $(TMTSF)_2PF_6$ one.

To summarize these recent findings, AO opens space for superconductivity in $(TMTSF)_2ClO_4$ by killing the SDW with the reconstruction of the electronic structure. AO also kills the tendency to form a $4k_F$ commensurate Mott-Hubbard gap by achieving an inter-stack electronic charge transfer.

### 7.3.2. Texture of the AO Phase

By its FS reconstruction AO controls the low temperature electronic properties of $TMTSF)_2ClO_4$. In particular the tuning between SDW and superconductivity which results from this electronic structure modification is revealed using the slow kinetics required to orient the anions. The slow kinetics is due to the fact that AO occurs at a temperature $T_{AO}$ = 24 K much lower than the barrier height (~240 K) necessary to overcome to change the orientation of the $ClO_4$. The achievement of the AO phase thus requires a very slow cooling rate. In that case relaxed $(TMTSF)_2ClO_4$ becomes a superconductor with a $T_S \sim$ 1.2 K [1]. For faster cooling rates the AO is incomplete and $(TMTSF)_2ClO_4$ undergoes a SDW order, due to the FS nesting, at a $T_{SDW}$ increasing with the cooling rate and capped to ~ 6.5 K for the most rapidly cooled (quenched) samples [2]. Although it is often stated that a very slow cooling rate achieves a perfect AO and that a very rapid cooling rate leaves the anions disordered, high resolution X-ray measurements [164] show that this assertion is incorrect: there is no true long range AO even in the best relaxed samples and a local AO is always present even in the best quenched samples. Also the cooling rate modifies progressively the fraction of ordered anions and the texture of the AO pattern, as we shall see below.

It is interesting to remark that, when $T_{AO}$ is low (let say below 40 K), the cooling rate affects the achievement of the AO process for anions, such as $ClO_4$ and $NO_3$ [165], weakly coupled to the TMTSF, but has no real effect for the $BF_4$ anion [149] more strongly linked to the donors (see Section 7.2.1). This means that the use of a phenomenological model of a static double well of potential, external to the anion, to describe the kinetics of ordering is oversimplified especially if the ordering process involves a sizeable interaction with the TMTSF stack able to modify substantially the shape of this potential in order to lower the potential height to overcome during the ordering process.

Let us now discuss more quantitatively the texture of the AO domains. First, superconductivity occurs in $ClO_4$ordered domains. Figure 12 shows that, in function of the cooling rate, the fraction of ordered $ClO_4$ [164] is identical to the fraction of the superconducting volume determined by the measurement of Meissner effect [166].

Second, the profile of the (0, 1/2, 0) super-lattice reflections evolves with the cooling rate [164], which means that the topology of the AO domain pattern changes. The superlattice reflections of relaxed samples have a Lorentzian square profile (LS in Figure 13b) and those of quench samples, a Gaussian profile (G in Figure 13a). The texture of relaxed samples (right part of Figure 13b) is made of a collection of domains where the $ClO_4$ is ordered in one of its two opposite orientations (I and II in Figure 13b). Due to the intricate pattern of the domain walls and to the random distribution of domain walls there is along a given direction (straight line in Figure 13b) a large distribution of domain size, which Fourier transform gives a LS function [167]. The texture of quenched samples (right part of Figure 13a) is made of AO domains of about the same size and shape which are embedded in a matrix of disordered $ClO_4$ (in grey in Figure 13a). With a small (*i.e.*, Gaussian) distribution of domain sizes, the Fourier transform of the shape of ordered domains is a G function [168]. For intermediate cooling rates the profile of the (0, 1/2, 0) superlattice reflections evolves from a LS to a G [164]. This means that non ordered $ClO_4$regions begin to develop between type I and II ordered domains.

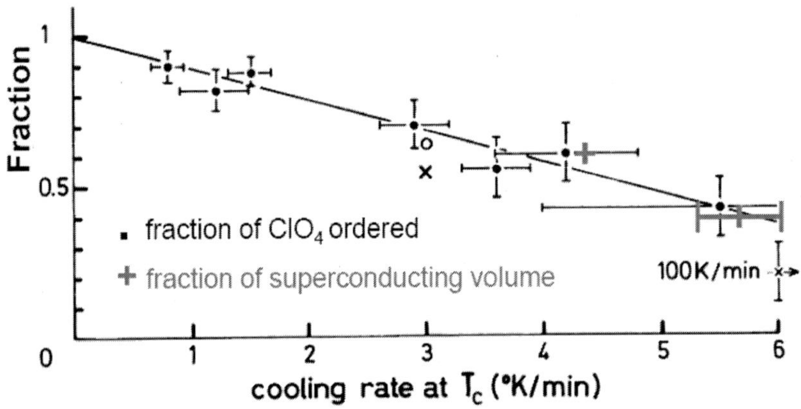


**Figure 12.** Fraction of ordered $ClO_4$ (black symbols defined in [164]) and fraction of superconducting volume (red crosses) in $(TMTSF)_2ClO_4$ in function of the cooling rate. The fraction of superconducting volume has been determined from the data of [166] assuming that the relaxed sample is completely superconducting.

The change of topology, and the presence of additional regions of non ordered $ClO_4$, has to be considered to interpret the cooling rate dependence of the superconductivity critical temperature, $T_S$. For intermediate cooling rates, let say between 0.5 K/min and 5 K/min, the reduction of the fraction of AO superconducting domains, given in Figure 12, is mostly due to the increase of the average distance, $l$, between the AO domains (by a factor 7) while average domain size L decreases more slowly (by a factor 2). For intermediate cooling rates $T_S$, after a rapid drop, saturates. In presence of such a texture superconductivity is achieved through Josephson coupling between AO superconducting domains through the non ordered regions. The quantitative evolution of $T_S$ has been calculated [169] using the dependence of L and $l$ with the cooling rate. It is found that the rate of decrease of $T_S$ saturates in agreement with the experimental finding. Finally, for larger cooling rate, when L becomes smaller than the superconductivity coherence length, $\xi_0$, $T_S$ rapidly drops to zero.

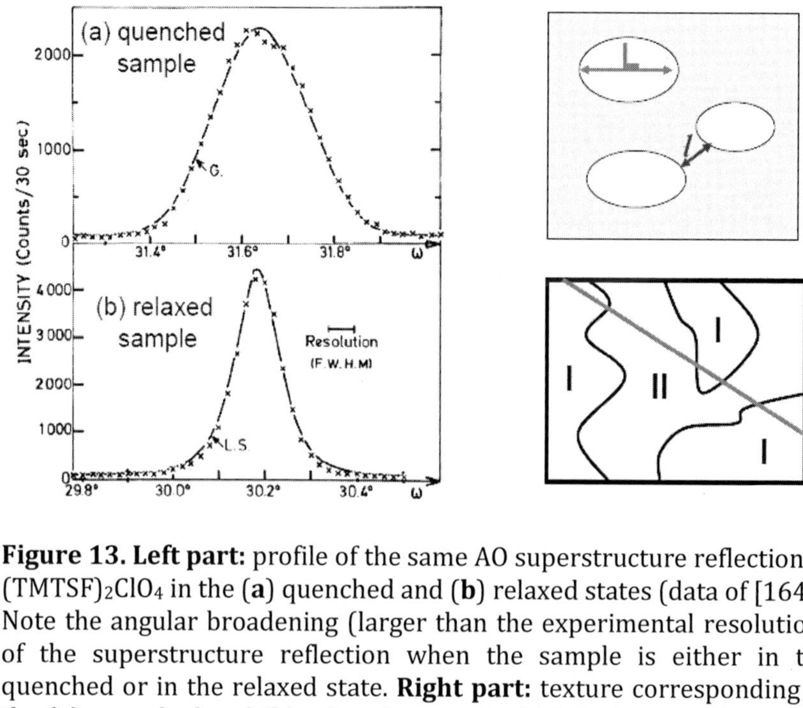


**Figure 13. Left part:** profile of the same AO superstructure reflection of $(TMTSF)_2ClO_4$ in the (**a**) quenched and (**b**) relaxed states (data of [164]). Note the angular broadening (larger than the experimental resolution) of the superstructure reflection when the sample is either in the quenched or in the relaxed state. **Right part:** texture corresponding to the (**a**) quenched and (**b**) relaxed states. In (a) L is the typical size of a domain of ordered $ClO_4$, $l$ is a typical inter-domain distance and the grey background represents non ordered $ClO_4$. In (b) domains of types I and II with opposite $ClO_4$orientations are indicated. Along the line drawn in (b) there is a large distribution of domain sizes.

In the solid solution $(TMTSF)_2(ClO_4)_{1-x}(ReO_4)_x$, the profile of the superlattice reflections remains LS in the domain of existence of the (0, 1/2, 0) local order [170]. However the probability μ to cross a domain wall per unit length [167] increases considerably with $x$. As $\mu^{-1}$ amounts to the average distance between $ReO_4$ substituent for$x$ = 5% and 7% [171], $ReO_4$ appears to be very efficient to break the (0, 1/2, 0) AO spatial coherence. This can be easily understood since $ReO_4$ develops different interactions with TMTSF (interaction 2 in Figure 9) than does $ClO_4$ (interaction 1 in Figure 9) [172]. With this texture it is found, by the measurement [173,174] of a quasi-linear decrease of *T*s with x increasing, that $ReO_4$ acts as non magnetic point defects on the superconductivity of $(TMTSF)_2ClO_4$. This behavior, interpreted as due to the reduction of the Cooper pair lifetime with the increasing number of defects, is taken as an evidence of the presence of nodes in the superconducting gap. However a microscopic theory linking this pair breaking effect at the local perturbations of the electronic structure due to the fragmentation of the (0, 1/2, 0) AO domain pattern with the $ReO_4$ is lacking.

An important aspect of textured $(TMTSF)_2ClO_4$ is that the electronic parameters should vary spatially because the local order around the "ordered" and "non ordered" $ClO_4$ is different. This is illustrated by the observation of a significant change of lattice parameters with the cooling rate. In particular the c parameter increases (decreases) in the quenched (relaxed) state with respect to the dependence extrapolated from the thermal variations above$T_{AO}$ [175,176]. Similarly the $\gamma$ angle increases (decreases) in the relaxed (quenched) state. The decrease of c, direction along which TMTSF and $ClO_4$ layers alternate, has to be related to the contraction of the methyl group cavity when the ordered $ClO_4$ establishes its H-bonds in the relaxed state. The opening of $\gamma$ corresponds to a shear deformation of the TMTSF layer when molecule A adopts the boat shape. In addition, high resolution measurements [164] show that the growth of the AO order parameter upon cooling below $T_{AO}$ is accompanied by erratic angular deformations of the lattice, consisting mostly in sudden rotations of the c direction probably due to the establishment of the H-bond network with newly ordered $ClO_4$. Each angular jump, accompanied by a narrowing of the angular distribution of c directions, realizes an improvement of the crystalline order. Differently, the decrease of the AO order parameter upon heating is accompanied by the increase of the angular distribution of c directions due to the disorder induced by the successive breaks of the H-bonds. This shows that the AO process, via the progressive setting of the H-bond network, is accompanied by considerable elastic constraints.

## 7.4. The AO/CO Transitions of the SCN Salts

The early discovery [94] of the 2nd order AO transition of $(TMTTF)_2$ SCN was a very important step to assess the importance of electron-electron repulsions in organic conductors. Indeed the stabilization of the (0, 1/2, 1/2) superstructure below $T_{AO}$ = 160 K was the first realization of a MI transition [66] not accompanied by the opening of a gap in the spin degrees of freedom. In other words the AO of $(TMTTF)_2$ SCN achieved the spin-charge decoupling expected from the stabilization of a $4k_F$ periodicity. Recently the structural refinement of the AO structure [103] was able to prove from the variation of the intra-molecular distances that the AO transition is accompanied by a $4k_F$ CDW of an amplitude of 0.15 electrons, with an excess of hole on the TMTTF towards which the SCN points and an excess of electron in the opposite situation. This transition is also accompanied by a reduction of the amplitude of the stack dimerization. This result is expected by the repulsive coupling between the $4k_F$ CDW and $4k_F$ BOW order parameters (this is exactly the opposite of the effect described in Section 6.2 and Annex A.3 where the establishment of the SP tetramerization reduced the amplitude of the CO). At low temperature $(TMTTF)_2$ SCN undergoes at $T_N$ = 7 K [94] an AF transition which stabilizes the $q_{AF}$ = (1/2, 1/4, ?) commensurate magnetic order [177].

$(TMTSF)_2SCN$ also undergoes a MI transition at ~90K which does not affect the spin degrees of freedom and which is also associated to a structural modulation [178]. Thus a "$4k_F$" charge localization should accompany this MI transition as in $(TMTTF)_2SCN$. However the structural modulation surprisingly corresponds to the establishment of a triply incommensurate short range order. The wave vector of this modulation (0.48 ± 0.015, 0.65 ± 0.01, 0.1 ± 0.02) is quite different to the one, (0, 1/2, 1/2), stabilized at the CO/AO transition of $(TMTTF)_2SCN$. It is however interesting to remark that the wave vector components of the incommensurate modulation of $(TMTTF)_2SCN$ approximately verify the relationship $\mathbf{qd_1} = 0$, *i.e.*, $q_a - q_b + q_c = 0$ (mod. 1) derived in [60], for the setting of CO via direct interactions (S...NCS) between anions and donors. As the SCN anion can only take two orientations, the incommensurability of the modulation is certainly related to the probability of occupancy of one of the two orientations of the anion. For such a wave the incommensurability observed in diffraction experiments results from a statistical average because locally the modulation must be commensurate. The perfection of such a wave is altered (*i.e.*, the harmonic content of the modulation is reduced) if there is a broad distribution of commensurate domain sizes. In addition, if visualizing the modulation from the origin of the lattice there are increasing fluctuations on the position of successive domain walls, or equivalently cumulative fluctuations of domain size, the long range order is lost. The Fourier transform of such a modulation pattern leads to broadened satellite reflections. Similar features are observed for the incommensurate CO modulation of $(DMtTTF)_2ClO_4$ [102].

## 7.5. The (1/2, 0, 0) AO Transition of the $NO_3$ Salts

Both $(TMTSF)_2NO_3$ and $(TMTTF)_2NO_3$ undergo an AO transition at $T_{AO}$ = 41 K [143] and 50 K [157] respectively and which stabilizes the same (1/2, 0, 0) superstructure. Even if this superstructure establishes the $2k_F$ periodicity in stack direction, the AO transition has a weak influence on the electronic properties of the donor stack.

The (1/2, 0, 0) superstructure of $(TMTSF)_2NO_3$ has been refined [165]. No substantial stack tetramerization and differentiation between the individual TMTSF are observed below $T_{AO}$ (the doubling of stack periodicity should, in principle, differentiate two molecules per unit cell). This means that the $NO_3$ anion orders without significantly perturbing its methyl group cavity. Such a result is expected because of the small size of the $NO_3$ anion. As the organic array is not perturbed by the AO process the electronic structure remains basically unchanged below $T_{AO}$and $(TMTSF)_2NO_3$ does not become semi-metallic as previously predicted [144,145]. Its FS remains quasi-1D below $T_{AO}$ [179] and available for a nesting instability. Indeed $(TMTSF)_2NO_3$ stabilizes a SDW ground state at$T_{SDW}$ = 9 K whose modulation wave vector, determined by NMR [180], is close to the wave vector of the SDW phase of $(TMTSF)_2PF_6$ [54,55].

$(TMTTF)_2NO_3$ undergoes a (1/2, 0, 0) AO transition below which the TMTTF stack is probably weakly tetramerized because a weak spin gap, less than $k_BT_{AO}$ [66,118], opens, resulting in the ground state of $(TMTTF)_2NO_3$ being non-magnetic at the difference of the magnetic SDW ground state of $(TMTSF)_2NO_3$.

# 8. CONCLUSIONS

We have reviewed the very rich panel of structural instabilities exhibited by the Bechgaard and Fabre salts. We have presented a coherent synthesis of the latest experimental developments and interpreted more quantitatively older results in a unified picture. This allows us to correlate the structural instabilities with the electronic, magnetic and superconducting properties of these salts.

The main conclusion is that structural degrees of freedom have a decisive influence on the physical properties the Bechgaard and Fabre salts; an influence which is underestimated in the present literature.

Structural instabilities are not considered in the interpretation of the generic electronic phase diagram [52] giving the evolution of the crossovers and of the ground states exhibited by the octahedral anion salts in function of a generalized coordinate, which can be the size of the anion or the pressure although there is not a direct correspondence between these variables [17]. The current interpretation [51] describes the evolution of 1D instabilities and ground states by taking into account the variation of electronic interactions and couplings with pressure, nature of the donor and size of the anion. However we have shown in this review that the orientation disorder of the anions and of the methyl groups should influence the quasi-1D electronic structure in a manner which is not considered until now. The generic phase diagram accounts for the Mott-Hubbard charge localization, the crossover from a 1D Luttinger liquid to 2D and 3D Fermi liquids, the FS nesting instability leading to the density wave ground state and, with the deterioration of the nesting, the restoration of the metallic state which achieves superconductivity. However BOW, CO and ferroelectric intra-chain instabilities involving intra-stack electronic charge redistributions and inter-chain Coulomb couplings are not included in this description. In particular structural degrees of freedom which take into account the reaction of the lattice at electronic charge redistributions should directly control the development of charge instabilities and set the inter-chain coupling via a generalized electron-phonon coupling including the anion shift. Lattice degrees of freedom play also a role in SP, AF and SDW transitions via "magneto-elastic" couplings which are quite important in the Fabre and Bechgaard salts [25,128,181]).

In this review we have also shown that anions set direct and/or indirect inter-chain coupling mechanisms allowing stabilizing CO, ferroelectric and SP ground states. Also the blockade of the anions in the methyl group cavities can be taken as responsible for the suppression of the CO and SP phase transitions under pressure and of the $2k_F$ BOW instability of $(TMTSF)_2PF_6$ at low temperature. In addition, the adiabaticity of the phonon field controlling the lattice dynamics in reaction at the charge redistribution must be considered. We provide here arguments showing that non-adiabaticity is relevant to explain the SP/AF phase diagram of Fabre salts.

The generic phase diagram does not include salts with non centro-symmetrical anions such as those with $ReO_4$or $FSO_3$ which become superconductors at larger $T_S$ than $T_S$ of salts with octahedral anions. In these salts, AO transitions, involving important charge redistributions, activate CDW and BOW instabilities of the organic stack. These charge displacements are accompanied by substantial deformations of the lattice structure. The modification of the electronic structure which thus results drives noticeable change of electronic properties. In this manner non centro-symmetrical anions can tune novel ground state competitions. Also little is known regarding the general pressure evolution of the AO ground states which, for salts with $ReO_4$ and $FSO_3$ anions stabilize new metallic super-structures which exhibit superconductivity, some of them being probably non-centrosymmetric when a ferroelectric dipolar anion order is achieved.

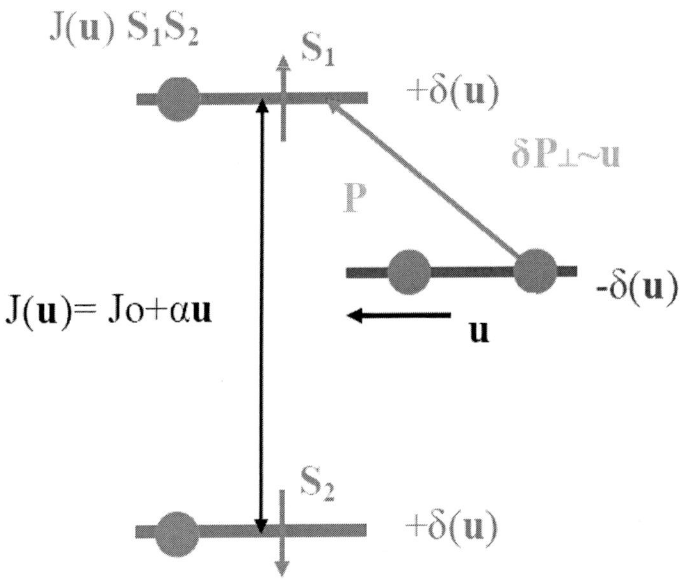


**Figure 14.** CO (of amplitude ±δ) and AF exchange coupling $JS_1S_2$ function of a generalized distortion parameter **u** of the organic stack. The drawing represents the extreme δ = 0.5 situation where the HOMO of the donor is alternatively occupied by one electron (ionic donor $D^+$) and two electrons (neutral donor $D^0$). In the drawing the pink dots represent the electrons. The holes, corresponding to the absence of one electron on $D^+$, bear the spins $\mathbf{S_1}$and $\mathbf{S_2}$. The exchange interaction J(**u**) is tuned by the position, **u**, of the neutral molecule which controls the intra-stack overlap of the HOMO's. The figure shows also that the molecular shift **u** changes the orientation of the electric dipole **P** of the $D^0D^+$ dimer.

A beautiful illustration of the interplay between SDW and superconductivity is shown by $(TMTSF)_2ClO_4$. In this salt we have shown that this competition is controlled by the texture of the AO pattern. This is a simple case where superconductivity and SDW can be simply associated to structurally different regions. Texture effects are also important in the competition between SDW and superconductivity in pressurized $(TMTSF)_2PF_6$ [182]. In a general way it is interesting to remark that, similarly to pressurized $(TMTSF)_2PF_6$, the highest $T_S$ in pressurized $(TMTSF)_2ReO_4$ [151,153] and $FSO_3$ [159,160] is achieved in the de-mixing region between different AO phases.

Finally we conclude on the possible interplay between ferroelectricity and (antiferro-) magnetism in the Fabre salts which could give rise to a multiferroic behavior [183] where the action on the dielectric polarization reacts on the magnetic polarization and *vice-versa*. In particular, CO and the associated lattice deformation could control via a sizeable spin-phonon coupling the magnitude of the exchange interaction and the direction of the dielectric polarization as schematically illustrated by Figure 14. In addition, dipolar anions located between donor stacks could add a new degree of freedom at the ferroelectricity. Multiferroicity is now well documented in inorganic materials (see for example [184,185]). Its manifestation in spin-charge decoupled organic materials such as the ones discussed in this review deserve special attention in the future.

## ACKNOWLEDGMENTS

The work reported here is due to wide collaborations which have been quoted in the references. For the most recent achievements outlined in this review, useful discussions with E. Canadell, S. Charfi-Kaddour, C. Coulon, M. de Souza, M. Dressel, G. Giovannetti, S. Haddad, P. Foury-Leylekian, M. Lang, M. Poirier, E. Rose and H. Sawa are acknowledged. The author is also very grateful to C. Bourbonnais for a general discussion on the content of the paper.

# REFERENCES AND NOTES


1. Jérome, D.; Schulz, H.J. Organic conductors and superconductors. *Adv. Phys.* **1982**, *31*, 299–490.

2. Ishiguro, T.; Yamaji, K.; Saito, G. *Organic Super-Conduc*

3. Toyota, N.; Lang, M.; Műller, J. *Low-Dimensional Molecular Metals*; Springer-Verlag: Berlin/Heidelberg, Germany, 2007; Volume 154.

4. Henriques, R.T.; Alcácer, L.; Pouget, J.P.; Jérome, D. Electrical conductivity and x-ray diffuse scattering study of organic conductors (perylene)$_2$M(mnt)$_2$, (M=Pt, Pd, Au). *J. Phys. C: Solid State Phys.* **1984**, *17*, 5197–5208.

5. Rovira, C.; Veciana, J.; Ribera, E.; Tarrès, J.; Canadell, E.; Rousseau, R.; Mas, M.; Molins, E.; Almeida, M.; Henriques, R.T.; *et al.* An organic spin-ladder molecular material. *Angew. Chem. Int. Ed.* **1997**, *36*, 2324–2326.

6. Canadell, E.; Rachidi, I. E.-I.; Ravy, S.; Pouget, J.-P.; Brossard, L.; Legros, J.P. On the band electronic structure of X[M(dmit)$_2$]$_2$ (X = TTF, $(CH_3)_4N$; M = Ni, Pd) molecular conductors and superconductors. *J. Phys. France* **1989**, *50*, 2967–2981.

7. Pouget, J.-P.; Ravy, S. Structural aspects of the bechgaard salts and related compounds. *J. Phys. I France* **1996**, *6*, 1501–1525.

8. Pouget, J.-P. Structural features of the Bechgaard salts and related compounds. *J. Phys. IV France* **2000**, *10*, Pr3-43–Pr3-56.

9. Thorup, N.; Rindorf, G.; Soling, H.; Bechgaard, K. The structure of Di(2,3,6,7-tetramethyl-1,4,5,8-tetraselenafulvalenium) Hexafluorophosphate, (TMTSF)$_2$PF$_6$, the First Superconducting Organic Solid. *Acta Cryst. B* **1981**, *37*, 1236–1240.

10. Pashkin, A.; Dressel, M.; Ebbinghaus, S.G.; Hanfland, M.; Kuntscher, C.A. Pressure-induced structural phase transition in the Bechgaard-Fabre salts. *Synth. Met.* **2009**, *159*, 2097–2100.

11. Gallois, B.; Gaultier, J.; Hauw, C.; Lamcharfi, T.; Filhol, A. Neutron Low-Temperature (4 and 20K) and X-ray High-Pressure ($6.5 \times 10^2$ and $9.8 \times 10^2$ MPa) Structures of the Organic Superconductor Di(2,3,6,7-tetramethyl-1,4,5,8-tetraselenafulvalenium)

Hexafluorophosphate, $(TMTSF)_2PF_6$. *Acta Cryst. B* **1986**, *42*, 564–575.

12. Emery, V.J.; Bruinsma, R.; Barisic, S. Electron-electron umklapp scattering in organic superconductors.*Phys. Rev. Lett.* **1982**, *48*, 1039–1043.

13. Ishibashi, S.; Manuel, A.A.; Kohyama, M. *Ab initio* pseudopotential calculation for $(TMTSF)_2ClO_4$. *J. Phys. Cond. Matt.* **1999**, *11*, 2279–2283.

14. Liautard, B.; Peytavin, S.; Brun, G.; Maurin, M. Corrélations structurales dans la série $(TMTTF)_2X$. *J. Phys. (Paris)* **1982**, *43*, 1453–1459.

15. Liautard, B.; Peytavin, S.; Brun, G.; Maurin, M. Structural studies and physical properties in the organic conductors series $(TMTTF)_2X$ and $(TMTSF)_2X$. *J. Phys. (Paris)* **1983**, *44*, C3-951–C3-956.

16. Kistenmacher, T.J. Cavity size versus anion size in $(TMTSF)_2X$ salts: Possible implications for the uniqueness of $(TMTSF)_2ClO_4$. *Solid State Commun.* **1984**, *50*, 729–733.

17. This assertion is incorrect for TMTSF and TMTSF salts because the anion modifies mainly the c parameter [14,15] while pressure leads to a relative decrease of the parameter a twice larger than for parameters b and c [10,11].

18. Kistenmacher, T.J. Anion-donor coupling in $(TMTSF)_2X$ salts: Symmetry considerations. *Solid State Commun.* **1984**, *51*, 931–934.

19. Beno, M.A.; Blackman, G.S.; Leung, P.C.W.; Williams, J.M. Hydrogen bond formation and anion ordering in superconducting $(TMTSF)_2ClO_4$ and $(TMTSF)_2AsF_6$. *Solid State Commun.* **1983**, *48*, 99–103.

20. Granier, T.; Gallois, B.; Fritsch, A.; Ducasse, L.; Coulon, C. 135K Crystallographic and Electronic Structure of $(TMTTF)_2SbF_6$. In *Lower-Dimensional Systems and Molecular Electronics*; Metzger, R.M., Day, P., Papavassiliou, G.C., Eds.; Plenum Press: New York, NY, USA, 1990; Volume 248, pp. 163–168.

21. Pouget, J.-P.; Foury-Leylekian, P.; Alemany, P.; Canadell, E. Charge ordering in low dimensional organic conductors: structural aspects. *Phys. Status Solidi B* **2012**, *249*, 937–942.

22. Jacobsen, C.S.; Tanner, D.B.; Bechgaard, K. Optical and infrared properties of tetramethyltetraselenafulvalene [$(TMTSF)_2X$] and tetramethyltetrathiafulvalene [$(TMTTF)_2X$] compounds.*Phys. Rev. B* **1983**, *28*, 7019–7032.

23. de Souza, M.; Hofmann, D.; Foury-Leylekian, P.; Moradpour, A.; Pouget, J.-P.; Lang, M. Exploring the charge-ordering transition in $(TMTTF)_2X$ via thermal expansion measurements. *Phys. B* **2010**, *405*, S92–S94.

24. de Souza, M.; Foury-Leylekian, P.; Moradpour, A.; Pouget, J.-P.; Lang, M. Evidence for lattice effects at the charge-ordering transition in $(TMTTF)_2X$. *Phys. Rev. Lett.* **2008**, *101*, 216403.

25. Lang, M.; Műller, J.; Steglich, F.; Brűhl, A.; Wolf, B.; Dressel, M. Anomalous thermal expansion behaviour in $(TMTTF)_2PF_6$ and $(TMTSF)_2PF_6$ around 100K: evidence for a strong spin-phonon interaction? *J. Phys. IV France* **2004**, *114*, 111–112.

26. Bechgaard, K.; Jacobsen, C.S.; Mortensen, K.; Pedersen, H.J.; Thorup, N. The properties of five highly conducting salts: $(TMTSF)_2X$, $X=PF_6^-$, $AsF_6^-$, $SbF_6^-$, $BF_4^-$ and $NO_3^-$ derived from tetra-methyltetraselenafulvalene (TMTSF). *Solid State Commun.* **1980**, *33*, 1119–1125.

27. Yu, W.; Zhang, F.; Zamborszky, F.; Alavi, B.; Baur, A.; Merlic, C.A.; Brown, S.E. Electron-lattice coupling and broken symmetries of the molecular salt $(TMTTF)_2SbF_6$. *Phys. Rev. B* **2004**, *70*, 121101 (R).

28. Furukawa, K.; Hara, T.; Nakamura, T. Deuteration effect and possible origin of the charge-ordering transition of $(TMTTF)_2X$. *J. Phys. Soc. Jpn.* **2005**, *74*, 3288–3294.

29. Williams, J.M.; Beno, M.A.; Sullivan, J.C.; Banovetz, L.M.; Braam, J.M.; Blackman, G.S.; Carlson, C.D.; Greer, D.L.; Loesing, D.M.; Carneiro, K. The design of organic metals based on TMTSF and TMTTF: novel structural implications and predictions. *J. Phys. (Paris)* **1983**, *44*, C3-941–C3-949.

30. Scott, J.C.; Pedersen, H.J.; Bechgaard, K. Proton NMR in the organic conductor tetramethyltetraselenafulvalenium hexafluorophosphate. *Phys. Rev. B* **1981**, *24*, 475–477.

31. McBrierty, V.J.; Douglass, D.C.; Wudl, F.; Aharon-Shalom, E. Nuclear resonance and relaxation in ditetramethyltetraselenafulvalenium salts. *Phys. Rev. B* **1982**, *26*, 4805–4809.

32. Scott, J.C.; Engler, E.M.; Clark, W.G.; Murayama, C.; Bechgaard, K.; Pedersen, H.J. NMR studies of $(TMTSF)_2PF_6$. *Mol. Cryst. Liq. Cryst.* **1982**, *79*, 61–65.

33. Stein, P.C.; Moradpour, A.; Jérome, D. Nuclear relaxation in tetramethyltetraselenafulvalene salts $(TMTSF)_2X$, ($X=PF_6$, $ClO_4$). *J. Phys. Lett. (Paris)* **1985**, *46*, L241–L247.

34. Foury-Leylekian, P.; Petit, S.; André, G.; de Souza, M.; Lang, M.; Ressouche, E.; Moradpour, A.; Pouget, J.-P*. *Low temperature structural ordering in the $(TMTSF)_2$ $PF_6$ and $AsF_6$ Bechgaard salts. In preparation, 2012.**; Laboratoire de Physique des Solides, Université Paris-sud: 91405 Orsay, France.

35. Aoyagi, B.; Kato, K.; Ota, A.; Yamochi, A.; Saito, G.; Suematsu, H.; Sakata, M.; Takata, M. Direct Observation of Bonding and Charge Ordering in $(EDO\text{-}TTF)_2PF_6$. *Angew. Chem. Int. Ed.* **2004**, *43*, 3670–3676.

36. Pouget, J.-P. Structural Instabilities. In *Highly Conducting Quasi One Dimensional Organic Crystals*; Conwell, E.M., Ed.; Academic Press: New York, NY, USA, 1988; Semiconductors and Semimetals; Volume 27, pp. 87–215.

37. Shirane, G.; Shapiro, S.M.; Comès, R.; Garito, A.F.; Heeger, A.J. Phonon dispersion and Kohn anomaly in tetrathiafulvalene-tetracyanoquinodimethane (TTF-TCNQ). *Phys. Rev. B* **1976**, *14*, 2325–2334.

38. Pouget, J.P. Structural Instabilities of One-Dimensional Conductors. In *Low-Dimensional Conductors and Superconductors*; Jérome, D., Caron, L.G., Eds.; Plenum Press: New York, NY, USA, 1987; Volume 155, pp. 17–45.

39. Ribault, M.; Pouget, J.-P.; Jérome, D.; Bechgaard, K. Superconductivity and Absence of a Kohn Anomaly in the Quasi One

Dimensional Organic Conductor: $(TMTSF)_2AsF_6$. *J. Phys. Lett. (Paris)* **1980**, *41*, L607–L610.

40. Pouget, J.-P.; Moret, R.; Comès, R.; Bechgaard, K.; Fabre, J.M.; Giral, L. X-Ray Diffuse Scattering Study of Some $(TMTSF)_2X$ and $(TMTTF)_2X$ Salts. *Mol. Cryst. Liq. Cryst.* **1982**, *79*, 129–143.

41. Slater, J.C. Magnetic Effect and the Hartree-Fock Equation. *Phys. Rev.* **1951**, *82*, 538–541.

42. Pouget, J.-P. X-ray diffuse scattering as precursor of incommensurate Peierls transitions in one-dimensional organic charge transfer conductors. *Z. Kristallogr.* **2004**, *219*, 711–718.

43. Cao, N.; Timusk, T.; Bechgaard, K. Unconventional Electrodynamic Response of the Quasi-One-Dimensional Organic Conductor $(TMTSF)_2ClO_4$. *J. Phys. I France* **1996**, *6*, 1719–1726.

44. Dressel, M.; Schwartz, A.; Grüner, G.; Degiorgi, L. Deviations from Drude Response in Low-Dimensional Metals: Electrodynamics of the Metallic State of $(TMTSF)_2PF_6$. *Phys. Rev. Lett.* **1996**, *77*, 398–401.

45. Schwartz, A.; Dressel, M.; Grüner, G.; Vescoli, V.; Degiorgi, L.; Giamarchi, T. On-chain electrodynamics of metallic $(TMTSF)_2X$ salts: observation of Tomonaga-Luttinger liquid response. *Phys. Rev. B* **1998**, *58*, 1261–1271.

46. Wzietek, P.; Creuzet, F.; Bourbonnais, C.; Jérome, D.; Bechgaard, K.; Batail, P. Nuclear-relaxation and electronic correlations in quasi-one-dimensional organic conductors. 2. Experiments. *J. Phys. I France* **1993**,*3*, 171–201.

47. Chaikin, P.M.; Tiedje, T.; Bloch, A.N. Sound velocity measurements in $(TMTSF)_2PF_6$. *Solid State Commun.* **1982**, *41*, 739–742.

48. Shi, X.D.; Chiang, L.; Upasani, R.; Chaikin, P.M. Sound velocity studies of Bechgaard salts $(TMTSF)_2PF_6$and$(TMTSF)_2ClO_4$. In *Advanced Organic Solid State Materials*; Chiang, L.Y., Chaikin, P.M., Cowan, D.O., Eds.; Materials Research Society: Pittsburgh, PA, USA, 1990; pp. 239–244.

49. Fertey, P.; Poirier, M.; Batail, P. Microwave transport approach to the coherence of interchain hopping in $(TMTSF)_2PF_6$. *Eur. Phys. J. B* **1999**, *10*, 305–309.

50. Fertey, P.; Poirier, M.; Batail, P. H-T behaviors of the transverse microwave conductivity of $(TMTSF)_2$ X crystals. *Synth. Met.* **1999**, *103*, 2076–2077.

51. Bourbonnais, C. Organic Superconductors: Reduced Dimensionality and Correlation Effects. *Synth. Met.***1997**, *84*, 19–24.

52. Bourbonnais, C.; Jérome, D. Electronic confinement in organic metals. *Science* **1998**, *281*, 1155–1156.

53. Giamarchi, T. Theoretical Framework for Quasi-One Dimensional Systems. *Chem. Rev.* **2004**, *104*, 5037–5055.

54. Delrieu, J.M.; Roger, M.; Toffano, Z.; Moradpour, A.; Bechgaard, K. NMR proton lineshape in $(TMTSF)_2X$- incommensurability of nesting wave vector and order parameter. *J. Phys. (Paris)* **1986**, *47*, 839–861.

55. Takahashi, T.; Maniva, Y.; Kawamura, H.; Saito, G. Determination of SDW characteristics in $(TMTSF)_2$ $PF_6$by $^1H$ NMR analysis. *J. Phys. Soc. Jpn.* **1986**, *55*, 1364–1373.

56. Ducasse, L.; Abderrabba, M.; Gallois, B. Temperature dependence of the Fermi surface topography in the $(TMTSF)_2X$ and $(TMTTF)_2X$ families. *J. Phys. C: Solid State Phys.* **1985**, *18*, L947–L951.

57. Vaca, P.; Coulon, C. Magnetic phase-transition in Bechgaard salts and related-compounds - electronic localization and competing interchain couplings. *Phase Transitions* **1991**, *30*, 49–62.

58. Moret, R.; Pouget, J.P.; Comès, R.; Bechgaard, K. X-ray Scattering evidence for anion ordering and structural distortions in the low-temperature phase of Di (Tetramethyltetraselenafulvalenium) perrhenate [$(TMTSF)_2ReO_4$]. *Phys. Rev. Lett.* **1982**, *49*, 1008–1012.

59. Rindorf, G.; Soling, H.; Thorup, N. Di(4,4',5,5'-tetramethyl-2,2'-bi-&,3-diselenolyliden)ium Perrhenate, $C_{20}H_{24}Se^{+}_{8}.ReO_4$, $(TMTSF)_2ReO_4$. Detailed superstructure at 120 K. *Acta Cryst. C* **1984**, *40*, 1137–1139.

60. By the same argument one obtains a 3D $4k_F$ CDW or CO if the alternation of charge rich and charge poor molecules along a given stack is fixed by uniform shifts of anions separated by **a** and $\mathbf{d_i} = \mathbf{a} \pm \mathbf{b} - \mathbf{c}$ (± corresponding to the two coupling process considered in the text and in Figure 9). $\mathbf{qa} = \mathbf{qd_i} = 0$ (mod. 2π) leads to $q_a=0$ and $q_a \pm q_b + qc = 0$ (mod. 1). These relationships leads to $\pm q_b + q_c = 0$ (mod.1) which has two solutions: $q_b=qc=0$ and $q_b = qc = 1/2$. They corresponds respectively to the ferroelectric $q_F = (0, 0, 0)$ and antiferroelectric $q_{AF} = (0, 1/2, 1/2)$ CO patterns.

61. Pouget, J.-P.; Shirane, G.; Bechgaard, K.; Fabre, J.M. X-ray evidence of a structural phase transition in di-tetramethyl-tetraselenafulvalenium perchlorate [$(TMTSF)_2$ $ClO_4$], pristine and slightly doped. *Phys. Rev. B***1983**, *27*, 5203–5206.

62. Pouget, J.-P.; Moret, R.; Comès, R.; Shirane, G.; Bechgaard, K.; Fabre, J.M. X-ray evidence of competing orderings in $(TMTSF)_2ClO_4$ and related alloys. *J. Phys. (Paris)* **1983**, *44*, C3-969–C3-975.

63. Liu, Q.; Ravy, S.; Pouget, J.P.; Johannsen, I. X-ray investigation of the tetramethyldithiadiselenafulvalene $(TMDTDSF)_2X$ series of organic conductors. II. Influence of the orientational disorder on the structural instabilities. *J. Phys. I France* **1993**, *3*, 821–837.

64. Liu, Q.; Ravy, S.; Pouget, J.P.; Johannsen, I.; Bechgaard, K. X-ray investigation of the tetramethyldithiadiselenafulvalene $(TMDTDSF)_2X$ series of organic conductors. I. Study of the orientational disorder of the TMDTDST molecule. *J. Phys. I France* **1993**, *3*, 803–819.

65. Pouget, J.-P.; Ravy, S. X-Ray evidence of charge density wave modulations in the magnetic phases of $(TMTSF)_2PF_6$ and $(TMTTF)_2Br$. *Synth. Met.* **1997**, *85*, 1523–1528.

66. Coulon, C.; Delhaes, P.; Flandrois, S.; Lagnier, R.; Bonjour, E.; Fabre, J.M. A new survey of the physical properties of the $(TMTTF)_2X$ series. Role of the counterion ordering. *J. Phys. (Paris)* **1982**, *43*, 1059–1067.

67. Coulon, C.; Delhaes, P.; Amiell, J.; Manceau, J.P.; Fabre, J.M.; Giral, L. Effet of doping $(TMTSF)_2ClO_4$ with TMTTF I. Ambient pressure result: A competition between the different possible ground states. *J. Phys. (Paris)* **1982**, *43*, 1721–1729.

68. Foury-Leylekian, P.; Le Bolloc'h, D.; Hennion, B.; Ravy, S.; Moradpour, A.; Pouget, J.-P. Neutron-scattering evidence for a spin-Peierls ground state in $(TMTTF)_2 PF_6$. *Phys. Rev. B* **2004**, *70*, 180405 (R).

69. Dumoulin, L.; Bourbonnais, C. Theory of lattice and electronic fluctuations in weakly localized spin-peierls systems. *J. Phys. I France* **1996**, *6*, 1727–1744.

70. Liu, Q.; Ravy, S.; Pouget, J.-P.; Coulon, C.; Bourbonnais, C. Structural fluctuations and spin-Peierls transitions revisited. *Synth. Met.* **1993**, *56*, 1840–1845.

71. Dumm, M.; Loidl, A.; Alavi, B.; Starkey, K.P.; Montgomery, L.K.; Dressel, M. Comprehensive ESR studies of the antiferromagnetic ground states in the one dimensional spin systems $(TMTSF)_2PF_6$, $(TMTSF)_2AsF_6$and $(TMTTF)_2Br$. *Phys. Rev. B* **2000**, *62*, 6512–6520, and references there in.

72. Pouget, J.-P. Microscopic interactions in $CuGeO_3$ and organic Spin-Peierls systems deduced from their pretransitional fluctuations. *Eur. Phys. J. B* **2001**, *20*, 321–333, (Erratum Eur. Phys. J. B 2001, 24, 415).

73. Pouget, J.-P. Bond and charge ordering in low-dimensional organic conductors. *Phys. B* **2012**, *407*, 1762–1770.

74. Creuzet, F.; Bourbonnais, C.; Caron, L.G.; Jérome, D.; Bechgaard, K. A $^{13}C$ NMR study of the interplay between the spin-Peierls and antiferromagnetic ground states in $(TMTTF)_2PF_6$ under pressure. *Synth. Met.***1987**, *19*, 289–294.

75. Nakamura, T.; Nobutoki, T.; Kobayashi, Y.; Takahashi, T.; Saito, G. $^1H$-NMR investigation of the SDW wave-number in $(TMTTF)_2$ Br. *Synth. Met.* **1995**, *70*, 1293–1294.

76. Seo, H.; Merino, J.; Yoshioka, H.; Ogata, M. Theoretical aspects of charge ordering in molecular conductors.*J. Phys. Soc. Jpn.* **2006**, *75*, 051009.

77. Zorina, L.; Simonov, S.; Mézière, C.; Canadell, E.; Suh, S.; Brown, S.E.; Foury-Leylekian, P.; Fertey, P.; Pouget, J.P.; Batail, P. Charge ordering, symmetry and electronic structures issues and Wigner crystal structure of the quarter-filled band Mott insulators and high

pressure metalsδ-(EDT-TTF-$CONMe_2$)$_2$X, X=Br and $AsF_6$. *J. Mat. Chem.* **2009**, *19*, 6980–6994.

78. Tomic, S.; Biskup, N.; Babic, S.D.; Maki, K. Commensurate spin-density–wave state in $(TMTTF)_2$Br- single particle and collective charge dynamics. *Europhys. Lett.* **1994**, *26*, 295–301.

79. Ashikawa, A.; Matsunaga, N.; Nomura, K.; Nakamura, T.; Takahashi, T.; Saito, G. Pressure and Magnetic Field Dependence of SDW transition in $(TMTTF)_2$Br. *Phys. Stat. Sol. B* **2001**, *223*, 539–543.

80. Fujiyama, S.; Nakamura, T. NMR studies of the localized states of $(TMTTF)_2$Br. *J. Phys. Chem. Solids* **2002**,*63*, 1259–1261.

81. Javadi, H.H.S.; Laversanne, R.; Epstein, A.J. Microwave conductivity and dielectric constant of tetramethyltetrathiafulvalene salts [$(TMTTF)_2$X, X=SCN, $ReO_4$, $SbF_6$. *Phys. Rev. B* **1988**, *37*, 4280–4283.

82. Nad, F.; Monceau, P.; Fabre, J.M. Low frequency dielectric permittivity of quasi-one-dimensional conductor $(TMTTF)_2$ Br. *Eur. Phys. J. B* **1998**, *3*, 301–306.

83. Clark, W.G.; Hanson, M.E.; Wong, W.H.; Alavi, B. Evidence that the SDW transition in $(TMTSF)_2$ $PF_6$ is first order. *Phys. B* **1994**, *194–196*, 285–286.

84. Kagoshima, S.; Saso, Y.; Maesato, M.; Kondo, R.; Hasegawa, T. Low temperature diffuse X-ray studies of charge-density waves coexisting with spin-density waves in the organic conductors $(TMTSF)_2PF_6$ and $(TMTSF)_2AsF_6$. *Solid State Commun.* **1999**, *110*, 479–483.

85. Gufan; Yu, M.; Larin, E.S. Theory of phase transitions described by two order parameters. *Sov. Phys. Solid State* **1980**, *22*, 270–275.

86. Takahashi, T.; Maniwa, Y.; Kawamura, H.; Saito, G. Determination of SDW characteristics in $(TMTSF)_2$ $PF_6$by $^1$H-NMR analysis. *Phys. B* **1986**, *143*, 417–421.

87. Takahashi, T.; Maniwa, Y.; Kawamura, H.; Murata, K.; Saito, G. NMR analysis of electronic properties in organic superconductors $(TMTSF)_2PF_6$ and β-$(BEDT\text{-}TTF)_2I_3$. *Synth. Met.* **1987**, *19*, 225–230.

88. Nickel, J.C.; Duprat, R.; Bourbonnais, C.; Dupuis, N. Triplet Superconductivity Pairing and Density Wave Instabilities in Organic Conductors. *Phys. Rev. Lett.* **2005**, *95*, 247001.

89. Nickel, J.C.; Duprat, R.; Bourbonnais, C.; Dupuis, N. Superconductivity pairing and density wave instabilities in quasi-one-dimensional conductors. *Phys. Rev. B* **2006**, *73*, 165126.

90. Kuroki, K. Pairing symmetry considerations in organic superconductors. *J. Phys. Soc. Jpn.* **2006**, *75*, 051013.

91. Ng, H.K.; Timusk, T.; Bechgaard, K. Far infra-red studies of bis (tetramethyltetraselenafulvalene) hexafluoroantimonate [$(TMTSF)_2$ $SbF_6$]: Coexistence of metallic and semiconducting states. *Phys. Rev. B***1984**, *30*, 5842–5846.

92. Grüner, G. The dynamics of spin-density waves. *Rev. Mod. Phys.* **1994**, *66*, 1–24.

93. Laversanne, R.; Coulon, C.; Gallois, B.; Pouget, J.P.; Moret, R. Structural and electrical properties of $(TMTTF)_2$ $MF_6$ salts (M=P, As, Sb). Role of the anions. *J. Phys. Lett.(Paris)* **1984**, *45*, L393–L399.

94. Coulon, C.; Maaroufi, A.; Amiell, J.; Dupart, E.; Flandrois, S.; Delhaes, P.; Moret, R.; Pouget, J.P.; Morand, J.P. Antiferromagnetic and structural instabilities in tetramethyltetrathiafulvalene thiocyanate [$(TMTTF)_2SCN$]. *Phys. Rev. B* **1982**, *26*, 6322–6325.

95. Coulon, C.; Parkin, S.S.P.; Laversanne, R. Structureless transition and strong localization effects in bis-tetramethyltetrathiafulvalenium salts [$(TMTTF)_2$ X]. *Phys. Rev. B* **1985**, *31*, 3583–3587.

96. Brown, S.E.; Javadi, H.H.S.; Laversanne, R. Sound Velocity Measurements in $(TMTTF)_2$ X Salts, X= $ReO_4$, $AsF_6$, $SbF_6$. In *Advanced Organic Solid State Materials*; Chiang, L.Y., Chaikin, P.M., Cowan, D.O., Eds.; Materials Research Society: Pittsburgh, PA, USA, 1990; pp. 245–250.

97. Chow, D.S.; Zamborsky, F.; Alavi, B.; Tantillo, D.J.; Baur, A.; Merlic, C.A.; Brown, S.E. Charge ordering in the TMTTF family of molecular conductors. *Phys. Rev. Lett.* **2000**, *85*, 1698–1701.

98. Dumm, M.; Salameh, B.; Abaker, M.; Montgomery, L.K.; Dressel, M. Magnetic and optical studies of spin and charge ordering in $(TMTTF)_2 AsF_6$. *J. Phys. IV (France)* **2004**, *114*, 57–60.

99. Dumm, M.; Abaker, M.; Dressel, M.; Montgomery, L.K. Charge order in $(TMTTF)_2PF_6$ investigated by infrared spectroscopy. *J. Low. Temp. Phys.* **2006**, *142*, 609–612.

100. Monceau, P.; Nad, F.; Brazovski, S. Ferroelectric Mott-Hubbard phase of organic $(TMTTF)_2$ X conductors.*Phys. Rev. Lett.* **2001**, *86*, 4080–4083.

101. Foury-Leylekian, P.; Ravy, S.; Pouget, J.-P. Stuctural properties of strongly correlated quasi-1D organic systems. *Phys. B* **2002**, *312–313*, 574–577.

102. Ravy, S.; Foury-Leylekian, P.; Le Bolloc'h, D.; Pouget, J.-P.; Fabre, J.M.; Prado, R.J.; Lagarde, P. Structural instability and electronic localization in the 2/1 salts: The case of the Fabre and the $(DMtTTF)_2ClO_4$ salts. *J. Phys. (Paris) IV* **2004**, *114*, 81–85.

103. Nogami, Y.; Nakamura, T. X-ray observation of $2k_F$ and $4k_F$ charge orderings in $(TMTTF)_2$ $ReO_4$ and $(TMTTF)_2SCN$ associated with anion orderings. *J. Phys. (Paris) IV* **2002**, *12*, Pr9-145–Pr9-148.

104. Coulon, C.; Lalet, G.; Pouget, J.-P.; Foury-Leylekian, P.; Moradpour, A.; Fabre, J.M. Anisotropic conductivity and charge ordering in $(TMTTF)_2X$ salts probed by ESR. *Phys. Rev. B* **2007**, *76*, 085126.

105. Foury-Leylekian, P.; Petit, S.; Andre, G.; Moradpour, A.; Pouget, J.P. Neutron scattering evidence for a lattice displacement at the charge ordering transition of $(TMTTF)_2PF_6$. *Phys. B* **2010**, *405*, S95–S97.

106. Granier, T.; Gallois, B.; Ducasse, L.; Fritsch, A.; Filhol, A. 4K crystallographic and electronic structures of $(TMTTF)_2X$ salts ($X^-$: $PF_6^-$; $AsF_6^-$). *Synth. Met.* **1988**, *24*, 343–356.

107. Sawa, H. Direct Observation of Electronic State of $(TMTTF)_2PF_6$ Using Synchrotron Radiation X-ray Analysis. In Proceedings of International Research School and Workshop on Electronic Crystals; ECRYS, Cargèse, France, 15–27 August 2011.

108. In this respect the analysis performed in [105] and which considers only the shift of the anion sublattice is incomplete. As the neutron scattering length is similar for all the nucleus of $(TMTTF)_2PF_6$, the 3 contributions given in the text have to be treated on equal footing in the structural refinement. Note that in deuterated samples the neutron scattering could be particularly sensitive to the deformation of the methyl group cavity at $T_{CO}$

109. Dressel, M.; Dumm, M.; Knoblauch, T.; Masimo, M. Comprehensive optical investigation of Charge Order in Organic chain compounds $(TMTTF)_2X$. *Crystals* **2012**. in press.

110. Rose, E.; Dressel, M. Coupling between molecular chains and anions in $(TMTTF)_2X$ salts. *Phys. B* **2012**, *407*, 1787–1892.

111. Riera, J.; Poilblanc, D. Influence of the anion potential on the charge ordering in quasi-one-dimensional charge-transfer salts. *Phys. Rev. B* **2001**, *63*, 241102(R).

112. Pouget, J.-P.; Foury-Leylekian, P.; Le Bolloc'h, D.; Hennion, B.; Ravy, S.; Coulon, C.; Cardoso, V.; Moradpour, A. Neutron-Scattering Evidence for a Spin-Peierls ground State in $(TMTTF)_2PF_6$. *J. Low. Temp. Phys.* **2006**, *142*, 147–152.

113. Kőhler, B.; Rose, E.; Dumm, M.; Untereiner, G.; Dressel, M. Comprehensive transport study of anisotropy and ordering phenomena in quasi-one-dimensional $(TMTTF)_2X$ salts (X=$PF_6$, $AsF_6$, $SbF_6$, $BF_4$, $ClO_4$, $ReO_4$.*Phys. Rev. B* **2011**, *84*, 035124.

114. Nad, F.; Monceau, P.; Nakamura, T.; Furukawa, K. The effect of deuteration on the transition into a charge ordered state of $(TMTTF)_2X$ salts. *J. Phys.: Condens. Matter* **2005**, *35*, L399–L406.

115. Moret, R.; Pouget, J.P.; Comès, R.; Bechgaard, K. X-ray study of the anion ordering transition in di(tetramethyltetraselenafulvalen)-ium perchlorate $(TMTSF)_2ClO_4$: quenching and irradiation effects. *J. Phys. (Paris)* **1985**, *46*, 1521–1732.

116. Zamborszky, F.; Yu, W.; Raas, W.; Brown, S.E.; Alavi, B.; Merlic, C.A.; Baur, A. Competition and coexistence of bond and charge orders in $(TMTTF)_2AsF_6$. *Phys. Rev. B* **2002**, *66*, 081103(R).

117. Nad, F.; Monceau, P. Dielectric Response of the Charge Ordered State in Quasi-One Dimensional Organic Conductors. *J. Phys. Soc. Jpn.* **2006**, *75*, 051005.

118. Coulon, C.; Foury-Leylekian, P.; Fabre, J.M.; Pouget, J.-P. Electronic instabilities and irradiation effects in the $(TMTTF)_2X$ series. Centre de Recherche Paul Pascal, Université Bordeaux I, France, 2012; Unpublished work.

119. Brazovskii, S. Ferroelectricity and Charge Ordering in Quasi-1D Organics Conductors. In *The Physics of Organic Superconductors and Conductors*; Lebed, A., Ed.; Springer-Verlag: Berlin, Germany, 2008; pp. 313–355.

120. Kirova, N.; Brazovskii, S. Ferroelectricity: From organic conductors to electronic polymers. *Phys. B* **2009**,*404*, 382–384.

121. Yamamoto, K.; Iwai, S.; Boyko, S.; Kashiwazaki, A.; Hiramatsu, F.; Okabe, C.; Nishi, N.; Yakushi, K. Strong optical nonlinearity and its ultrafast response associated with electron ferroelectricity in an organic conductor. *J. Phys. Soc. Jpn.* **2008**, *77*, 074709.

122. Alemany, P.; Pouget, J.-P.; Canadell, E. Essential role of anions in the charge ordering transition of $\alpha$-$(BEDT\text{-}TTF)_2I_3$. *Phys. Rev. B* **2012**, *85*, 195118.

123. Pouget, J.-P.; Khanna, S.K.; Denoyer, F.; Comès, R.; Garito, A.F.; Heeger, A.J. X-ray Observation of $2k_F$ and $4k_F$ Scatterings in Tetrathiafulvalene-Tetracyanoquinodimethane (TTF-TCNQ). *Phys. Rev. Lett.* **1976**, *37*, 437–440.

124. Nogami, Y.; Moret, R.; Pouget, J.-P.; Yamamoto, Y.; Oshima, K.; Hiraki, K.; Kanoda, K. X-ray structural study of $4k_F$ Superlattice in $(DMDCNQI)_2Ag$. *Synth. Met.* **1999**, *102*, 1778.

125. Foury-Leylekian, P.; Auban-Senzier, P.; Coulon, C.; Jeannin, O.; Fourmigué, M.; Pasquier, C.; Pouget, J.-P. Phase diagram of the correlated quarter-filled-band organic salt series $(o\text{-}DMTTF)_2X$ (X=Cl, Br, I) . *Phys. Rev. B* **1994**, *50*, 7131–7139.

126. Zamborszky, F.; Yu, W.; Raas, W.; Brown, S.E.; Alavi, B.; Merlic, C.A.; Baur, A.; Lefebvre, S.; Wzieteck, P. Influence of charge order on the ground state of TMTTF conductors. *J. Phys. (Paris) IV* **2002**, *12*, Pr9-139–Pr9-144.

127. In contrast $^{75}$As NMR $T_1^{-1}$ does not exhibit a divergence at $T_{CO}$ but only a change of slope. This result could be understood if $AsF_6$ plays a weaker role in the CO transition (through process 1) than in the SP transition (through process 2).

128. de Souza, M.; Brűhl, A.; Műller, J.; Foury-Leylekian, P.; Moradpour, A.; Pouget, J.-P.; Lang, M. Thermodynamic studies at the charge ordering and spin-Peierls transitions in $(TMTTF)_2X$. *Phys. B* **2009**,*404*, 494–498.

129. Dumm, M.; Loidl, A.; Fravel, B.W.; Starkey, K.P.; Montgomery, L.K.; Dressel, M. Electron spin resonance studies on the organic linear chain compounds $(TMTTF)_2X$ (C= S, Se; X= $PF_6$, $AsF_6$, $ClO_4$, Br). *Phys. Rev. B*2001, 61, 511–521.

130. Salameh, B.; Yasin, S.; Dumm, M.; Untereiner, G.; Montgomery, L.; Dressel, M. Spin dynamics of the organic linear chain compounds $(TMTTF)_2X$ (X = $SbF_6$, $AsF_6$, $BF_4$, $ReO_4$, and SCN). *Phys. Rev. B* **2011**, *83*, 205126.

131. Foury-Leylekian, P.; Petit, S.; Coulon, C.; Hennion, B.; Moradpour, A.; Pouget, J.-P. Inelastic neutron scattering investigation of the magnetic excitations in the spin-Peierls ground state of $(TMTTF)_2PF_6$. *Phys. B*2009, *404*, 537–540.

132. Langlois, A.; Poirier, M.; Bourbonnais, C.; Foury-Leylekian, P.; Moradpour, A.; Pouget, J.-P. Microwave dielectric study of spin-Peierls and charge ordering transitions in (TMTTF)2$PF_6$ salts. *Phys. Rev. B* **2010**, *81*, 125101.

133. Brown, S.E.; Clark, W.G.; Zamborszky, F.; Klemme, B.J.; Kriza, G.; Alavi, B.; Merlic, C.; Kuhns, P.; Moulton, W. $^{13}$C NMR measurements of the high-magnetic-field, low-temperature phases of $(TMTTF)_2PF_6$. *Phys. Rev. Lett.* **1998**, *80*, 5429–5432.

134. Brown, S.E.; Clark, W.G.; Alavi, B.; Hall, D.; Naughton, M.J.; Tantillo, D.J.; Merlic, C.A. High–field magnetization of the spin-Peierls compound $(TMTTF)_2PF_6$. *Phys. Rev. B* **1999**, *60*, 6270–6272.

135. Hirsch, J.E.; Scalapino, D.J. $2p_F$ and $4p_F$ instabilities in the one-quarter-filled-band Hubbard model. *Phys. Rev. B* **1983**, *27*, 7169–7185.

136. Hirsch, J.E.; Scalapino, D.J. $2p_F$ and $4p_F$ instabilities in the one dimensional Hubbard model. *Phys. Rev. B***1984**, *29*, 5554–5561.

137. Ilakovac, V.; Ravy, S.; Pouget, J.P.; Lenoir, C.; Boubekeur, K.; Batail, P.; Dolanski Babic, S.; Biskup, N.; Korin-Hamzic, B.; Tomic, S.; Bourbonnais, C. Enhanced charge localization in the organic alloys $[(TMTSF)_{1-x}(TMTTF)_x]_2ReO_4$. *Phys. Rev. B* **2011**, *84*, 195134.

138. The assertion done in [137] that the inversion symmetry is broken on the anion sites at RT above the AO transition is incorrect. Only inversion centers located on the organic stacks between the donors are removed by the chemical CO.

139. Chow, D.S.; Wzietek, P.; Fogliatti, D.; Alavi, B.; Tantilo, D.J.; Merlic, C.A; Brown, S.E. Singular behavior in the pressure-tuned competition between spin-Peierls and antiferromagnetic ground states of $(TMTTF)_2PF_6$.*Phys. Rev. Lett.* **1998**, *81*, 3984–3987.

140. Rose, E.; Loose, C.; Kortus, J.; Pashkin, A.; Kuntscher, C.A.; Ebbinghaus, S.G.; Hanfland, M.; Lissner, F.; Schleid, Th.; Dressel, M. Pressure dependence of Crystal Structure and Electronic Band Structure of $(TMTTF)_2PF_6$. *J. Phys.: Condens. Matter* **2012**. in press.

141. Le Pévelen, D.; Gaultier, J.; Barrens, Y.; Chasseau, D.; Castet, F.; Ducasse, L. Temperature and pressure dependencies of the crystal structure of the organic superconductor $(TMTSF)_2ClO_4$. *Eur. Phys. J. B* **2001**, *19*, 363–373.

142. Jacobsen, C.S.; Pedersen, H.J.; Mortensen, K.; Rindorf, G.; Thorup, N.; Torrance, J.B.; Bechgaard, K. An unusual metal-insulator transition: bis(tetramethyltetraselenafulvalenium)-perrhenate $(TMTSF)_2ReO_4$. *J. Phys. C: Solid State Phys.* **1982**, *15*, 2651–2663.

143. Pouget, J.-P.; Moret, R.; Comès, R.; Bechgaard, K. X-ray diffuse scattering study of superstructure formation in tetramethyltetraselenafulvalenium perrhenate $(TMTSF)_2ReO_4$ and nitrate $(TMTSF)_2NO_3$. *J. Phys. Lett. (Paris)* **1981**, *42*, L543–L546.

144. Grant, P.M. Broken-Symmetry Band Structure of Ditetramethyltetradelenafulvalene-X $[(TMTSF)_2X]$. *Phys. Rev. Lett.* **1983**, *50*, 1005–1008.

145. Grant, P.M. Electronic structure of the 2:1 charge transfer salts of TMTCF. *J. Phys. (Paris)* **1983**, *44*, C3-847–C3-857.

146. Emery, V.J. Some basic questions in organic superconductivity. *J. Phys. (Paris)* **1983**, *44*, C3-977–C3-982.

147. Brazovslii, S.; Yakovenko, V. On the theory of phase transitions in organic superconductors. *J. Phys. Lett.(Paris)* **1985**, *46*, L111–L116.

148. Bruinsma, R.; Emery, V.J. Theory of anion ordering in TMTSF compounds. *J. Phys. (Paris)* **1983**, *44*, C3-1115–C3-1120.

149. Emge, T.J.; Wang, M.H.; Beno, M.A.; Williams, J.M.; Whangbo, M.H.; Evain, M. Effect of anion ordering on the H-anion interaction and band electronic structure of $(TMTSF)_2BF_4$ at 20K. *J. Am. Chem. Soc.* **1986**, *108*, 8215–8223.

150. Ducasse, L.; Abderrabba, M.; Gallois, B.; Chasseau, D. Influence of temperature on the band structures of $(TMTSF)_2X$ and $(TMTTF)_2X$ ($X^-$= tetrahedral anion). *Synth. Met.* **1987**, *19*, 327–332.

151. Tomic, S.; Jérome, D.; Bechgaard, K. Influence of the cooling rate on the ground state of the organic conductor $(TMTSF)_2ReO_4$. *J. Phys. C: Solid State Phys.* **1984**, L11–L16.

152. Moret, R.; Ravy, S.; Pouget, J.P.; Comès, R.; Bechgaard, K. Anion-Ordering Phase Diagram of di(tetramethyltetraselenafulvalenium)-Perrhenate, $(TMTSF)_2ReO_4$. *Phys. Rev. Lett.* **1986**, *57*, 1915–1918.

153. Jo, Y.J.; Kang, H.; Kang, W. Review of the P-T phase diagram of the organic superconductor $(TMTSF)_2ReO_4$. *Synth. Met.* **2001**, *120*, 1043–1044.

154. Nakamura, T.; Furukawa, K.; Hara, T. $^{13}C$ NMR Analyses of Successive Charge Ordering in $(TMTTF)_2ReO_4$. *J. Phys. Soc. Jpn.* **2006**, *75*, 013707.

155. Nogami, Y.; Ito, T.; Yamamoto, K.; Irie, N.; Horita, S.; Kambe, T.; Nagao, N.; Oshima, K.; Ikeda, N.; Nakamura, T. X-ray structural study of charge and anion orderings of TMTTF salts. *J. Phys. (Paris) IV* **2005**,*131*, 39–42.

156. Ravy, S.; Pouget, J.-P.; Moret, R.; Wudl, F. Successive phase transitions in the organic conductor $(TMTSF)_2PF_2O_2$. *J. Phys. I France* **1991**, *1*, 703–720.

157. Moret, R.; Pouget, J.-P.; Comès, R.; Bechgaard, K. Structural phase transitions in the $(TMTSF)_2X$ and $(TMTTF)_2X$ series: A survey with some new results. *J. Phys. (Paris)* **1983**, *44*, C3-957–C3-962.

158. Yamaura, J.; Jo, Y.J.; Kang, H.; Chung, O.H.; Kang, W. X-ray study at low temperature and under pressure in $(TMTSF)_2FSO_3$. **2005**. Unpublished work.

159. Lacoe, R.C.; Wolf, S.A.; Chaikin, P.M.; Wudl, F.; Aharon-Shalom, E. Metal-insulator transitions and superconductivity in ditetramethyltetraselenafulvalenium fluorosulfonate [$(TMTSF)_2$ $FSO_3$]. *Phys. Rev. B***1983**, *27*, 1947–1950.

160. Jo, Y.J.; Choi, E.S.; Kang, H.; Kang, W.; Seo, I.S.; Chung, O.H. Refinement of the pressure-temperature phase diagram of the organic superconductor with asymmetric anion $(TMTSF)_2FSO_3$. *Phys. Rev. B* **2003**, *67*, 014516.

161. Takahashi, T. Charge fluctuation, charge ordering, and zero-gap state in organic conductors. *Phys. B* **2012**,*407*, 1757–1761.

162. Alemany, P.; Canadell, E.; Pouget, J.-P. *Ab-initio* electronic structure of $(TMTSF)_2ClO_4$. Institut de Ciència de Materials de Barcelona: Bellaterra, Spain, 2012; Unpublished work.

163. Nagai, Y.; Nakamura, H.; Machida, M. Superconducting gap function in the organic superconductor $(TMTSF)_2ClO_4$ with anion ordering; First-principles calculations and quasiclassical analysis for angle-resolved heat capacity. *Phys. Rev. B* **2011**, *83*, 104523.

164. Pouget, J.-P.; Kagoshima, S.; Tamegai, T.; Nogami, Y.; Kubo, K.; Nakajima, T.; Bechgaard, K. High resolution x-ray scattering study of the anion ordering phase transition of $(TMTSF)_2ClO_4$. *J. Phys. Soc. Jpn.***1990**, *59*, 2036–2053.

165. Barrens, Y.; Gaultier, J.; Brachetti, S.; Guionneau, P.; Chasseau, D.; Fabre, J.M. Low temperature structural properties of the $(TMTSF)_2NO_3$ salt. *Synth. Met.* **1999**, *103*, 2042–2043.

166. Matsunaga, N.; Ishikawa, A.; Hoshikawa, A.; Nomura, K.; Takasaki, S.; Yamada, J.; Nakatsuji, S.; Anzai, H. Anion disorder and two-dimensionality in the superconducting and SDW states of $(TMTSF)_2ClO_4$. *J. Low Temp. Phys.* **1999**, *117*, 1735–1739.

167. In the case of a random distribution of domain sizes, the probability that two sites separated by x belong to the same domain is given by exp-$\mu$x, where $\mu$ is the constant probability to cross a domain wall per unit length. The 3D Fourier transform of this probability function, assuming $\mu$ isotropic, is a Lorentzian squared function, whose HWHM $\Delta q$ is linearly related to $\mu$ by $\Delta q \approx 0.64\ \mu$. $\mu$ is connected to the linear density of domain walls along the direction of measurement of $\Delta q$. $\mu$ should also depend upon the geometry the domain pattern.

168. The average domain size L in a given direction is inversely proportional to the HWHM $\Delta q$ of the reflection (expressed in $Å^{-1}$), measured in this direction. In a general manner the profile of the reflection is the Fourier transform of the average shape of the ordered domain. L is given by the relationship $L \approx C(d)/\Delta q$, where the constant C(d) depends upon the dimension, d, of independent directions involved in the Fourier transform. For a segment of length L (or along one edge of a parallelepiped): $C(1) \approx 2.79$; for a disc of diameter L: $C(2) \approx 3.2$, and for a sphere of diameter L: $C(3) \approx 3.72$. L depends weakly of the shape of the domain. In [164], L is obtained using C(1).

169. Haddad, S.; Charfi-Kaddour, S.; Pouget, J.-P. Inhomogeneous superconductivity in organic conductors: the role of disorder and magnetic field. *J. Phys.: Condens. Matter* **2011**, *23*, 464205.

170. Ilakovac, V.; Ravy, S.; Boubekeur, K.; Lenoir, C.; Batail, P.; Pouget, J.-P. Substitutional disorder and anion ordering transition in the $(TMTSF)_2(ReO_4)_{1-x}(ClO_4)_x$ solid solution. *Phys. Rev. B* **1997**, *56*, 13878–13887.

171. $\mu^{-1}$, which is quasi-isotropic, amounts to 400 Å in relaxed samples and decreases rapidly to 25Å / 20 Å for *x*= 5% / 7% of $ReO_4$. Interestingly for these last two concentrations, $\mu^{-1}$ amounts to the average distance between $ReO_4$ substituent (24 Å and 21 Å respectively).

172. In this respect one observes at low temperature for *x* = 7% of $ReO_4$ a coexistence of (0, 1/2, 0) AO and (1/2, 1/2, 1/2) AO domains in close contact and whose diffraction pattern is a superimposition of LS shaped superlattice reflections [170]. As there is a random distribution of the two species of anion, (1/2, 1/2, 1/2) AO domains are nucleated in regions of excess of $ReO_4$ (with respect to the

average concentration) while (0, 1/2, 0) AO domains are nucleated in regions deficient in $ReO_4$.

173. Joo, N.; Auban-Senzier, P.; Pasquier, C.R.; Monod, P.; Jérome, D.; Bechgaard, K. Supression of superconductivity by non-magnetic disorder in the organic superconductor $(TMTSF)_2(ClO_4)_{1-x}(ReO_4)_x$.*Eur. Phys. J. B* **2004**, *40*, 43–48.

174. Joo, N.; Auban-Senzier, P.; Pasquier, C.R.; Jérome, D.; Bechgaard, K. Impurity-controlled superconductivity/spin density wave interplay in the organic superconductor $(TMTSF)_2ClO_4$. *Europhys. Lett.* **2005**, *72*, 645–651.

175. Fertey, P.; Sayetat, F.; Muller, J.; Pouget, J.-P.; Lenoir, C.; Batail, P. Thermal behaviour of $(TMTSF)_2ClO_4$lattice parameters in the range 10K-300K. *Physica C* **1994**, *235–240*, 2459–2460.

176. Fertey, P.; Canadell, E.; Pouget, J.-P.; Sayetat, F.; Lenoir, C.; Batail, P; Muller, J. Cooling rate dependence of the lattice parameters of $(TMTSF)_2ClO_4$. *Synth. Met.* **1995**, *70*, 761–762.

177. Nakamura, T.; Kinami, R.; Takahashi, T.; Saito, G. $^1H$ NMR Study of the Magnetic Structure in $(TMTTF)_2SCN$. *Synth. Met.* **1997**, *86*, 2053–2054.

178. Parkin, S.S.P.; Coulon, C.; Moret, R.; Pouget, J.-P. Incommensurate structural modulation and electronic localization in bis-tetramethyltetraselenafulvalene thiocyanate. *Phys. Rev. B* **1987**, *36*, 2246–2250.

179. Kang, W.; Chung, O.H. Quasi-one-dimensional Fermi surface of $(TMTSF)_2NO_3$. *Phys. Rev. B* **2009**, *79*, 045115.

180. Hiraki, K.; Nemoto, T.; Takahashi, T.; Kang, K.; Jo, Y.; Kang, W.; Cung, O.-H. NMR studies of the exotic members of the Bechgaard salts, $NO_3$ and $FSO_3$ salts. *Synth. Met.* **2003**, *135–136*, 691–692. [Google Scholar]

181. Poirier, M.; Langlois, A.; Bourbonnais, C.; Foury-Leylekian, P.; Moradpour, A.; Pouget, J.-P. Magneto-elastic coupling in the spin-Peierls ground state of hydrogenated and deuterated $(TMTTF)_2PF_6$ salts. Departement de Physique, Université de Sherbrooke: Shergrooke, Canada, 2012; Unpublished work.

182. Pasquier, C.R.; Kang, N.; Salameh, B.; Auban-Senzier, P.; Jérome, D.; Brazovskii, S. Evolution of the spin-density wave-superconductivity texture in the organic superconductor $(TMTSF)_2PF_6$ under pressure. *Phys. B* **2012**, *407*, 1806–1809, and references there in.

183. Giovannetti, G.; Kumar, S.; Pouget, J.-P.; Cappone, M. Magnetic and charge orderings in predicting multiferroicity in $TMTTF_2$-$PF_6$ organic crystals. **2012**. Unpublished work.

184. Khomskii, D.I. Multiferroics: Different ways to combine magnetism and ferroelectricity. *J. Magn. Magn. Mater.* **2006**, *306*, 1–8.

185. van den Brink, J.; Khomskii, D.I. Multiferroicity due to charge ordering. *J. Phys.: Condens. Matter* **2008**, *20*, 434217.

186. Holakovský, J. A New Type of Ferroelectric Phase Transition. *Phys. Stat. Sol. B* **1973**, *56*, 615–619.

187. Imry, Y. On the statistical mechanics of coupled order parameters. *J. Phys. C: Solid State Phys.* **1975**, *8*, 567–577.

# INDEX